JOHNSON/EVINRUDE Outboards
1956-70 REPAIR MANUAL
1½ - 40 HORSEPOWER, 1 AND 2 CYLINDER

SELOC®

Managing Partners	Dean F. Morgantini, S.A.E.
	Barry L. Beck
Executive Editor	Kevin M. G. Maher, A.S.E.
Manager-Marine/Recreation	James R. Marotta, A.S.E.
Production Specialist	Melinda Possinger
Authors	Joan and Clarence Coles

Manufactured in USA
© 1997 Seloc Publications
104 Willowbrook Lane
West Chester, PA 19382
ISBN 0-89330-007-1
7890123456 4321098765

www.selocmarine.com
1-866-SELOC55

Other titles
Brought to you by

CHILTON™ MARINE

Title	Part #
Chrysler Outboards, All Engines, 1962-84	018-7
Force Outboards, All Engines, 1984-96	024-1
Honda Outboards, All Engines 1988-98	1200
Johnson/Evinrude Outboards, 1-2 Cyl, 1956-70	007-1
Johnson/Evinrude Outboards, 1-2 Cyl, 1971-89	008-X
Johnson/Evinrude Outboards, 1-2 Cyl, 1990-95	026-8
Johnson/Evinrude Outboards, 3, 4 & 6 Cyl, 1973-91	010-1
Johnson/Evinrude Outboards, 3-4 Cyl, 1958-72	009-8
Johnson/Evinrude Outboards, 4, 6 & 8 Cyl, 1992-96	040-3
Kawasaki Personal Watercraft, 1973-91	032-2
Kawasaki Personal Watercraft, 1992-97	042-X
Marine Jet Drive, 1961-96	029-2
Mariner Outboards, 1-2 Cyl, 1977-89	015-2
Mariner Outboards, 3, 4 & 6 Cyl, 1977-89	016-0
Mercruiser Stern Drive, Type I, Alpha/MR, Bravo I & II, 1964-92	005-5
Mercruiser Stern Drive, Alpha I (Generation II), 1992-96	039-X
Mercruiser Stern Drive, Bravo I, II & III, 1992-96	046-2
Mercury Outboards, 1-2 Cyl, 1965-91	012-8
Mercury Outboards, 3-4 Cyl, 1965-92	013-6
Mercury Outboards, 6 Cyl, 1965-91	014-4
Mercury/Mariner Outboards, 1-2 Cyl, 1990-94	035-7
Mercury/Mariner Outboards, 3-4 Cyl, 1990-94	036-5
Mercury/Mariner Outboards, 6 Cyl, 1990-94	037-3
Mercury/Mariner Outboards, All Engines, 1995-99	1416
OMC Cobra Stern Drive, Cobra, King Cobra, Cobra SX, 1985-95	025-X
OMC Stern Drive, 1964-86	004-7
Polaris Personal Watercraft, 1992-97	045-4
Sea Doo/Bombardier Personal Watercraft, 1988-91	033-0
Sea Doo/Bombardier Personal Watercraft, 1992-97	043-8
Suzuki Outboards, All Engines, 1985-99	1600
Volvo/Penta Stern Drives 1968-91	011-X
Volvo/Penta Stern Drives, Volvo Engines, 1992-93	038-1
Volvo/Penta Stern Drives, GM and Ford Engines, 1992-95	041-1
Yamaha Outboards, 1-2 Cyl, 1984-91	021-7
Yamaha Outboards, 3 Cyl, 1984-91	022-5
Yamaha Outboards, 4 & 6 Cyl, 1984-91	023-3
Yamaha Outboards, All Engines, 1992-98	1706
Yamaha Personal Watercraft, 1987-91	034-9
Yamaha Personal Watercraft, 1992-97	044-6
Yanmar Inboard Diesels, 1975-98	7400

SAFETY NOTICE

Proper service and repair procedures are vital to the safe, reliable operation of all marine engines, as well as the personal safety of those performing repairs. This manual outlines procedures for servicing and repairing outboards using safe, effective methods. The procedures contain many NOTES, CAUTIONS and WARNINGS which should be followed, along with standard procedures, to eliminate the possibility of personal injury or improper service which could damage the vessel or compromise its safety.

It is important to note that repair procedures and techniques, tools and parts for servicing marine engines, as well as the skill and experience of the individual performing the work, vary widely. It is not possible to anticipate all of the conceivable ways or conditions under which these engines may be serviced, or to provide cautions as to all possible hazards that may result. Standard and accepted safety precautions and equipment should be used during cutting, grinding, chiseling, prying, or any other process that can cause material removal or projectiles.

Some procedures require the use of tools specially designed for a specific purpose. Before substituting another tool or procedure, you must be completely satisfied that neither your personal safety, nor the performance of the marine engine, will be compromised.

Although information in this manual is based on industry sources and is complete as possible at the time of publication, the possibility exists that some vehicle manufacturers made later changes which could not be included here. While striving for total accuracy, Chilton Marine cannot assume responsibility for any errors, changes or omissions that may occur in the compilation of this data.

PART NUMBERS

Part numbers listed in this reference are not recommendations by Chilton Marine for any product brand name. They are references that can be used with interchange manuals and aftermarket supplier catalogs to locate each brand supplier's discrete part number.

SPECIAL TOOLS

Special tools are recommended by the marine manufacturer to perform a specific task. Use has been kept to a minimum, but, where absolutely necessary, they are referred to in the text by the part number of the tool manufacturer. These tools can be purchased, under the appropriate part number, from your local dealer or regional distributor, or an equivalent tool can be purchased locally from a tool supplier or parts outlet. Before substituting any tool for the one recommended, read the SAFETY NOTICE at the top of this page.

ACKNOWLEDGMENTS

Chilton Marine expresses sincere appreciation to Outboard Marine Corporation for their assistance in the production of this manual.

ALL RIGHTS RESERVED

No part of this publication may be reproduced, transmitted or stored in any form or by any means, electronic or mechanical, including photocopy, recording, or by information storage or retrieval system, without prior written permission from the publisher.

FOREWORD

This is a comprehensive tune-up and repair manual for Johnson/Evinrude outboards manufactured between 1956 and 1970. Competition, high-performance, and commercial units (including aftermarket equipment), are not covered. The book has been designed and written for the professional mechanic, the do-it-yourselfer, and the student developing his mechanical skills.

Professional Mechanics will find it to be an additional tool for use in their daily work on outboard units because of the many special techniques described.

Boating enthusiasts interested in performing their own work and in keeping their unit operating in the most efficient manner will find the step-by-step illustrated procedures used throughout the manual extremely valuable. In fact, many have said this book almost equals an experienced mechanic looking over their shoulder giving them advice.

Students and Instructors have found the chapters divided into practical areas of interest and work. Technical trade schools, from Florida to Michigan and west to California, as well as the U.S. Navy and Coast Guard, have adopted these manuals as a standard classroom text.

Troubleshooting sections have been included in many chapters to assist the individual performing the work in quickly and accurately isolating problems to a specific area without unnecessary expense and time-consuming work. As an added aid and one of the unique features of this book, many worn parts are illustrated to identify and clarify when an item should be replaced.

Illustrations and procedural steps are so closely related and identified with matching numbers that, in most cases, captions are not used. Exploded drawings show internal parts and their interrelationship with the major component.

TABLE OF CONTENTS

1 SAFETY

INTRODUCTION	1-1
CLEANING, WAXING, AND POLISHING	1-1
CONTROLLING CORROSION	1-2
PROPELLERS	1-2
FUEL SYSTEM	1-7
LOADING	1-9
HORSEPOWER	1-10
FLOTATION	1-10
EMERGENCY EQUIPMENT	1-12
COMPASS	1-15
STEERING	1-17
ANCHORS	1-17
MISCELLANEOUS EQUIPMENT	1-18
BOATING ACCIDENT REPORTS	1-19
NAVIGATION	1-19

2 TUNING

INTRODUCTION	2-1
TUNE-UP SEQUENCE	2-2
COMPRESSION CHECK	2-3
SPARK PLUG INSPECTION	2-4
IGNITION SYSTEM	2-4
SYNCHRONIZING	2-5
BATTERY SERVICE	2-5
CARBURETOR ADJUSTMENTS	2-7
FUEL PUMPS	2-9
STARTER AND SOLENOID	2-10
INTERNAL WIRING HARNESS	2-11
WATER PUMP CHECK	2-12
PROPELLER	2-13
LOWER UNIT	2-14
BOAT TESTING	2-15

3 POWERHEAD

INTRODUCTION	3-1
Theory of Operation	3-1
CHAPTER ORGANIZATION	3-4
POWERHEAD DISASSEMBLING	3-5
HEAD SERVICE	3-5
REED SERVICE	3-6
Description	3-6
Reed Valve Adjustment	3-8
Cleaning and Service	3-9
BYPASS COVERS	3-10
EXHAUST COVER	3-11
Cleaning	3-11
TOP SEAL	3-12
Removal	3-12
BOTTOM SEAL	3-13
Inspection	3-14
CENTERING PINS	3-15
MAIN BEARING BOLTS AND CRANKCASE SIDE BOLTS	3-15
CRANKCASE COVER	3-16
Removal	3-16
Cleaning and Inspecting	3-16
CONNECTING RODS AND PISTONS	3-16
Removal	3-17
Disassembly	3-18
Rod Inspection and Service	3-21
Piston and Ring Inspection and Service	3-22
Assembling	3-24
CRANKSHAFT	3-27
Removal	3-27
Cleaning and Inspection	3-27
Assembling	3-28
CYLINDER BLOCK SERVICE	3-28
Honing Procedures	3-29
Assembling	3-30
Piston and Rod Assembly Installation	3-30
Crankshaft Installation Large Horsepower Engines 15 hp to 40 hp	3-33
Crankshaft Installation Small Horsepower Engines 1.5 hp, 5.0 hp, 5.5 hp, 6.0 hp, 9.5 hp	3-35
Crankshaft Installation Small Horsepower Engines 3.0 hp, 4.0 hp, 7.5 hp	3-37
Crankcase Cover Installation	3-38
Main Bearing Bolt and Crankcase Side Bolt Installation	3-39
Bottom Seal Installation 15 hp to 40 hp Engines	3-39
Exhaust Cover and Bypass Cover Installation	3-40
Reed Box Installation	3-40
Head Installation	3-41
BREAK-IN PROCEDURES	3-41
EXPLODED DRAWINGS	3-42 - 3-50

4 FUEL

INTRODUCTION	4-1
GENERAL CARBURETION INFORMATION	4-1
FUEL SYSTEM TROUBLESHOOTING	4-4
Fuel Pump Tests	4-6
Fuel Line Test	4-7
Testing with Pressure Tank	4-8
Rough Engine Idle	4-10
Excessive Fuel Consumption	4-11
CARBURETORS	4-12
TYPE I CARBURETOR	4-13
Removal and Disassembling	4-13
Cleaning and Inspecting	4-15
Assembling	4-16
CHOKE SYSTEM SERVICE	4-23
Heat/Electric Choke	4-23
All Electric Choke	4-25
Water Choke	4-26
TYPE II CARBURETOR	4-28
Disassembling	4-28
Cleaning and Inspecting	4-31
Assembling	4-33
ASSEMBLING CHOKES TO TYPE II CARBURETORS	4-38
Adjustments	4-42
TYPE III CARBURETORS	4-43
Removal	4-43
Cleaning and Inspecting	4-45
Assembling	4-47
Adjustments	4-50
FUEL PUMP SERVICE	4-51
Troubleshooting	4-52
Removal and Repair	4-52
Cleaning and Inspecting	4-54
Assembling and Installation	4-54
FUEL TANK AND LINE SERVICE	4-57
Disassembling	4-58
Cleaning and Inspecting	4-61
Assembling	4-61
LATE MODEL FUEL TANK SERVICE	4-67

5 IGNITION

INTRODUCTION	5-1
SPARK PLUG EVALUATION	5-2
POLARITY CHECK	5-3
WIRING HARNESS	5-4
FLYWHEEL MAGNETO IGNITION	5-5
TROUBLESHOOTING	5-6
SERVICING FLYWHEEL MAGNETO IGNITION SYSTEM	5-13
Removal	5-13
Cleaning and Inspecting	5-19
Assembling	5-20
SYNCHRONIZATION FUEL AND IGNITION SYSTEMS	5-26
Primary Pickup Adjustments and Locations	5-26

6 ELECTRICAL

INTRODUCTION	6-1
BATTERIES	6-1
Marine Batteries	6-1
Battery Construction	6-2
Battery Location	6-2
Battery Service	6-2
Jumper Cables	6-5
Dual Battery Installation	6-5
GAUGES AND HORNS	6-7
Constant-Voltage System	6-7
SERVICE PROCEDURES	6-7
Temperature Gauges	6-7
Warning Lights	
Thermomelt Sticks	6-8
FUEL SYSTEM	6-8
Fuel Gauge	6-8
Fuel Gauge Hookup	6-8
Troubleshooting	6-9
TACHOMETER	6-10
HORNS	6-10
ELECTRICAL SYSTEM GENERAL INFORMATION	6-11
CHARGING CIRCUIT SERVICE	6-12
Troubleshooting	6-12
Generator Service	6-16
Armature Testing	6-17
Cleaning and Inspecting	6-18
Assembling	6-20
CHOKE CIRCUIT SERVICE	6-22
STARTER MOTOR CIRCUIT SERVICE	6-22
Circuit Description	6-22
Starter Motor Description	6-22
Troubleshooting	6-24
Testing	6-25
STARTER DRIVE GEAR SERVICE	6-26
Starter Removal	6-26
Drive Gear Disassembling	6-27
Cleaning and Inspecting	6-27
Assembling Type I Drive	6-28
Disassembling Type II	6-28
Cleaning and Inspecting	6-28
Assembling Type II Drive	6-28
DELCO-REMY SERVICE	6-29
Removal	6-29
Disassembling	6-29
Armature Testing	6-30
Cleaning and Inspecting	6-31
Assembling	6-32

6 ELECTRICAL (CONT)

- AUTOLITE STARTER MOTOR
  - SERVICE — 6-34
    - Removal — 6-34
    - Disassembling — 6-35
    - Armature Testing — 6-35
    - Cleaning and Inspecting — 6-37
    - Assembling — 6-37
- PRESTOLITE STARTER MOTOR
  - SERVICE — 6-39
    - Removal — 6-39
    - Disassembling — 6-40
    - Armature Testing — 6-40
    - Cleaning and Inspecting — 6-42
    - Assembling — 6-43
- STARTER MOTOR TESTING — 6-44
- STARTER MOTOR INSTALLATION — 6-44

7 ACCESSORIES

- INTRODUCTION — 7-1
- SHIFT BOXES — 7-1
  - Description — 7-1
- OLD-STYLE DOUBLE LEVER — 7-3
  - Troubleshooting — 7-3
  - Disassembling — 7-4
  - Cleaning and Inspection — 7-5
  - Assembling — 7-5
- NEW-STYLE SHIFT LEVER — 7-6
  - Troubleshooting — 7-6
  - Removal — 7-8
  - Disassembling — 7-8
  - Cleaning and Inspecting — 7-9
  - Assembling — 7-10
- ELECTRIC GEAR BOXES AND SINGLE LEVER CONTROL — 7-12
  - Troubleshooting — 7-12
  - Disassembling — 7-14
  - Cleaning and Inspecting — 7-15
  - Assembling — 7-16
- PUSH BUTTON SHIFT BOX SERVICE EVINRUDE UNITS ONLY — 7-18
  - Troubleshooting — 7-19
  - Disassembling — 7-21
  - Cleaning and Inspecting — 7-22
  - Assembling — 7-22
- CABLE END FITTING INSTALLATION AT THE ENGINE END — 7-24

8 LOWER UNIT

- DESCRIPTION — 8-1
  - Chapter Coverage — 8-1
  - Illustrations — 8-2
- TROUBLESHOOTING MANUAL SHIFT — 8-2
- PROPELLER REMOVAL — 8-7
- DRAINING LOWER UNIT — 8-8
- LOWER UNIT SERVICE
  - 1.5 hp to 4.0 hp -- NO SHIFT — 8-8
  - Lower Unit Removal — 8-9
  - Water Pump Removal — 8-9
  - Disassembling — 8-10
  - Cleaning and Inspecting — 8-11
  - Assembling — 8-13
  - Water Pump Installation — 8-14
  - Lower Unit Installation — 8-15
  - Filling the Lower Unit — 8-16
  - Propeller Installation — 8-16
- LOWER UNIT SERVICE MANUAL SHIFT -- 5 HP TO 25 HP — 8-17
  - Removal — 8-19
  - Water Pump Removal — 8-19
  - Disassembling — 8-20
  - Cleaning and Inspecting — 8-22
  - Assembling — 8-28
  - Water Pump Installation — 8-31
  - Lower Unit Installation — 8-33
- LOWER UNIT SERVICE MANUAL SHIFT -- 28 HP TO 40 HP — 8-36
  - Removal — 8-37
  - Water Pump Removal — 8-38
  - Disassembling — 8-38
  - Cleaning and Inspecting — 8-41
  - Assembling — 8-46
  - Lower Unit Installation — 8-50
- ELECTROMATIC LOWER UNIT — 8-53
  - Description — 8-53
  - Troubleshooting — 8-53
  - Removal — 8-56
  - Disassembling — 8-57
  - Cleaning and Inspecting — 8-65
  - Assembling — 8-66
  - Water Pump Installation — 8-72
  - Lower Unit Installation — 8-74

9 HAND STARTERS

- INTRODUCTION — 9-1
  - Operation — 9-2
- TYPE I STARTER CYLINDER WITH PINION GEAR 5 HP and 6 HP ENGINES — 9-3
  - Starter Rope Replacement — 9-4
    - Removal — 9-4
    - Installation — 9-4
  - Starter Removal — 9-5
  - Disassembling — 9-7
  - Cleaning and Inspecting — 9-7
  - Assembling — 9-7

9 HAND STARTERS (CONT)

TYPE I STARTER
CYLINDER WITH PINION GEAR
ALL 9.5 HP ENGINES

Starter Rope Replacement	9-11
Removal	9-11
Installation	9-12
Starter Removal	9-12
Cleaning and Inspecting	9-14
Assembling	9-14
Installation	9-15

TYPE II STARTER
COIL SPRING WITH SWING ARM DRIVE GEAR
3 HP 1968
4 HP 1969-70

Removal	9-17
Disassembling	9-17
Cleaning and Inspecting	9-20
Assembling	9-20

TYPE III STARTER
MOUNTED ATOP FLYWHEEL MODEL WITH RETURN SPRINGS
28 HP 1962-63
30 HP 1956
35 HP 1957-59
40 HP 1960-63

Removal	9-24
Cleaning and Inspecting	9-26
Assembling	9-27
Rope Installation	9-29
Starter Installation	9-30

TYPE III STARTER
MOUNTED ATOP FLYWHEEL MODEL WITH NO RETURN SPRINGS
28 HP 1964
33 HP 1965-70
40 HP 1964-70

Removal	9-32
Cleaning and Inspecting	9-34
Assembling	9-34
Rope Installation	9-37
Starter Installation	9-39

TYPE III STARTER
MOUNTED ATOP FLYWHEEL MODEL WITH ONE NYLON PAWL
3 HP 1956-68
5.5 HP 1956-64
7.5 HP 1956-58
10 HP 1956-63
15 HP 1956
18 HP 1956-70
20 HP 1966-70
25 HP 1969-70   9-39

Removal	9-40
Disassembling	9-40
Cleaning and Inspecting	9-42
Assembling	9-43
Rope Installation	9-43
Starter Installation	9-46

10 MAINTENANCE

INTRODUCTION	10-1
ENGINE SERIAL NUMBERS	10-2
FIBERGLASS HULLS	10-3
ALUMINUM HULLS	10-3
BELOW WATERLINE SERVICE	10-4
SUBMERGED ENGINE SERVICE	10-5
WINTER STORAGE	10-7
LOWER UNIT SERVICE	10-9
Propeller	10-9
BATTERY STORAGE	10-13
PRESEASON PREPARATION	10-13

APPENDIX

METRIC CONVERSION CHART	A-1
DRILL SIZE CONVERSION CHART	A-2
TORQUE SPECIFICATIONS	A-3
POWERHEAD SPECS	A-4 & A-5
TUNE-UP SPECS	A-6 to A10
GEAR OIL CAPACITIES	A-11
STARTER MOTOR SPECS	A-12
REGULATOR SPECS	A-12
GENERATOR SPECS	A-12
CONDENSER SPECS	A-13
STARTER ROPE SPECS	A-14

WIRE INDENTIFICATION DRAWINGS

20 hp and 25 hp -- 1971-72	A-15
33 hp with Generator -- 1965-67	A-16
33 hp with Generator -- 1968	A-17
33 hp with Generator -- 1969-70	A-18
35 hp -- 1957-59	A-19
40 hp Standard Shift with Generator -- 1960-66	A-20
40 hp Standard Shift with Generator -- 1967-68	A-21
40 hp Standard Shift with Generator -- 1969-70	A-22
40 hp Electric Shift with Generator -- 1961-66	A-23
40 hp Electric Shift with Generator -- 1967-68	A-24
40 hp Electric Shift with Generator -- 1969-70	A-25

1
SAFETY

1-1 INTRODUCTION

Your boat probably represents a sizeable investment for you. In order to protect this investment and to receive the maximum amount of enjoyment from your boat it must be cared for properly while being used and when it is out of the water. Always store your boat with the bow higher than the stern and be sure to remove the transom drain plug and the inner hull drain plugs. If you use any type of cover to protect your boat, plastic, canvas, whatever, be sure to allow for some movement of air through the hull. Proper ventilation will assure evaporation of any condensation that may form due to changes in temperature and humidity.

1-2 CLEANING, WAXING, AND POLISHING

An outboard boat should be washed with clear water after each use to remove surface dirt and any salt deposits from use in salt water. Regular rinsing will extend the time between waxing and polishing. It will also give you "pride of ownership", by having a sharp looking piece of equipment. Elbow grease, a mild detergent, and a brush will be required to remove stubborn dirt, oil, and other unsightly deposits.

Stay away from harsh abrasives or strong chemical cleaners. A white buffing compound can be used to restore the original gloss to a scratched, dull, or faded area. The finish of your boat should be thoroughly cleaned, buffed, and polished at least once each season. Take care when buffing or polishing with a marine cleaner not to overheat the surface you are working, because you will burn it.

A small outboard engine mounted on an aluminum boat should be removed from the boat and stored separately. Under all circumstances, any outboard engine must **ALWAYS** be stored with the powerhead higher than the lower unit and exhaust system. This position will prevent water trapped in the lower unit from draining back through the exhaust ports into the powerhead.

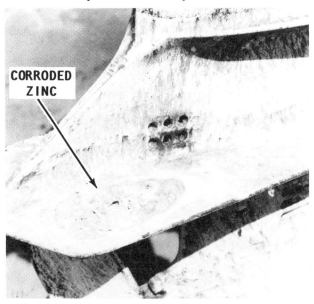

Lower unit badly corroded because the zinc was not replaced. Once the zinc is destroyed, more costly parts will be damaged. Attention to the zinc condition is extremely important during boat operation in salt water.

Whenever the boat is stored, for long or short periods, the bow should be slightly higher than the stern and the drain plug in the transom removed to ensure proper drainage of rain water.

1-2 SAFETY

A new zinc prior to installation. This inexpensive item will save corrosion on more valuable parts.

Most outboard engines have a flat area on the back side of the powerhead. When the engine is placed with the flat area on the powerhead and the lower unit resting on the floor, the engine will be in the proper altitude with the powerhead higher than the lower unit.

1-3 CONTROLLING CORROSION

Since man first started out on the water, corrosion on his craft has been his enemy. The first form was merely rot in the wood and then it was rust, followed by other forms of destructive corrosion in the more modern materials. One defense against corrosion is to use similar metals throughout the boat. Even though this is difficult to do in designing a new boat, particularly the undersides, similar metals should be used whenever and wherever possible.

A second defense against corrosion is to insulate dissimilar metals. This can be done by using an exterior coating of Sea Skin or by insulating them with plastic or rubber gaskets.

Using Zinc

The proper amount of zinc attached to a boat is extremely important. The use of too much zinc can cause wood burning by placing the metals close together and they become "hot". On the other hand, using too small a zinc plate will cause more rapid deterioration of the metal you are trying to protect. If in doubt, consider the fact that is is far better to replace the zincs than to replace planking or other expensive metal parts from having an excess of zinc.

When installing zinc plates, there are two routes available. One is to install many different zincs on all metal parts and thus run the risk of wood burning. Another route, is to use one large zinc on the transom of the boat and then connect this zinc to every underwater metal part through internal bonding. Of the two choices, the one zinc on the transom is the better way to go.

Small outboard engines have a zinc plate attached to the cavitation plate. Therefore, the zinc remains with the engine at all times.

1-4 PROPELLERS

As you know, the propeller is actually what moves the boat through the water. This is how it is done. The propeller operates in water in much the manner as a wood screw does in wood. The propeller "bites" into the water as it rotates. Water passes between the blades and out to the rear in the shape of a cone. The propeller "biting" through the water in much the same manner as a wood auger is what propels the boat.

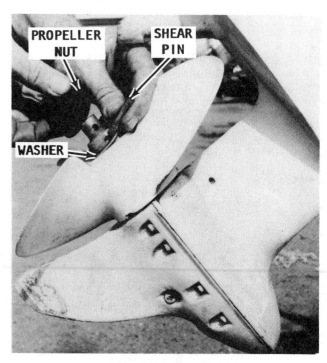

Propeller and associated parts in order, washer, shear-pin, and nut, ready for installation.

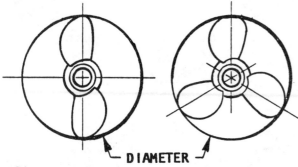

Diameter and pitch are the two basic dimensions of a propeller. The diameter is measured across the circumference of a circle scribed by the propeller blades, as shown.

Arrangement of propeller and associated parts, in order, for a small horsepower engine.

Diameter and Pitch

Only two dimensions of the propeller are of real interest to the boat owner: the diameter and the pitch. These two dimensions are stamped on the propeller hub and always appear in the same order: the diameter first and then the pitch. For instance, the number 15-19 stamped on the hub, would mean the propeller had a diameter of 15 inches with a pitch of 19.

The diameter is the measured distance from the tip of one blade to the tip of the other as shown in the accompanying illustration.

The pitch of a propeller is the angle at which the blades are attached to the hub. This figure is expressed in inches of water travel for each revolution of the propeller. In our example of a 15-19 propeller, the propeller should travel 19 inches through the water each time it revolves. If the propeller action was perfect and there was no slippage, then the pitch multiplied by the propeller rpms would be the boat speed.

Most outboard manufacturers equip their units with a standard propeller with a diameter and pitch they consider to be best suited to the engine and the boat. Such a propeller allows the engine to run as near to the rated rpm and horsepower (at full throttle) as possible for the boat design.

The blade area of the propeller determines its load-carrying capacity. A two-blade propeller is used for high-speed running under very light loads.

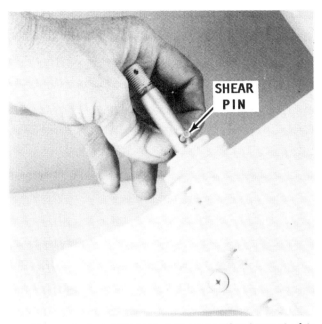

Shear-pin installed behind the propeller instead of in front of the propeller.

A four-blade propeller is installed in boats intended to operate at low speeds under very heavy loads such as tugs, barges, or large houseboats. The three-blade propeller is the happy medium covering the wide range between the high performance units and the load carrying workhorses.

Propeller Selection

There is no standard propeller that will do the proper job in very many cases. The list of sizes and weights of boats is almost endless. This fact coupled with the many boat-engine combinations makes the propeller selection for a specific purpose a difficult job. In fact, in many cases the propeller is changed after a few test runs. Proper selection is aided through the use of charts set up for various engines and boats. These charts should be studied and understood when buying a propeller. However, bear in mind, the charts are based on average boats

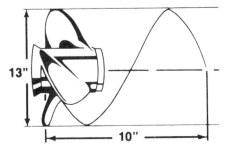
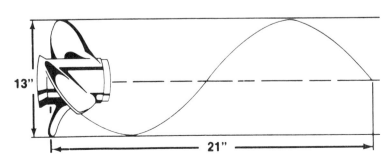

Diagram to explain the pitch dimension of a propeller. The pitch is the theoretical distance a propeller would travel through the water if there was no slippage.

with average loads, therefore, it may be necessary to make a change in size or pitch, in order to obtain the desired results for the hull design or load condition.

A wide range of pitch is available for each of the larger horsepower engines. The choice available for the smaller engines, up to about 25 hp, is restricted to one or two sizes. Remember, a low pitch takes a smaller bite of the water than the high pitch propeller. This means the low pitch propeller will travel less distance through the water per revolution. The low pitch will require less horsepower and will allow the engine to run faster and more efficiently.

It stands to reason, and it's true, that the high pitch propeller will require more horsepower, but will give faster boat speed if the engine is allowed to turn to its rated rpm.

If a higher-pitched propeller is installed on a boat, in an effort to get more speed, extra horsepower will be required. If the extra power is not available, the rpms will be reduced to a less efficient level and the actual boat speed will be less than if the lower-pitched propeller had been left installed.

All engine manufacturers design their units to operate with full throttle at, or slightly above, the rated rpm. If you run your engine at the rated rpm, you will increase spark plug life, receive better fuel economy, and obtain the best performance from your boat and engine. Therefore, take time to make the proper propeller selection for the rated rpm of your engine at full throttle with what you consider to be an average load. Your boat will then be correctly balanced between engine and propeller throughout the entire speed range.

A reliable tachometer must be used to measure engine speed at full throttle to ensure the engine will achieve full horsepower and operate efficiently and safely. To test for the correct propeller, make your run in a body of smooth water with the lower unit in forward gear at full throttle. Observe the tachometer at full throttle. **NEVER** run the engine at a high rpm when a flush attachment is installed. If the reading is above the manufacturer's recommended operating range, you must try propellers of greater pitch, until you find the one that allows the engine to operate continually within the recommended full throttle range.

If the engine is unable to deliver top performance and you feel it is properly tuned, then the propeller may not be to blame. Operating conditions have a marked effect on performance. For instance, an engine will lose rpm when run in very cold water. It will also lose rpm when run in salt water as compared with fresh water. A hot, low-barometer day will also cause your engine to lose power.

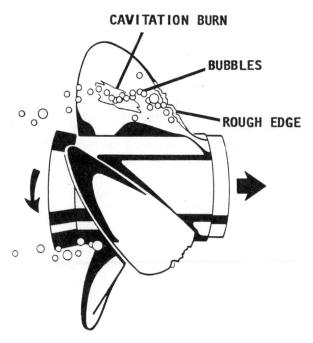

Cavitation (air bubbles) formed at the propeller. Manufacturers are constantly fighting this problem, as explained in the text.

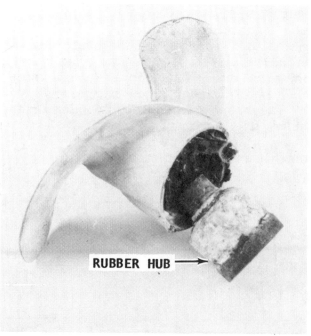

Example of a damaged propeller. This unit should have been replaced long before this amount of damage was sustained.

Ventilation

Ventilation is the forming of voids in the water just ahead of the propeller blades. Marine propulsion designers are constantly fighting the battle against the formation of these voids due to excessive blade tip speed and engine wear. The voids may be filled with air or water vapor, or they may actually be a partial vacuum. Ventilation may be caused by installing a piece of equipment too close to the lower unit, such as the knot indicator pickup, depth sounder, or bait tank pickup.

Vibration

Your propeller should be checked regularly to be sure all blades are in good condition. If any of the blades become bent or nicked, this condition will set up vibrations in the drive unit and the motor. If the vibration becomes very serious it will cause a loss of power, efficiency, and boat performance. If the vibration is allowed to continue over a period of time it can have a damaging effect on many of the operating parts.

Vibration in boats can never be completely eliminated, but it can be reduced by keeping all parts in good working condition and through proper maintenance and lubrication. Vibration can also be reduced in some cases by increasing the number of blades. For this reason, many racers use two-blade props and luxury cruisers have four- and five-blade props installed.

Shock Absorbers

The shock absorber in the propeller plays a very important role in protecting the shafting, gears, and engine against the shock of a blow, should the propeller strike an underwater object. The shock absorber allows the propeller to stop rotating at the instant of impact while the power train continues turning.

How much impact the propeller is able to withstand before causing the clutch hub to slip is calculated to be more than the force needed to propel the boat, but less than the amount that could damage any part of the power train. Under normal propulsion loads of moving the boat through the water, the hub will not slip. However, it will slip if the propeller strikes an object with a force that would be great enough to stop any part of the power train.

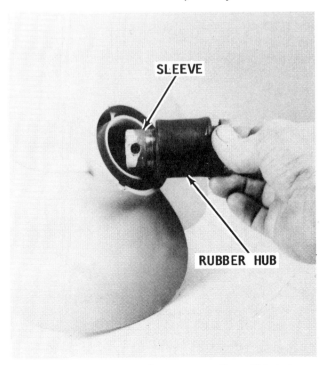

Rubber hub removed from a propeller. This hub was removed because the hub was slipping in the propeller.

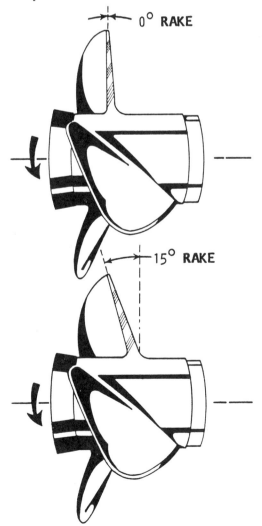

Illustration depicting the rake of a propeller, as explained in the text.

If the power train was to absorb an impact great enough to stop rotation, even for an instant, something would have to give and be damaged. If a propeller is subjected to repeated striking of underwater objects, it would eventually slip on its clutch hub under normal loads. If the propeller would start to slip, a new hub and shock absorber would have to be installed.

Propeller Rake

If a propeller blade is examined on a cut extending directly through the center of the hub, and if the blade is set vertical to the propeller hub, as shown in the accompanying illustration, the propeller is said to have a zero degree (0°) rake. As the blade slants back, the rake increases. Standard propellers have a rake angle from 0° to 15°.

A higher rake angle generally improves propeller performance in a cavitating or ventilating situation. On lighter, faster boats, higher rake often will increase performance by holding the bow of the boat higher.

Progressive Pitch

Progressive pitch is a blade design innovation that improves performance when forward and rotational speed is high and/or the propeller breaks the surface of the water.

Progressive pitch starts low at the leading edge and progressively increases to the trailing edge, as shown in the accompanying illustration. The average pitch over the entire blade is the number assigned to that propeller. In the illustration of the progressive pitch, the average pitch assigned to the propeller would be 21.

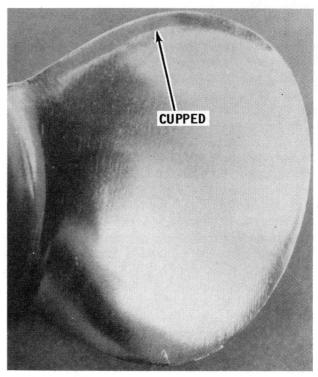

Propeller with a "cupped" leading edge. "Cupping" gives the propeller a better "hold" in the water.

Cupping

If the propeller is cast with a edge curl inward on the trailing edge, the blade is said to have a cup. In most cases, cupped blades improve performance. The cup helps the blades to **"HOLD"** and not break loose, when operating in a cavitating or ventilating situation. This action permits the engine to be trimmed out further, or to be mounted higher on the transom. This is especially true on high-performance boats. Either of these two adjustments will usually add to higher speed.

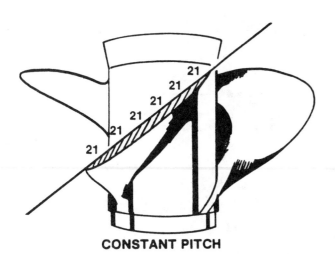
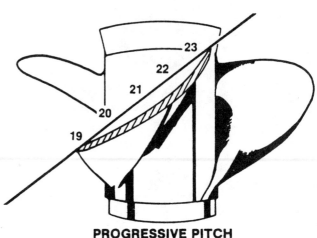

Comparison of a constant and progressive pitch propeller. Notice how the pitch of the progressive pitch propeller, right, changes to give the blade more thrust and therefore, the boat more speed.

The cup has the effect of adding to the propeller pitch. Cupping usually will reduce full-throttle engine speed about 150 to 300 rpm below the same pitch propeller without a cup to the blade. A propeller repair shop is able to increase or decrease the cup on the blades. This change, as explained, will alter engine rpm to meet specific operating demands. Cups are rapidly becoming standard on propellers.

In order for a cup to be the most effective, the cup should be completely concave (hollowed) and finished with a sharp corner. If the cup has any convex rounding, the effectiveness of the cup will be reduced.

Rotation

Propellers are manufactured as right-hand rotation (RH), and as left-hand rotation (LH). The standard propeller for outboards is RH rotation.

A right-hand propeller can easily be identified by observing it as shown in the accompanying illustration. Observe how the blade slants from the lower left toward the upper right. The left-hand propeller slants in the opposite direction, from upper left to lower right, as shown.

When the propeller is observed rotating from astern the boat, it will be rotating clockwise when the engine is in forward gear. The left-hand propeller will rotate counterclockwise.

1-5 FUEL SYSTEM

With Built-in Fuel Tank

All parts of the fuel system should be selected and installed to provide maximum service and protection against leakage. Reinforced flexible sections should be installed in fuel lines where there is a lot of motion, such as at the engine connection. The flaring of copper tubing should be annealed after it is formed as a protection against hardening. **CAUTION:** Compression fittings should **NOT** be used because they are so easily overtightened, which places them under a strain and subjects them to fatigue. Such conditions will cause the fitting to leak after it is connected a second time.

The capacity of the fuel filter must be large enough to handle the demands of the engine as specified by the engine manufacturer.

A manually-operated valve should be installed if anti-siphon protection is not provided. This valve should be installed in the fuel line as close to the gas tank as possible. Such a valve will maintain anti-siphon protection between the tank and the engine.

Fuel tanks should be mounted in dry, well ventilated places. Ideally, the fuel tanks should be installed above the cockpit floors, where any leakage will be quickly detected. In order to obtain maximum circulation of air around fuel tanks, the tank should not come in contact with the boat hull except through the necessary supports. The supporting surfaces and hold-downs must fasten the tank firmly and they should be insulated from the tank surfaces. This insulation material should be non-abrasive and non-absorbent material. Fuel tanks installed in the forward portion of the boat should be especially well secured and protected because shock loads in this area can be as high as 20 to 25 g's.

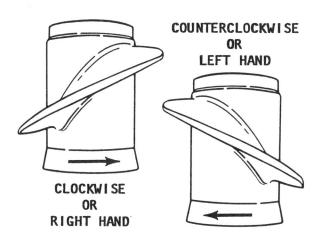

Right- and left-hand propellers showing how the angle of the blades is reversed. Right-hand propellers are by far the most popular.

A three-position valve permits fuel to be drawn from either tank or to be shut off completely. Such an arrangement prevents accidental siphoning of fuel from the tank.

Taking On Fuel

The fuel tank of your boat should be kept full to prevent water from entering the system through condensation caused by temperature changes. Water droplets forming is one of the greatest enemies of the fuel system. By keeping the tank full, the air space in the tank is kept to an absolute minimum and there is no room for moisture to form. It is a good practice not to store fuel in the tank over an extended period, say for six months. Today, fuels contain ingredients that change into gums when stored for any length of time. These gums and varnish products will cause carburetor problems and poor spark plug performance. An additive (Sta-Bil) is available and can be used to prevent gums and varnish from forming.

Static Electricity

In very simple terms, static electricity is called frictional electricity. It is generated by two dissimilar materials moving over each other. One form is gasoline flowing through a pipe or into the air. Another form is when you brush your hair or walk across a synthetic carpet and then touch a metal object. All of these actions cause an electrical charge. In most cases, static electricity is generated during very dry weather conditions, but when you are filling the fuel tank on your boat it can happen at any time.

Fuel Tank Grounding

One area of protection against the build-up of static electricity is to have the fuel

An OMC fuel tank equipped with a quick-disconnect fitting. This type of arrangement is handy when the tank must be removed from the boat to obtain fuel.

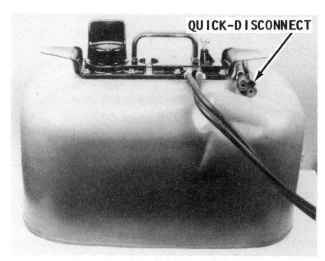

Old style pressure-type tank showing the fuel line to the engine and quick-disconnect fitting.

tank properly grounded (also known as bonding). A direct metal-to-metal contact from the fuel hose nozzle to the water in which the boat is floating. If the fill pipe is made of metal, and the fuel nozzle makes a good contact with the deck plate, then a good ground is made.

As an economy measure, some boats use rubber or plastic filler pipes because of compound bends in the pipe. Such a fill line does not give any kind of ground and if your boat has this type of installation and you do

Adding fuel to a six-gallon OMC fuel tank. Some fuel must be in the tank before oil is added to prevent the oil from accumulating on the tank bottom.

not want to replace the filler pipe with a metal one, then it is possible to connect the deck fitting to the tank with a copper wire. The wire should be 8 gauge or larger.

The fuel line from the tank to the engine should provide a continuous metal-to-metal contact for proper grounding. If any part of this line is plastic or other non-metallic material, then a copper wire must be connected to bridge the non-metal material. The power train provides a ground through the engine and drive shaft, to the propeller in the water.

Fiberglass fuel tanks pose problems of their own. One method of grounding is to run a copper wire around the tank from the fill pipe to the fuel line. However, such a wire does not ground the fuel in the tank. Manufacturers should imbed a wire in the fiberglass and it should be connected to the intake and the outlet fittings. This wire would avoid corrosion which could occur if a wire passed through the fuel. **CAUTION: It is not advisable to use a fiberglass fuel tank if a grounding wire was not installed.**

Anything you can feel as a "shock" is enough to set off an explosion. Did you know that under certain atmospheric conditions you can cause a static explosion yourself, particularly if you are wearing synthetic clothing. It is almost a certainty you could cause a static spark if you are **NOT** wearing insulated rubber-soled shoes.

As soon as the deck fitting is opened, fumes are released to the air. Therefore, to be safe you should ground yourself before opening the fill pipe deck fitting. One way to ground yourself is to dip your hand in the water overside to discharge the electricity in your body before opening the filler cap. Another method is to touch the engine block or any metal fitting on the dock which goes down into the water.

1-6 LOADING

In order to receive maximum enjoyment, with safety and performance, from your boat, take care not to exceed the load capacity given by the manufacturer. A plate attached to the hull indicates the U.S. Coast Guard capacity information in pounds for persons and gear. If the plate states the maximum person capacity to be 750 pounds and you assume each person to weigh an average of 150 lbs., then the boat could carry five persons safely. If you add another 250 lbs. for motor and gear, and the maximum weight capacity for persons and gear is 1,000 lbs. or more, then the five persons and gear would be within the limit.

Try to load the boat evenly port and starboard. If you place more weight on one side than on the other, the boat will list to the heavy side and make steering difficult. You will also get better performance by placing heavy supplies aft of the center to keep the bow light for more efficient planing.

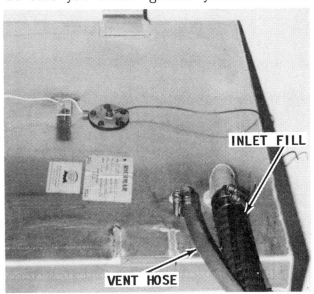

A fuel tank properly grounded to prevent static electricity. Static electricity could be extremely dangerous when taking on fuel.

U.S. Coast Guard plate affixed to all new boats. When the blanks are filled in, the plate will indicate the Coast Guard's recommendations for persons, gear, and horsepower to ensure safe operation of the boat. These recommendations should not be exceeded, as explained in the text.

1-10 SAFETY

Clarification

Much confusion arises from the terms, certification, requirements, approval, regulations, etc. Perhaps the following may clarify a couple of these points.

1- The Coast Guard does not approve boats in the same manner as they "Approve" life jackets. The Coast Guard applies a formula to inform the public of what is safe for a particular craft.

2- If a boat has to meet a particular regulation, it must have a Coast Guard certification plate. The public has been led to believe this indicates approval of the Coast Guard. Not so.

3- The certification plate means a willingness of the manufacturer to meet the Coast Guard regulations for that particular craft. The manufacturer may recall a boat if it fails to meet the Coast Guard requirements.

4- The Coast Guard certification plate, see accompanying illustration, may or may not be metal. The plate is a regulation for the manufacturer. It is only a warning plate and the public does not have to adhere to the restrictions set forth on it. Again, the plate sets forth information as to the Coast Guard's opinion for safety on that particular boat.

5- Coast Guard Approved equipment is equipment which has been approved by the Commandant of the U.S. Coast Guard and has been determined to be in compliance with Coast Guard specifications and regulations relating to the materials, construction, and performance of such equipment.

1-7 HORSEPOWER

The maximum horsepower engine for each individual boat should not be increased by any great amount without checking requirements from the Coast Guard in your area. The Coast Guard determines horsepower requirements based on the length, beam, and depth of the hull. **TAKE CARE NOT** to exceed the maximum horsepower listed on the plate or the warranty and possibly the insurance on the boat may become void.

1-8 FLOTATION

If your boat is less than 20 ft. overall, a Coast Guard or BIA (Boating Industry of America) now changed to NMMA (National Marine Manufacturers Association) requirement is that the boat must have buoyant material built into the hull (usually foam) to keep it from sinking if it should become swamped. Coast Guard requirements are mandatory but the NMMA is voluntary.

"Kept from sinking" is defined as the ability of the flotation material to keep the boat from sinking when filled with water

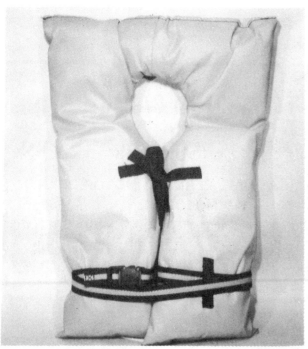

Type I PFD Coast Guard approved life jacket. This type flotation device provides the greatest amount of buoyancy. **NEVER** *use them for cushions or other purposes.*

A Type IV PFD cushion device intended to be thrown to a person in the water. If air can be squeezed out of the cushion it is no longer fit for service as a PFD.

and with passengers clinging to the hull. One restriction is that the total weight of the motor, passengers, and equipment aboard does not exceed the maximum load capacity listed on the plate.

Life Preservers —Personal Flotation Devices (PFDs)

The Coast Guard requires at least one Coast Guard approved life-saving device be carried on board all motorboats for each person on board. Devices approved are identified by a tag indicating Coast Guard approval. Such devices may be life preservers, buoyant vests, ring buoys, or buoyant cushions. Cushions used for seating are serviceable if air cannot be squeezed out of it. Once air is released when the cushion is squeezed, it is no longer fit as a flotation device. New foam cushions dipped in a rubberized material are almost indestructible.

Life preservers have been classified by the U.S. Coast Guard into five distinct categories. PFDs presently acceptable on recreational boats fall into one of these five designations. All PFDs **MUST** be U.S. Coast Guard approved, in good and serviceable condition, and of an appropriate size for the persons who intend to wear them. Wearable PFDs **MUST** be readily accessible and throwable devices **MUST** be immediately available for use.

Type I PFD has the greatest required buoyancy and is designed to turn most **UNCONSCIOUS** persons in the water from a face down position to a vertical or slightly backward position. The adult size device provides a minimum buoyancy of 22 pounds and the child size provides a minimum buoyancy of 11 pounds. The Type I PFD provides the greatest protection to its wearer and is most effective for all waters and conditions.

Type II PFD is designed to turn its wearer in a vertical or slightly backward position in the water. The turning action is not as pronounced as with a Type I. The device will not turn as many different type persons under the same conditions as the Type I. An adult size device provides a minimum buoyancy of 15½ pounds, the medium child size provides a minimum of 11 pounds, and the infant and small child sizes provide a minimum buoyancy of 7 pounds.

Type III PFD is designed to permit the wearer to place himself (herself) in a vertical or slightly backward position. The Type III device has the same buoyancy as the Type II PFD but it has little or no turning ability. Many of the Type III PFD are designed to be particularly useful when water skiing, sailing, hunting, fishing, or engaging in other water sports. Several of this type will also provide increased hypothermia protection.

Type IV PFD is designed to be thrown to a person in the water and grasped and held by the user until rescued. It is **NOT** designed to be worn. The most common Type IV PFD is a ring buoy or a buoyant cushion.

Type V PFD is any PFD approved for restricted use.

Coast Guard regulations state, in general terms, that on all boats less than 16 ft. overall, one Type I, II, III, or IV device shall be carried on board for each person in the boat. On boats over 26 ft., one Type I, II, or III device shall be carried on board for each person in the boat **plus** one Type IV device.

It is an accepted fact that most boating people own life preservers, but too few actually wear them. There is little or no excuse for not wearing one because the modern comfortable designs available today do not subtract from an individual's boating pleasure. Make a life jacket available to

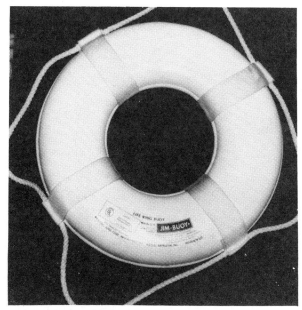

Type IV PFD ring buoy designed to be thrown. On ocean cruisers, this type device usually has a weighted pole with flag, attached to the buoy.

1-12 SAFETY

your crew and advise each member to wear it. If you are a crew member ask your skipper to issue you one, especially when boating in rough weather, cold water, or when running at high speed. Naturally, a life jacket should be a must for non-swimmers any time they are out on the water in a boat.

1-9 EMERGENCY EQUIPMENT

Visual Distress Signals
The Regulation

Since January 1, 1981, Coast Guard Regulations require all recreation boats when used on coastal waters, which includes the Great Lakes, the territorial seas and those waters directly connected to the Great Lakes and the territorial seas, up to a point where the waters are less than two miles wide, and boats owned in the United States when operating on the high seas to be equipped with visual distress signals.

The only exceptions are during daytime (sunrise to sunset) for:

Recreational boats less than 16 ft. (5 meters) in length.

Boats participating in organized events such as races, regattas or marine parades.

Open sailboats not equipped with propulsion machinery and less than 26 ft. (8 meters) in length.

Manually propelled boats.

The above listed boats need to carry night signals when used on these waters at night.

Pyrotechnic visual distress signaling devices **MUST** be Coast Guard Approved, in serviceable condition and stowed to be readily accessible. If they are marked with a date showing the serviceable life, this date must not have passed. Launchers, produced before Jan. 1, 1981, intended for use with approved signals are not required to be Coast Guard Approved.

USCG Approved pyrotechnic visual distress signals and associated devices include:

Pyrotechnic red flares, hand held or aerial.

Pyrotechnic orange smoke, hand held or floating.

Launchers for aerial red meteors or parachute flares.

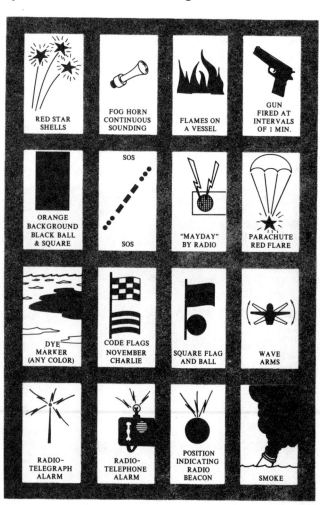

Internationally accepted distress signals.

Moisture-protected flares should be carried on board for use as a distress signal.

None-pyrotechnic visual distress signaling devices must carry the manufacturer's certification that they meet Coast Guard requirements. They must be in serviceable condition and stowed so as to be readily accessible.

This group includes:

Orange distress flag at least 3 x 3 feet with a black square and ball on an orange background.

Electric distress light -- not a flashlight but an approved electric distress light which **MUST** automatically flash the international **SOS** distress signal (. . . - - - . . .) four to six times each minute.

Types and Quantities

The following variety and combination of devices may be carried in order to meet the requirements.

1- Three hand-held red flares (day and night).

2- One electric distress light (night only).

3- One hand-held red flare and two parachute flares (day and night).

4- One hand-held orange smoke signal, two floating orange smoke signals (day) and one electric distress light (day and night).

If young children are frequently aboard your boat, careful selection and proper stowage of visual distress signals becomes especially important. If you elect to carry pyrotechnic devices, you should select those in tough packaging and not easy to ignite should the devices fall into the hands of children.

Coast Guard Approved pyrotechnic devices carry an expiration date. This date can **NOT** exceed 42 months from the date of manufacture and at such time the device can no longer be counted toward the minimum requirements.

SPECIAL WORDS

In some states the launchers for meteors and parachute flares may be considered a firearm. Therefore, check with your state authorities before acquiring such a launcher.

First Aid Kits

The first-aid kit is similar to an insurance policy or life jacket. You hope you don't have to use it but if needed, you want it there. It is only natural to overlook this essential item because, let's face it, who likes to think of unpleasantness when planning to have only a good time. However, the prudent skipper is prepared ahead of time, and is thus able to handle the emergency without a lot of fuss.

Good commercial first-aid kits are available such as the Johnson and Johnson "Marine First-Aid Kit". With a very modest expenditure, a well-stocked and adequate kit can be prepared at home.

Any kit should include instruments, supplies, and a set of instructions for their use. Instruments should be protected in a water-tight case and should include: scissors, tweezers, tourniquet, thermometer, safety

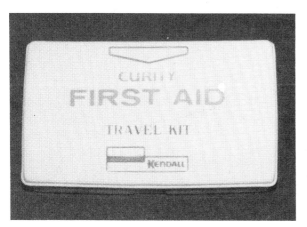

An adequately stocked first-aid kit should be on board for the safety of crew and guests.

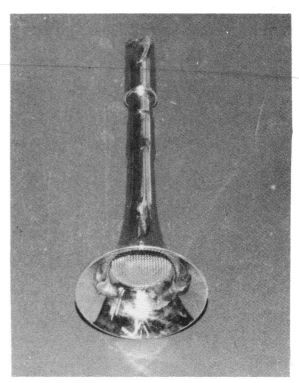

A sounding device should mounted close to the helmsman for use in sounding an emergency alarm.

pins, eye-washing cup, and a hot water bottle. The supplies in the kit should include: assorted bandages in addition to the various sizes of "band-aids", adhesive tape, absorbent cotton, applicators, petroleum jelly, antiseptic (liquid and ointment), local ointment, aspirin, eye ointment, antihistamine, ammonia inhalent, sea-sickness pills, antacid pills, and a laxative. You may want to consult your family physician about including antibiotics. Be sure your kit contains a first-aid manual because even though you have taken the Red Cross course, you may be the patient and have to rely on an untrained crew for care.

Fire Extinguishers

All fire extinguishers must bear Underwriters Laboratory (UL) "Marine Type" approved labels. With the UL certification, the extinguisher does not have to have a Coast Guard approval number. The Coast Guard classifies fire extinguishers according to their size and type.

Type B-I or B-II Designed for extinguishing flammable liquids. This type extinguisher is required on all motorboats.

The Coast Guard considers a boat having one or more of the following conditions as a "boat of closed construction" subject to fire extinguisher regulations.

1- Inboard engine or engines.
2- Closed compartments under thwarts and seats wherein portable fuel tanks may be stored.
3- Double bottoms not sealed to the hull or which are not completely filled with flotation materials.
4- Closed living spaces.
5- Closed stowage compartments in which combustible or flammable material is stored.
6- Permanently installed fuel tanks.

Detailed classification of fire extinguishers is by agent and size:

B-I contains 1-1/4 gallons foam, or 4 pounds carbon dioxide, or 2 pounds dry chemical agent, or 2-1/2 pounds Halon.

B-II contains 2-1/2 gallons foam, or 15 pounds carbo dioxide, or 10 pounds dry chemical agent, or 10 pounds Halon.

The class of motorboat dictates how many fire extinguishers are required on board. One B-II unit can be substituted for two B-I extinguishers. When the engine compartment of a motorboat is equipped with a fixed (built-in) extinguishing system, one less portable B-I unit is required.

Dry chemical fire extinguishers without

A suitable fire extinguisher should be mounted close to the helmsman for emergency use.

At least one gallon of emergency fuel should be kept on board in an approved container.

gauges or indicating devices must be weighed and tagged every 6 months. If the gross weight of a carbon dioxide (CO_2) fire extinguisher is reduced by more than 10% of the net weight, the extinguisher is not acceptable and must be recharged.

READ labels on fire extinguishers. If the extinguisher is U.L. listed, it is approved for marine use.

DOUBLE the number of fire extinguishers recommended by the Coast Guard, because their requirements are a bare **MINIMUM** for safe operation. Your boat, family, and crew, must certainly be worth much more than "bare minimum".

1-10 COMPASS

Selection

The safety of the boat and her crew may depend on her compass. In many areas weather conditions can change so rapidly that within minutes a skipper may find himself "socked-in" by a fog bank, a rain squall, or just poor visibility. Under these conditions, he may have no other means of keeping to his desired course except with the compass. When crossing an open body of water, his compass may be the only means of making an accurate landfall.

During thick weather when you can neither see nor hear the expected aids to navigation, attempting to run out the time on a given course can disrupt the pleasure of the cruise. The skipper gains little comfort in a chain of soundings that does not match those given on the chart for the expected area. Any stranding, even for a short time, can be an unnerving experience.

A pilot will not knowingly accept a cheap parachute. A good boater should not accept a bargain in lifejackets, fire extinguishers, or compass. Take the time and spend the few extra dollars to purchase a compass to fit your expected needs. Regardless of what the salesman may tell you, postpone buying until you have had the chance to check more than one make and model.

Lift each compass, tilt and turn it, simulating expected motions of the boat. The compass card should have a smooth and stable reaction.

The card of a good quality compass will come to rest without oscillations about the lubber's line. Reasonable movement in your hand, comparable to the rolling and pitching

The compass is a delicate instrument and deserves respect. It should be mounted securely and in position where it can be easily observed by the helmsman.

of the boat, should not materially affect the reading.

Installation

Proper installation of the compass does not happen by accident. Make a critical check of the proposed location to be sure compass placement will permit the helmsman to use it with comfort and accuracy. First, the compass should be placed directly in front of the helmsman and in such a position that it can be viewed without body stress as he sits or stands in a posture of relaxed alertness. The compass should be in the helmsman's zone of comfort. If the compass is too far away, he may have to bend forward to watch it; too close and he must rear backward for relief.

Do not hesitate to spend a few extra dollars for a good reliable compass. If in doubt, seek advice from fellow boaters.

1-16 SAFETY

Second, give some thought to comfort in heavy weather and poor visibilty conditions during the day and night. In some cases, the compass position may be partially determined by the location of the wheel, shift lever, and throttle handle.

Third, inspect the compass site to be sure the instrument will be at least two feet from any engine indicators, bilge vapor detectors, magnetic instruments, or any steel or iron objects. If the compass cannot be placed at least two feet (six feet would be better) from one of these influences, then either the compass or the other object must be moved, if first order accuracy is to be expected.

Once the compass location appears to be satisfactory, give the compass a test before installation. Hidden influences may be concealed under the cabin top, forward of the cabin aft bulkhead, within the cockpit ceiling, or in a wood-covered stanchion.

Move the compass around in the area of the proposed location. Keep an eye on the card. A magnetic influence is the only thing that will make the card turn. You can quickly find any such influence with the compass. If the influence can not be moved away or replaced by one of non-magnetic material, test to determine whether it is merely magnetic, a small piece of iron or steel, or some magnetized steel. Bring the north pole of the compass near the object, then shift and bring the south pole near it. Both the north and south poles will be attracted if the compass is demagnetized. If the object attracts one pole and repels the other, then the compass is magnetized. If your compass needs to be demagnetized, take it to a shop equipped to do the job **PROPERLY.**

After you have moved the compass around in the proposed mounting area, hold it down or tape it in position. Test everything you feel might affect the compass and cause a deviation from a true reading. Rotate the wheel from hard over to hard over. Switch on and off all the lights, radios, radio direction finder, radio telephone, depth finder and the shipboard intercom, if one is installed. Sound the electric whistle, turn on the windshield wipers, start the engine (with water circulating through the engine), work the throttle, and move the gear shift lever. If the boat has an auxiliary generator, start it.

If the card moves during any one of these tests, the compass should be relocated. Naturally, if something like the windshield wipers cause a slight deviation, it may be necessary for you to make a different deviation table to use only when certain pieces of equipment is operating. Bear in mind, following a course that is only off a degree or two for several hours can make considerable difference at the end, putting you on a reef, rock, or shoal.

"Innocent" objects close to the compass, such as diet coke in an aluminum can, may cause serious problems and lead to disaster, as these three photos and the accompanying text illustrate.

Check to be sure the intended compass site is solid. Vibration will increase pivot wear.

Now, you are ready to mount the compass. To prevent an error on all courses, the line through the lubber line and the compass card pivot must be exactly parallel to the keel of the boat. You can establish the fore-and-aft line of the boat with a stout cord or string. Use care to transfer this line to the compass site. If necessary, shim the base of the compass until the stile-type lubber line (the one affixed to the case and not gimbaled) is vertical when the boat is on an even keel. Drill the holes and mount the compass.

Magnetic Items After Installation

Many times an owner will install an expensive stereo system in the cabin of his boat. It is not uncommon for the speakers to be mounted on the aft bulkhead up against the overhead (ceiling). In almost every case, this position places one of the speakers in very close proximity to the compass, mounted above the ceiling.

As we all know, a magnet is used in the operation of the speaker. Therefore, it is very likely that the speaker, mounted almost under the compass in the cabin will have a very pronounced affect on the compass accuracy.

Consider the following test and the accompanying photographs as prove of the statements made.

First, the compass was read as 190 degrees while the boat was secure in her slip.

Next a full can of diet coke in an **aluminum** can was placed on one side and the compass read as 204 degrees, a good 14 degrees off.

Next, the full can was moved to the opposite side of the compass and again a reading was observed. This time as 189 degrees, 11 degrees off from the original reading.

Finally the contents of the can were consumed, the can placed on both sides of the compass with **NO** affect on the compass reading.

Two very important conclusions can be drawn from these tests.

1- Something must have been in the contents of the can to affect the compass so drastically.

2- Keep even "innocent" things clear of the compass to avoid any possible error in the boat's heading.

REMEMBER, a boat moving through the water at 10 knots on a compass error of just 5 degrees will be almost 1.5 miles off course in only **ONE** hour. At night, or in thick weather, this could very possibly put the boat on a reef, rock, or shoal, with disastrous results.

1-11 STEERING

USCG or BIA certification of a steering system means that all materials, equipment, and installation of the steering parts meet or exceed specific standards for strength, type, and maneuverability. Avoid sharp bends when routing the cable. Check to be sure the pulleys turn freely and all fittings are secure.

1-12 ANCHORS

One of the most important pieces of equipment in the boat next to the power plant is the ground tackle carried. The engine makes the boat go and the anchor and its line are what hold it in place when the boat is not secured to a dock or on the beach.

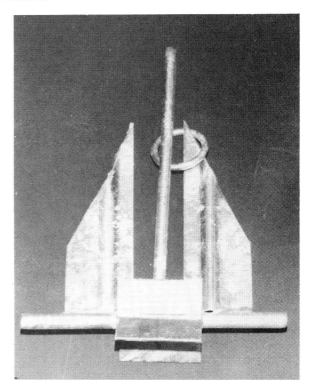

*The weight of the anchor **MUST** be adequate to secure the boat without dragging.*

1-18 SAFETY

The anchor must be of suitable size, type, and weight to give the skipper peace of mind when his boat is at anchor. Under certain conditions, a second, smaller, lighter anchor may help to keep the boat in a favorable position during a non-emergency daytime situation.

In order for the anchor to hold properly, a piece of chain must be attached to the anchor and then the nylon anchor line attached to the chain. The amount of chain should equal or exceed the length of the boat. Such a piece of chain will ensure that the anchor stock will lay in an approximate horizontal position and permit the flutes to dig into the bottom and hold.

1-13 MISCELLANEOUS EQUIPMENT

In addition to the equipment you are legally required to carry in the boat and those previously mentioned, some extra items will add to your boating pleasure and safety. Practical suggestions would include: a bailing device (bucket, pump, etc.), boat hook, fenders, spare propeller, spare engine parts, tools, an auxiliary means of propulsion (paddle or oars), spare can of gasoline, flashlight, and extra warm clothing. The area of your boating activity, weather conditions, length of stay aboard your boat, and the specific purpose will all contribute to the kind and amount of stores you put aboard. When it comes to personal gear, heed the advice of veteran boaters who say, "Decide on how little you think you can get by with, then cut it in half".

Bilge Pumps

Automatic bilge pumps should be equipped with an overriding manual switch. They should also have an indicator in the operator's position to advise the helmsman when the pump is operating. Select a pump that will stabilize its temperature within the manufacturer's specified limits when it is operated continuously. The pump motor should be a sealed or arcless type, suitable for a marine atmosphere. Place the bilge pump inlets so excess bilge water can be removed at all normal boat trims. The intakes should be properly screened to prevent the pump from sucking up debris from the bilge. Intake tubing should be of a high quality and stiff enough to resist kinking and not collapse under maximum pump suction condition if the intake becomes blocked.

To test operation of the bilge pump, operate the pump switch. If the motor does not run, disconnect the leads to the motor. Connect a voltmeter to the leads and see if voltage is indicated. If voltage is not indicated, then the problem must be in a blown fuse, defective switch, or some other area of the electrical system.

If the meter indicates voltage is present at the leads, then remove, disassemble, and inspect the bilge pump. Clean it, reassemble, connect the leads, and operate the switch again. If the motor still fails to run, the pump must be replaced.

To test the bilge pump switch, first disconnect the leads from the pump and connect them to a test light or ohmmeter. Next, hold the switch firmly against the mounting location in order to make a good ground. Now, tilt the opposite end of the switch upward until it is activated as indicated by the test light coming on or the ohmmeter showing continuity. Finally, lower the switch slowly toward the mounting

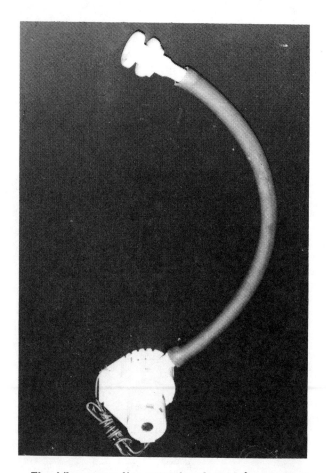

The bilge pump line must be cleaned frequently to ensure the entire bilge pump system will function properly in an emergency.

position until it is deactivated. Measure the distance between the point the switch was activated and the point it was deactivated. For proper service, the switch should deactivate between 1/2-inch and 1/4-inch from the planned mounting position. **CAUTION: The switch must never be mounted lower than the bilge pump pickup.**

1-14 BOATING ACCIDENT REPORTS

New federal and state regulations require an accident report to be filed with the nearest State boating authority within 48 hours if a person is lost, disappears, or is injured to the degree of needing medical treatment beyond first aid.

Accidents involving only property or equipment damage **MUST** be reported within 10 days, if the damage is in excess of $500.00. Some states require reporting of accidents with propery damage less then $500.00 or a total boat loss. A $1,000.00 **PENALTY** may be assessed for failure to submit the report.

WORD OF ADVICE

Take time to make a copy of the report to keep for your records or for the insurance company. Once the report is filed, the Coast Guard will not give out a copy, even to the person who filled the report.

The report must give details of the accident and include:

1- The date, time, and exact location of the occurrence.

2- The name of each person who died, was lost, or injured.

3- The number and name of the vessel.

4- The names and addresses of the owner and operator.

If the operator cannot file the report for any reason, each person on aboard **MUST** notify the authorities, or determine that the report has been filed.

1-15 NAVIGATION

Buoys

In the United States, a buoyage system is used as an assist to all boaters of all size craft to navigate our coastal waters and our navigable rivers in safety. When properly read and understood, these buoys and markers will permit the boater to cruise with comparative confidence that he will be able to avoid reefs, rocks, shoals, and other hazards.

In the spring of 1983, the Coast Guard began making modifications to U.S. aids to navigation in support of an agreement sponsored by the International Associaiton of Lighthouse Authorities (IALA) and signed by representatives from most of the maritime nations of the world. The primary purpose of the modifications is to improve safety by making buoyage systems around the world more alike and less confusing.

The modifications shown in the accompanying illustrations should be completed by the end of 1989.

Lights

The following information regarding lights required on boats between sunset and sunrise or during restricted visibility is taken directly from a U.S. Coast Guard publication dated 1984.

The terms **"PORT"** and **"STARBOARD"** are used to refer to the left and right side of the boat, when looking forward. One easy way to remember this basic fundamental is to consider the words "port" and "left" both have four letters and go together.

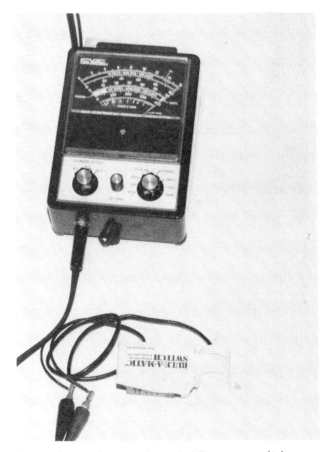

Hookup for testing an automatic bilge pump switch.

Waterway Rules

On the water, certain basic safe-operating practices must be followed. You should learn and practice them, for to **know**, is to be able to handle your boat with confidence and safety. Knowledge of what to do, and not do, will add a great deal to the enjoyment you will receive from your boating investment.

Rules of the Road

The best advice possible and a Coast Guard requirement for boats over 39' 4" (12 meters) since 1981, is to obtain an official copy of the "Rules of the Road", which includes Inland Waterways, Western Rivers, and the Great Lakes for study and ready reference.

The following two paragraphs give a **VERY** brief condensed and abbreviated -- almost a synopsis of the rules and should not be considered in any way as covering the entire subject.

Powered boats must yield the right-of-way to all boats without motors, except when being overtaken. When meeting another boat head-on, keep to starboard, unless you are too far to port to make this practical. When overtaking another boat, the right-of-way belongs to the boat being overtaken. If your boat is being passed, you must maintain course and speed.

When two boats approach at an angle and there is danger of collision, the boat to port must give way to the boat to starboard. Always keep to starboard in a narrow channel or canal. Boats underway must stay clear of vessels fishing with nets, lines, or trawls. (Fishing boats are not allowed to fish in channels or to obstruct navigation.)

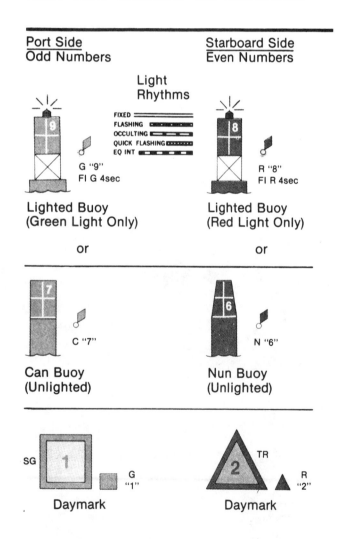

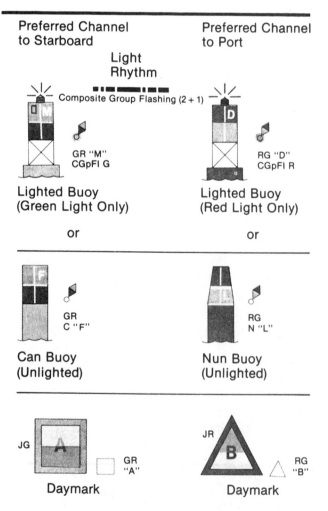

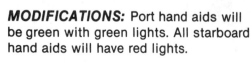

MODIFICATIONS: Port hand aids will be green with green lights. All starboard hand aids will have red lights.

MODIFICATIONS: Green will replace black. Light rhythm will be changed to Composite Gp Fl (2 + 1).

2
TUNING

2-1 INTRODUCTION

The efficiency, reliability, fuel economy and enjoyment available from engine performance are all directly dependent on having it tuned properly. The importance of performing service work in the sequence detailed in this chapter cannot be over emphasized. Before making any adjustments, check the Specifications in the Appendix. **NEVER** rely on memory when making critical adjustments.

Before beginning to tune any engine, check to be sure the engine has satisfactory compression. An engine with worn or broken piston rings, burned pistons, or badly scored cylinder walls, cannot be made to perform properly no matter how much time and expense is spent on the tune-up. Poor compression must be corrected or the tune-up will not give the desired results.

The opposite of poor compression would be to consider good compression as evidence of a satisfactory cylinder. However, this is not necessarily the case, when working on an outboard engine. As the professional mechanic has discovered, many times the compression check will indicate a satisfactory cylinder, but after the head is pulled and an inspection made, the cylinder will require service.

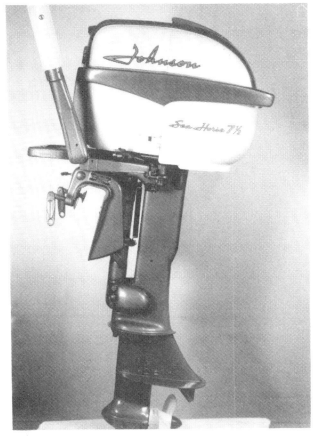

A clean exterior engine appearance reflects this owner's pride in his unit. Keeping the interior well lubricated and properly adjusted will give him the enjoyment deserved for his investment.

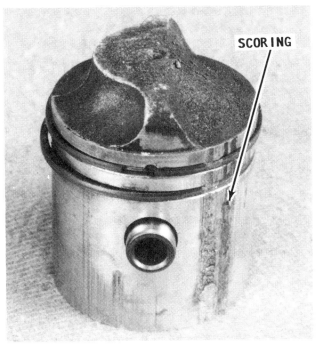

Damaged piston, probably caused by inaccurate fuel mixture, or improper point setting.

2-2 TUNING

A practical maintenance program that is followed throughout the year, is one of the best methods of ensuring the engine will give satisfactory performance at any time.

The extent of the engine tune-up is usually dependent on the time lapse since the last service. A complete tune-up of the entire engine would entail almost all of the work outlined in this manual. A logical sequence of steps will be presented in general terms. If additional information or detailed service work is required, the chapter containing the instructions will be referenced.

Each year higher compression ratios are built into modern outboard engines and the electrical systems become more complex, especially with electronic (capacitor discharge) units. Therefore, the need for reliable, authoratative, and detailed instructions becomes more critical. The information in this chapter and the referenced chapters fulfill that requirement.

2-2 TUNE-UP SEQUENCE

If twenty different mechanics were asked the question, "What constitutes a major and minor tune-up?", it is entirely possible twenty different answers would be given. As the terms are used in this manual and other Seloc outboard books, the following work is normally performed for a minor and major tune-up.

Minor Tune-up
Lubricate engine.
Drain and replace gear oil.
Adjust points.
Adjust carburetor.
Clean exterior surface of engine.
Tank test engine for fine adjustments.

Major Tune-up
Remove head.
Clean carbon from pistons and cylinders.
Clean and overhaul carburetor.
Clean and overhaul fuel pump.
Rebuild and adjust ignition system.
Lubricate engine.
Drain and replace gear oil.
Clean exterior surface of engine.
Tank test engine for fine adjustments.

During a major tune-up, a definite sequence of service work should be followed to return the engine to the maximum performance desired. This type of work should not be confused with attempting to locate problem areas of "why" the engine is not performing satisfactorily. This work is classified as "troubleshooting". In many cases, these two areas will overlap, because many times a minor or major tune-up will correct the malfunction and return the system to normal operation.

The following list is a suggested sequence of tasks to perform during the tune-up service work. The tasks are merely listed here. Generally procedures are given in subsequent sections of this chapter. For more detailed instructions, see the referenced chapter.

1- Perform a compression check of each cylinder. See Chapter 3.

2- Inspect the spark plugs to determine their condition. Test for adequate spark at the plug. See Chapter 5.

3- Start the engine in a body of water and check the water flow through the engine. See Chapter 8.

4- Check the gear oil in the lower unit. See Chapter 8.

The time, effort, and expense of a tune-up will not restore an engine to satisfactory performance, if the pistons are damaged.

A boat and lower unit covered with marine growth. Such a condition is a serious hinderance to satisfactory performance.

5- Check the carburetor adjustments and the need for an overhaul. See Chapter 4.
6- Check the fuel pump for adequate performance and delivery. See Chapter 4.
7- Make a general inspection of the ignition system. See Chapter 5.
8- Test the starter motor and the solenoid. See Chapter 6.
9- Check the internal wiring.
10- Check the synchronization. See Chapter 5.

2-3 COMPRESSION CHECK

A compression check is extremely important, because an engine with low or uneven compression between cylinders **CANNOT** be tuned to operate satisfactorily. Therefore, it is essential that any compression problem be corrected before proceeding with the tune-up procedure. See Chapter 3.

If the powerhead shows any indication of overheating, such as discolored or scorched paint, especially in the area of the top (No. 1) cylinder, inspect the cylinders visually thru the transfer ports for possible scoring. A more thorough inspection can be made if the head is removed. It is possible for a cylinder with satisfactory compression to be scored slightly. Also, check the water pump. The overheating condition may be caused by a faulty water pump.

An overheating condition may also be caused by running the engine out of the water. For unknown reasons, many operators have formed a bad habit of running a small engine without the lower unit being submerged. Such a practice will result in an overheated condition in a matter of seconds. It is interesting to note, the same operator would never operate or allow anyone else to run a large horsepower engine without water circulating through the lower unit for cooling. Bear-in-mind, the laws governing operation and damage to a large unit **ALL** apply equally as well to the small engine.

Checking Compression
Remove the spark plug wires. **ALWAYS** grasp the molded cap and pull it loose with a twisting motion to prevent damage to the connection. Remove the spark plugs and keep them in **ORDER** by cylinder for evaluation later. Ground the spark plug leads to the engine to render the ignition system inoperative while performing the compression check.

Insert a compression gauge into the No. 1, top, spark plug opening. Crank the engine with the starter, or pull on the starter cord, thru at least 4 complete strokes with the throttle at the wide-open position, or until the highest possible reading is observed on the gauge. Record the reading. Repeat the test and record the compression for each cylinder. A variation

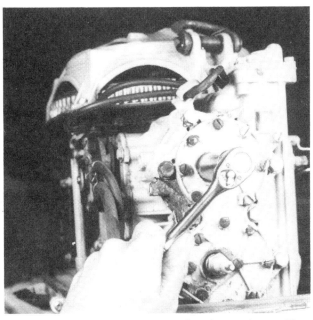

Removing the spark plugs for inspection. Worn plugs are one of the major contributing factors to poor engine performance.

A compression check should be taken in each cylinder before spending time and money on tune-up work. Without adequate compression, efforts in other areas to regain engine performance will be wasted.

Damaged spark plugs. Notice the broken electrode on the left plug. The broken part must be found and removed before returning the engine to service.

between cylinders is far more important than the actual readings. A variation of more than 5 psi between cylinders indicates the lower compression cylinder may be defective. The problem may be worn, broken, or sticking piston rings, scored pistons or worn cylinders. These problems may only be determined after the head has been removed. Removing the head on an outboard engine is not that big a deal and may save many hours of frustration and the cost of purchasing unnecessary parts to correct a faulty condition.

2-4 SPARK PLUG INSPECTION

Inspect each spark plug for badly worn electrodes, glazed, broken, blistered, or lead fouled insulators. Replace all of the plugs, if one shows signs of excessive wear.

Make an evaluation of the cylinder performance by comparing the spark condition with those shown in Chapter 5. Check each spark plug to be sure they are all of the same manufacturer and have the same heat range rating.

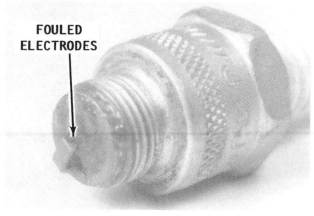

A foul spark plug. The condition of this plug indicates problems in the cylinder that should be corrected.

Inspect the threads in the spark plug opening of the head and clean the threads before installing the plug. If the threads are damaged, the head should be removed and and a Heli-coil insert installed. If an attempt is made to drill out the opening with the head in place, some of the filings may fall into the cylinder and cause damage to the cylinder wall during operation. Because the head is made of aluminum, the filings cannot be removed with a magnet.

When purchasing new spark plugs, **ALWAYS** ask the marine dealer if there has been a spark plug change for the engine being serviced.

Crank the engine through several revolutions to blow out any material which might have become dislodged during cleaning.

Install the spark plugs and tighten them to a torque value of 17 ft-lbs. **ALWAYS** use a new gasket and wipe the seats in the block clean. The gasket must be fully compressed on clean seats to complete the heat transfer process and to provide a gas tight seal in the cylinder. If the torque value is too high, the heat will dissipate too rapidly. Conversely, if the torque value is too low, heat will not dissipate fast enough.

2-5 IGNITION SYSTEM

Only one ignition system, a flywheel-magneto, is used on outboard engines covered in this manual. If the engine performance is less than expected, and the ignition is diagnosed as the problem area, refer to Chapter 5 for detailed service procedures. To properly synchronize the ignition system with the fuel system, see Chapter 5.

*Today, numerous type spark plugs are available for service. **ALWAYS** check with your local marine dealer to be sure you are purchasing the proper plugs for the engine being serviced.*

Worn ignition points are a common problem area contributing to poor engine performance.

Breaker Points

Rough or discolored contact surfaces is sufficient reason for replacement. The cam follower will usually have worn away by the time the points have become unsatisfactory for efficient service.

Check the resistance across the contacts. If the test indicates **ZERO** resistance, the points are serviceable. A slight resistance across the points will affect idle operation. A high resistance may cause the ignition system to malfunction and loss of spark. Therefore, if any resistance across the points is indicated, the point set should be replaced.

2-6 SYNCHRONIZING

The timing on small OMC (Johnson and Evinrude) outboard engines is controlled through adjustment of the points. If the points are adjusted too closely, the spark plugs will fire early; if adjusted with excessive gap, the plugs will fire too late, for efficient operation. Therefore, correct point adjustment and synchronization are essential for proper engine operation. An engine may be in apparent excellent mechanical condition, but perform poorly, unless the points and synchronization have been adjusted precisely, according to the Specifications in the Appendix. To synchronize the engine, see Chapter 5.

2-7 BATTERY SERVICE

Many owner/operators are not fully aware of the role a battery performs with a magneto ignition system outboard engine. To clarify: With a magneto ignition system, a battery is only used to crank the engine for starting purposes. Once the engine is running properly, the battery could very well be removed without affecting engine operation. Therefore, if the battery is completely dead, the engine may be hand started with a pull cord and operate efficiently.

If a battery is used for starting, inspect and service the battery, cables and connections. Check for signs of corrosion. Inspect the battery case for cracks or bulges, dirt, acid, and electrolyte leakage. Check the electrolyte level in each cell.

Fill each cell to the proper level with distilled water or water passed thru a demineralizer.

*The fuel and ignition systems on any engine **MUST** be properly synchronized before maximum performance can be obtained from the unit.*

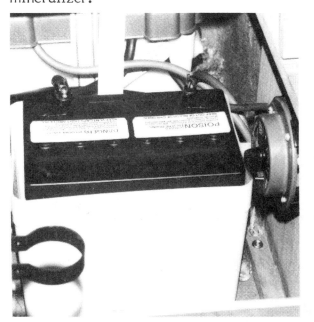

*The battery **MUST** be located near the engine in a well-ventilated area. It must be secured in such a manner that absolutely no movement is possible in any direction under the most violent action of the boat.*

2-6 TUNING

Clean the top of the battery. The top of a 12-volt battery should be kept especially clean of acid film and dirt, because of the high voltage between the battery terminals. For best results, first wash the battery with a diluted ammonia or baking soda solution to neutralize any acid present. Flush the solution off the battery with clean water. Keep the vent plugs tight to prevent the neutralizing solution or water from entering the cells.

Check to be sure the battery is fastened securely in position. The hold-down device should be tight enough to prevent any movement of the battery in the holder, but not so tight as to place a strain on the battery case.

If the battery posts or cable terminals are corroded, the cables should be cleaned separately with a baking soda solution and a wire brush. Apply a thin coating of Multipurpose Lubricant to the posts and cable clamps before making the connections. The lubricant will help to prevent corrosion.

If the battery has remained under-charged, check for high resistance in the charging circuit. If the battery appears to be using too much water, the battery may be defective, or it may be too small for the job.

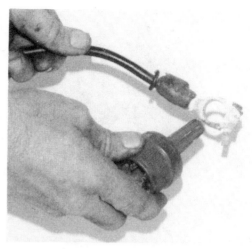

An inexpensive brush should be purchased and used to clean the battery terminals. Clean terminals will ensure a proper connection.

Jumper Cables

If booster batteries are used for starting an engine the jumper cables must be connected correctly and in the proper sequence to prevent damage to either battery, or the alternator diodes.

ALWAYS connect a cable from the positive terminals of the dead battery to the positive terminal of the good battery **FIRST**. **NEXT,** connect one end of the other cable to the negative terminals of the good battery and the other end of the **ENGINE** for a good ground. By making the ground connection on the engine, if there is an arc when you

A check of the electrolyte in the battery should be a regular task on the maintenance schedule on any boat.

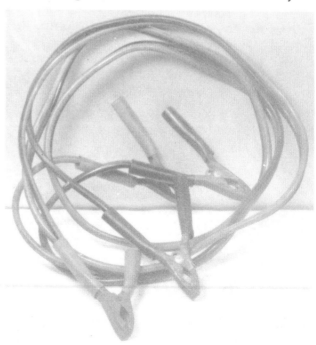

Common set of jumper cables for using a second battery to crank and start the engine. EXTREME care should be used when using a second battery, as explained in the text.

make the connection it will not be near the battery. An arc near the battery could cause an explosion, destroying the battery and causing serious personal injury.

If it is necessary to use a fast-charger on a dead battery, **ALWAYS** disconnect one of the boat cables from the battery first, to prevent burning out the diodes in the alternator.

NEVER use a fast charger as a booster to start the engine because the diodes in the generator will be **DAMAGED**.

Generator Charging

Normally a generating system is not standard equipment on the smaller horsepower engines, up to the 35 hp model. However, a generator kit may be purchased and installed on the 35 hp and 40 hp engines for battery charging while the engine is operating. A generator system is standard equipment on the 40 hp electric shift model.

When the battery is partially discharged, the ammeter should change from discharge to charge between 1500 to 1800 rpm for all models. If the battery is fully-charged, the rpm will be a little higher.

With the engine running, in gear, in the water, increase the throttle until the rpm is approximately 5200 rpm. The ammeter reading should meet the Alternator Specifications in the Appendix. With a fully-charged battery the ammeter reading will be a bit lower because of the self-regulating characteristics of the generating systems.

A 40 hp engine with a starter installed on the starbaord side and a generator on the port side. The generator, in kit form, is available from the local OMC dealer.

Pressure-type fuel tank used on many outboard installations until the late 1950's. The text explains how to update such a system with a fuel pump and modern fuel tank.

Before disconnecting the ammeter, remove the red harness lead connected to the positive battery terminal. For generating service, see Chapter 6.

2-8 CARBURETOR ADJUSTMENTS

Fuel and Fuel Tanks

Take time to check the fuel tank and all of the fuel lines, fittings, couplings, valves, flexible tank fill and vent. Turn on the fuel supply valve at the tank, if the engine is equipped with a self-contained fuel tank. If the gas was not drained at the end of the previous season, make a careful inspection for gum formation. When gasoline is allowed to stand for long periods of time, particularly in the presence of copper, gummy deposits form. This gum can clog the

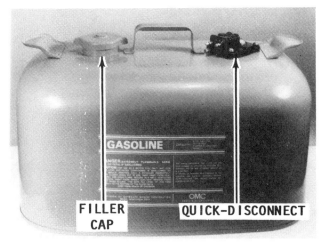

An ideal OMC fuel tank and fuel line arrangement. The tank is well secured and clean. The quick-disconnect device affords easy removal of the tank for filling.

filters, lines, and passageway in the carburetor.

If the condition of the fuel is in doubt, drain, clean, and fill the tank with fresh fuel.

Fuel pressure at the carburetor should be checked whenever a lack of fuel volume at the carburetor is suspected.

High-speed Adjustment

The high-speed needle valve is adjustable on most models covered in this manual through 1965. After 1965, the high-speed needle valve is fixed at the factory and is **NOT** adjustable. However, larger or smaller needle valves may be installed for different elevations. On all Johnson/Evinrude engines, the high-speed needle valve is the lower valve on the carburetor. The upper needle is always the idle adjustment.

A beginning "rough" adjustment for the high-speed needle valve is 3/4 turn out (counterclockwise) from the lightly seated (closed) position. **TAKE CARE** not to seat the valve firmly to prevent damage to the valve or the carburetor.

To make the high-speed adjustment:

a- Mount the engine in a test tank or body of water, preferably with a test wheel. Engines up to 40 hp may be operated in the high rpm range in a test tank without sustaining damage.

NEVER, AGAIN NEVER, operate the engine at high speed with a flush device attached. The engine, operating at high speed with such a device attached, would **RUN-A-WAY** from lack of a load on the propeller, causing extensive damage.

b- Connect a tachometer to the engine.

CAUTION: Water must circulate through the lower unit to the engine any time the engine is run to prevent damage to the water pump in the lower unit. Just five seconds without water will damage the water pump.

c- Start the engine and allow it to warm to operating temperature.

d- Shift the engine into forward gear.

e- With the engine running in forward gear, advance the throttle to the wide open position, and then very **SLOWLY** turn the high-speed needle valve inward (**CLOCKWISE**) until the engine begins to loose rpm. Now, **SLOWLY** rotate the needle valve outward (**COUNTERCLOCKWISE**) until the engine peaks out at the highest rpm.

If the high-speed needle valve adjustment is too lean, the low-speed adjustment will be affected. Under certain conditions it may be necessary to adjust the high-speed needle valve just a bit richer in order to obtain a satisfactory idle adjustment.

After the high-speed needle adjustment has been obtained, proceed with the idle adjustment as outlined in the next paragraphs.

Idle Adjustment

Due to local conditions, it may be necessary to adjust the carburetor while the engine is running in a test tank or with the boat in a body of water. For maximum performance, the idle mixture and the idle rpm should be adjusted under actual operating conditions.

Set the idle mixture screw at the specified number of turns open from a lightly seated position. In most cases this is from 1 to 1½ turns open from close.

A 35 hp unit showing location of the idle and high-speed adjustments. Notice the pressure-type fuel connector.

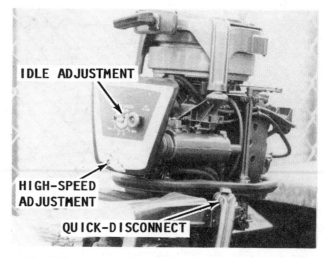

Small horsepower engine mounted in a test tank with the low- and high-speed adjustments indicated.

Start the engine and allow it to warm to operating temperature.

CAUTION: Water must circulate through the lower unit to the engine any time the engine is run to prevent damage to the water pump in the lower unit. Just five seconds without water will damage the water pump.

NEVER, AGAIN NEVER, operate the engine at high speed with a flush device attached. The engine, operating at high speed with such a device attached, would **RUN-A-WAY** from lack of a load on the propeller, causing extensive damage.

With the engine running in forward gear, slowly turn the idle mixture screw **COUNTERCLOCKWISE** until the affected cylinders start to load up or fire unevenly, due to an over-rich mixture. Slowly turn the idle mixture screw **CLOCKWISE** until the cylinders fire evenly and engine rpm increases. Continue to slowly turn the screw **CLOCKWISE** until too lean a mixture is obtained and the rpms fall off and the engine begins to misfire. Now, set the idle mixture screw one-quarter (1/4) turn out (counterclockwise) from the lean-out position. This adjustment will result in an approximate true setting. A too-lean setting is a major cause of hard starting a cold engine. It is better to have the adjustment on the rich side rather than on the lean side. Stating it another way, do not make the adjustment any leaner than necessary to obtain a smooth idle.

If the engine hesitates during acceleration after adjusting the idle mixture, the mixture is too lean. Enrich the mixture slightly, by turning the adjustment screw outward until the engine accelerates correctly.

With the engine running in forward gear, rotate the nylon idle adjustment screw, located on the portside of the powerhead until the powerhead idles at the recommended rpm. This idle adjustment screw is always exposed on the outside of the shroud.

Repairs and Adjustments

For detailed procedures to disassemble, clean, assemble, and adjust the carburetor, see the appropriate section in Chapter 4 for the carburetor type on the engine being serviced.

2-9 FUEL PUMPS

Many times a defective fuel pump diaphragm is mistakenly diagnosed as a problem in the ignition system. The most common problem is a tiny pin-hole in the diaphragm. Such a small hole will permit gas

A 3.0 hp engine with the low- and high-speed adjustments clearly visible.

A larger hp engine with the idle return stop screw indicated. This stop screw is adjustable and prevents the throttle from returning so far as to shut down the engine.

2-10 TUNING

Two of the many types of fuel pumps installed on OMC outboard engines. These fuel pumps cannot be rebuilt, as explained in the text.

to enter the crankcase and will foul the spark plug at idle-speed. During high-speed operation, gas quantity is limited, the plug is not fouled and will therefore fire in a satisfactory manner.

If the fuel pump fails to perform properly, an insufficient fuel supply will be delivered to the carburetor. This lack of fuel will cause the engine to run lean, lose rpm or cause piston scoring.

When a fuel pressure gauge is added to the system, it should be installed at the end of the fuel line leading to the upper carburetor. To ensure maximum performance, the fuel pressure must be 2 psi or more at full throttle.

This fuel pump is the only OMC pump that can be rebuilt. Replacement parts are available from the local OMC dealer.

Tune-up Task

Most fuel pumps are equipped with a fuel filter. The filter may be cleaned by first removing the cap, then the filter element, cleaning the parts and drying them with compressed air, and finally installing them in their original position.

A fuel pump pressure test should be made any time the engine fails to perform satisfactorily at high speed.

NEVER use liquid Neoprene on fuel line fittings. Always use Permatex when making fuel line connections. Permatex is available at almost all marine and hardware stores.

Only one Johnson/Evinrude fuel pump may be rebuilt, see accompanying illustration. All others pumps must be replace as a unit. For fuel pump service, see Chapter 4.

2-10 STARTER AND SOLENOID

Starter Motor Test

Check to be sure the battery has a 70-ampere rating and is fully charged. Would you believe, many starter motors are needlessly disassembled, when the battery is actually the culprit.

Lubricate the pinion gear and screw shaft with No. 10 oil.

Connect one lead of a voltmeter to the positive terminal of the starter motor.

Commercial additives, such as Sta-bil, may be used to keep the gasoline in the fuel tank fresh. Under favorable conditions, such additives will prevent the fuel from "souring" for up to twelve months.

INTERNAL WIRING 2-11

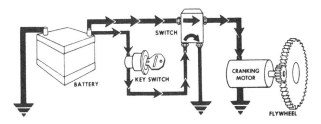

Functional diagram of a typical cranking circuit.

Connect the other meter lead to a good ground on the engine. Check the battery voltage under load by turning the ignition switch to the **START** position and observing the voltmeter reading.

If the reading is 9-1/2 volts or greater, and the starter motor fails to operate, repair or replace the starter motor. See Chapter 6.

Solenoid Test

An ohmmeter is the only instrument required to effectively test a solenoid. Test the ohmmeter by connecting the red and black leads together. Adjust the pointer to the right side of the scale.

On all Johnson/Evinrude engines, the case of the solenoid does **NOT** provide a suitable ground to the engine. Hundreds of solenoids have been discarded because of the erroneous belief the case is providing a ground and the unit should function when 12-volts is applied. Not so! One terminal of the solenoid is connected to a 12-volt source. The other terminal is connected via a white wire to a cutout switch on top of the engine. This cutout switch provides a safety to break the ground to the solenoid in the event the engine starts at a high rpm. Therefore, the solenoid ground is made and broken by the cutout switch.

NEVER connect the battery leads to the large terminals of the solenoid, or the test meter will be damaged. Connect each lead of the test meter to each of the large terminals on the solenoid.

Using battery jumper leads, connect the positive lead from the positive terminal of the battery to the the small **"S"** terminal of the solenoid. Connect the negative lead to the negative battery terminal and the **"I"** terminal of the solenoid. If the meter pointer hand moves into the **OK** block, the solenoid is serviceable. If the pointer fails to reach the **OK** block, the solenoid must be replaced.

2-11 INTERNAL WIRING HARNESS

An internal wiring harness is only used on the larger horsepower engines covered in this manual. If the engine is equipped with a wiring harness, the following checks and test will apply.

Check the internal wiring harness if problems have been encountered with any of

Typical Johnson/Evinrude starter motor installed.

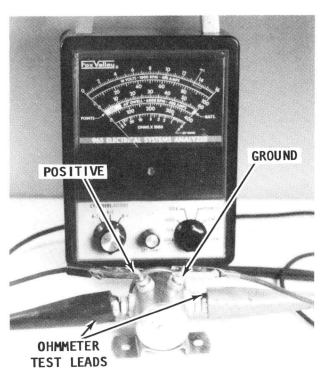

Proper hook-up of a voltmeter in preparation to testing a starter solenoid.

2-12 TUNING

the electrical components. Check for frayed or chafed insulation and/or loose connections between wires and terminal connections.

Check the harness connector for signs of corrosion. Inspect the electrical "prongs" to be sure they are not bent or broken. If the harness shows any evidence of the foregoing problems, the problem must be corrected before proceeding with any harness testing.

Verify that the "prongs" of the harness connector are clean and free of corrosion. Convince yourself that a good electrical connection is being made between the harness connector and the remote control harness.

Short Test (See the Wiring Diagram in the Appendix)

Disconnect the internal wiring harness from the electrical components. Use a magneto analyzer, set on Scale No. 3 and check for continuity between any of the wires in the harness. Use Scale No. 3 and check for continuity between any wire and a good ground. If continuity exists, the harness **MUST** be repaired or replaced.

Resistance Test (See the Wiring Diagram in the Appendix.)

Use a magneto analyzer, set on Scale No. 2. Clip the small red and black leads together. Turn the meter adjustment knob for Scale No. 2 until the meter pointer aligns with the set position on the left side of the **"OK"** block on Scale No. 2. Separate the small red and black leads. Use the Wiring Diagram in the Appendix, and check each wire for resistance between the harness connection and the terminal ends. If resistance exists (meter reading outside the **"OK"** block) the harness **MUST** be repaired or replaced.

2-12 WATER PUMP CHECK

FIRST A GOOD WORD: The water pump **MUST** be in very good condition for the engine to deliver satisfactory service. The pump performs an extremely important function by supplying enough water to properly cool the engine. Therefore, in most cases, it is advisable to replace the complete water pump assembly at least once a year, or anytime the lower unit is disassembled for service.

The terminals of a side-mounted electrical connector should be inspected and cleaned each season. This connector is exposed and vulnerable to dampness and corrosion.

Small horsepower engine operating in a test tank to check water circulation through the engine and discharge from the idle relief.

Using a flush attachment with a garden hose hook-up to clean the engine water circulation system with fresh water. This arrangement may also be used while operating the engine at idle speeds to make adjustments.

Sometimes during adjustment procedures, it is necessary to run the engine with a flush device attached to the lower unit. **NEVER** operate the engine over 1000 rpm with a flush device attached, because the engine may **"RUN-A-WAY"** due to the no-load condition on the propeller. A "run-a-way" engine could be severely damaged. As the name implies, the flush device is primarily used to flush the engine after use in salt water or contaminated fresh water. Regular use of the flush device will prevent salt or silt deposits from accumulating in the water passageway. During and immediately after flushing, keep the motor in an upright position until all of the water has drained from the drive shaft housing. This will prevent water from entering the power head by way of the drive shaft housing and the exhaust ports, during the flush. It will also prevent residual water from being trapped in the drive shaft housing and other passageways.

To test the water pump, the lower unit **MUST** be placed in a test tank or the boat moved into a body of water. The pump must now work to supply a volume to the engine.

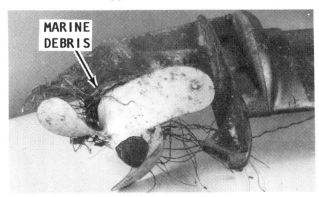

Small horsepower engine showing debris, including fish line, wrapped around the propeller shaft. Entangled fish line can cut through the oil seal behind the propeller causing loss of lubricant from the lower unit.

Water pump installed on a small horsepower lower unit.

Lack of adequate water supply from the water pump thru the engine will cause any number of power head failures, such as stuck rings, scored cylinder walls, burned pistons, etc.

2-13 PROPELLER

Check the propeller blades for nicks, cracks, or bent condition. If the propeller is damaged, the local marine dealer can make repairs or send it out to a shop specializing in such work.

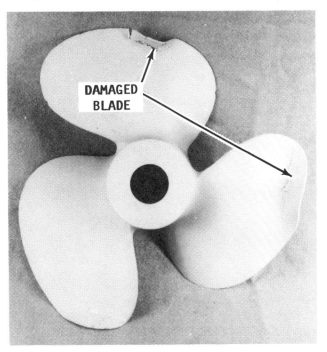

Example of a damaged propeller. This unit should have been replaced long before this amount of damage was sustained.

2-14 TUNING

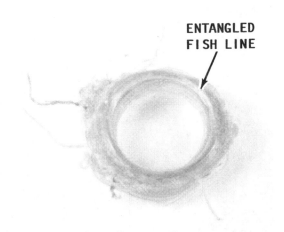

Considerable amount of fish line entangled around the propeller shaft. Some of the fish line actually melted, giving it the appearance of a washer.

Remove the cotter key, propeller nut, shear pin, and the propeller from the shaft. Check the propeller shaft seal to be sure it is not leaking. Check the area just forward of the seal to be sure a fish line is not wrapped around the shaft.

Operation At Recommended RPM

Check with the local OMC dealer, or a propeller shop for the recommended size and pitch for a particular size engine, boat, and intended operation. The correct propeller should be installed on the engine to enable operation at recommended rpm.

Two rpm ranges are usually given. The lower rpm is recommended for large, heavy slow boats, or for commercial applications. The higher rpm is recommended for light, fast boats. The wide rpm range will result in greater satisfaction because of maximum performance and greater fuel economy. If the engine speed is above the recommended rpm, try a higher pitch propeller or the same pitch cupped. See Chapter 1 for explanation of propeller terms, pitch, diameter, cupped, etc.

For a dual engine installation, the next higher pitch propeller may prove the most satisfactory condition for water skiing.

2-14 LOWER UNIT

NEVER remove the vent or filler plugs when the lower unit is hot. Expanded lubricant would be released through the plug hole. Check the lubricant level after the unit has been allowed to cool. Add only OMC approved gear lubricant. **NEVER** use regular automotive-type grease in the lower unit, because it expands and foams too

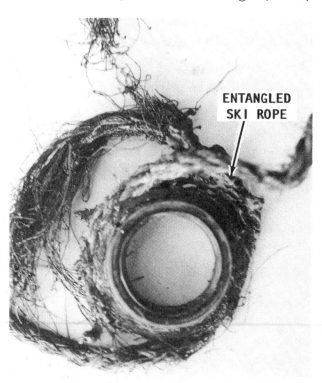

Ski rope entangled around the propeller shaft. Some of the rope has actually melted and fused together. In this case, the rope cut through the lower unit oil seal, allowing lubricant to escape.

New propeller ready for installation. On the smaller horsepower engines, a damaged propeller can cause excessive vibration, resulting in failure of more expensive parts. The modest cost of a new propeller is therefore justified. Rebuilding a small propeller is not economical when balanced against the minor difference of purchasing a new unit.

much. Outboard lower units do not have provisions to accommodate such expansion.

If the lubricant appears milky brown, indicating the presence of water, a check should be made to determine how the water entered. If large amounts of lubricant must be added to bring the lubricant up to the full mark, a thorough inspection should be made to find the cause of the lubricant loss.

Draining Lower Unit

The fill/drain plug on Johnson/Evinrude lower units may be located towards the bottom of the unit on the port side, starboard side, or on the leading edge of the lower unit. On many models a Phillips screw will be found very close to the fill/drain plug. **NEVER** remove this Phillips screw because the lower unit would then have to be disassembled in order to return the cradle for the shift dog back in place.

Remove the drain plug and then remove the vent plug located just above the anti-cavitation plate.

Filling Lower Unit

Position the drive unit approximately vertical and without a list to either port or starboard. Insert the lubricant tube into the **FILL/DRAIN** hole at the bottom plug hole, and inject lubricant until the excess begins to come out the **VENT** hole. Install the **VENT** plug first then replace the **FILL** plug with **NEW** gaskets. Check to be sure the gaskets are properly positioned to prevent water from entering the housing. Many times some of the gear lubricant is lost during installation of the plugs. Therefore, if the vent plug is removed again, and more lubricant added very **SLOWLY** using a small-spout oil can to allow air to pass out the opening, there is no doubt but what the unit will be filled to capacity.

For detailed lower unit service procedures, see Chapter 8. For lower unit lubrication capacities, see the Appendix.

Repairs and Adjustments

For detailed procedures to disassemble, clean, assemble, and adjust the carburetor, see the appropriate section in Chapter 4 for the carburetor type on the engine being serviced.

2-15 BOAT TESTING

Hook and Rocker

Before testing the boat, check the boat

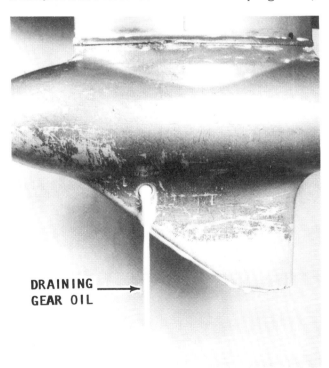

The gear oil in the lower unit should be checked on a daily basis during the season of operation. The oil should be drained and replenished with new oil every 100 hours of operation.

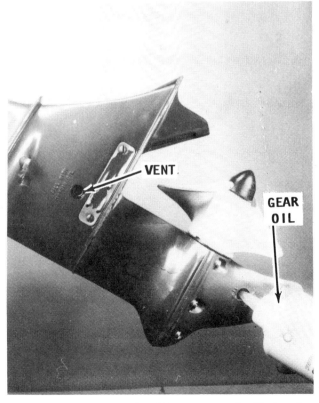

Filling the lower unit with new gear oil. Notice the unit is filled through the lower plug, but the upper plug MUST be removed to allow trapped air to escape.

bottom carefully for marine growth or evidence of a "hook" or a "rocker" in the bottom. Either one of these conditions will greatly reduce performance.

Performance

Mount the motor on the boat. Install the remote control cables and check for proper adjustment.

Make an effort to test the boat with what might be considered an average gross load. The boat should ride on an even keel, without a list to port or starboard. Adjust the motor tilt angle, if necessary, to permit the boat to ride slightly higher than the stern. If heavy supplies are stowed aft of the center, the bow will be light and the boat will "plane" more efficiently. For this test the boat must be operated in a body of water.

Check the engine rpm at full throttle. The rpm should be within the Specifications in the Appendix. All OMC engine model serial number identification plates indicate the horsepower rating and rpm range for the engine. If the rpm is not within specified range, a propeller change may be in order. A higher pitch propeller will decrease rpm, and a lower pitch propeller will increase rpm.

For maximum low speed engine performance, the idle mixture and the idle rpm should be readjusted under actual operating conditions.

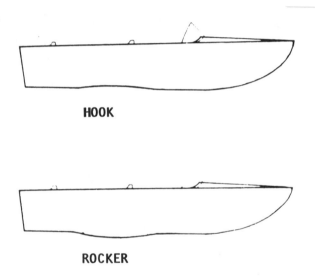

Boat performance will be drastically hampered, if the bottom is damaged.

Maximum engine performance can only be obtained through proper tuning using a tachometer.

3
POWERHEAD

3-1 INTRODUCTION

The carburetion and ignition principles of two-cycle engine operation **MUST** be understood in order to perform a proper tune-up on an outboard motor. Therefore, it would be well worth the time to study the principles of two-cycle engines as outlined in this section.

A Polaroid, or equivalent instant-type camera, is an extremely useful item providing the means of accurately recording the arrangement of parts and wire connections **BEFORE** the disassembly work begins. Such a record is most valuable during the assembly work.

Tags are handy to identify wires after they are disconnected to ensure they will be connected to the same terminal from which they were removed. These tags may also be used for parts where marks or other means of identification are not possible.

THEORY OF OPERATION

The two-cycle engine differs in several ways from a conventional four-cycle (automobile) engine.

1- The method by which the fuel-air mixture is delivered to the combustion chamber.
2- The complete lubrication system.
3- In most cases, the ignition system.
4- The frequency of the power stroke.

These differences will be discussed briefly and compared with four-cycle engine operation.

Intake/Exhaust

Two-cycle engines utilize an arrangement of port openings to admit fuel to the combustion chamber and to purge the exhaust gases after burning has been completed. The ports are located in a precise pattern in order for them to be open and closed off at an exact moment by the piston as it moves up and down in the cylinder. The exhaust port is located slightly higher than the fuel intake port. This arrangement opens the exhaust port first, as the piston starts downward, and therefore, the exhaust phase begins a fraction of a second before the intake phase.

Actually, the intake and exhaust ports are spaced so closely together that both open almost simultaneously. For this reason, the pistons of most two-cycle engines have a deflector-type top. This design of the piston top serves two purposes very effectively.

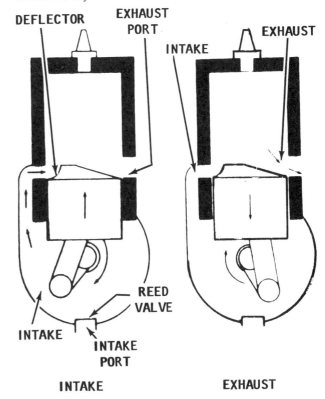

Drawing to depict the intake and exhaust cycles of a two-cycle engine.

3-2 POWERHEAD

First, it creates turbulence when the incoming charge of fuel enters the combustion chamber. This turbulence results in more complete burning of the fuel than if the piston top were flat. The second effect of the deflector-type piston crown is to force the exhaust gases from the cylinder more rapidly.

This system of intake and exhaust is in marked contrast to individual valve arrangement employed on four-cycle engines.

Lubrication

A two-cycle engine is lubricated by mixing oil with the fuel. Therefore, various parts are lubricated as the fuel mixture passes through the crankcase and the cylinder. Four-cycle engines have a crankcase containing oil. This oil is pumped through a circulating system and returned to the crankcase to begin the routing again.

Power Stroke

The combustion cycle of a two-cycle engine has four distinct phases.

1- Intake
2- Compression
3- Power
4- Exhaust

Three phases of the cycle are accomplished with each stroke of the piston, and the fourth phase, the power stroke occurs with each revolution of the crankshaft. Compare this system with a four-cycle engine. A stroke of the piston is required to accomplish each phase of the cycle and the power stroke occurs on every other revolution of the crankshaft. Stated another way, two revolutions of the four-cycle engine crankshaft are required to complete one full cycle, the four phases.

Physical Laws

The two-cycle engine is able to function because of two very simple physical laws.

One: Gases will flow from an area of high pressure to an area of lower pressure. A tire blowout is an example of this principle. The high-pressure air escapes rapidly if the tube is punctured.

Two: If a gas is compressed into a smaller area, the pressure increases, and if a gas expands into a larger area, the pressure is decreased.

If these two laws are kept in mind, the operation of the two-cycle engine will be easier understood.

Actual Operation

Beginning with the piston approaching top dead center on the compression stroke: The intake and exhaust ports are closed by the piston; the reed valve is open; the spark plug fires; the compressed fuel-air mixture

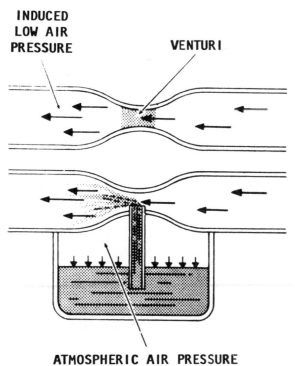

Air flow principle for a modern carburetor.

Adding OMC approved oil into the fuel tank.

is ignited; and the power stroke begins. The reed valve was open because as the piston moved upward, the crankcase volume increased, which reduced the crankcase pressure to less than the outside atmosphere.

As the piston moves downward on the power stroke, the combustion chamber is filled with burning gases. As the exhaust port is uncovered, the gases, which are under great pressure, escape rapidly through the exhaust ports. The piston continues its downward movement. Pressure within the crankcase increases, closing the reed valves against their seats.

The crankcase then becomes a sealed chamber. The fuel mixture is compressed ready for delivery the combustion chamber. As the piston continues to move downward the intake port is uncovered. Fresh fuel rushes through the intake port into the combustion chamber striking the top of the piston where it is deflected along the cylinder wall. The valve remains closed until the piston moves upward again.

When the piston begins to move upward on the compression stroke, the reed valve opens because the crankcase volume has been increased, reducing crankcase pressure to less than the outside atmosphere. The intake and exhaust ports are closed and the fresh fuel charge is compressed inside the combustion chamber.

Pressure in the crankcase decreases as the piston moves upward and a fresh charge of air flows through the carburetor picking up fuel. As the piston approaches top dead center, the spark plug ignites the air-fuel mixture, the power stroke begins and one complete cycle is completed.

Cross Fuel Flow Principle

OMC pistons are a deflector dome type. The design is necessary to deflect the fuel charge up and around the combustion chamber. The fresh fuel mixture enters the combustion chamber through the intake ports and flows across the top of the piston. The piston design contributes to clearing the combustion chamber, because the incoming fuel pushes the burned gases out the exhaust ports.

Loop Scavenging

Some powerheads have what is commonly known as a loop scavenging system. The piston dome is relatively flat on top with just a small amount of crown. Pressurized fuel in the crankcase is forced up through the skirt of the piston and out through irregular shaped openings cut in the skirt. After the fuel is forced out the piston skirt openings it is transferred upward through long deep grooves molded in the cylinder walls. The fuel then enters the com-

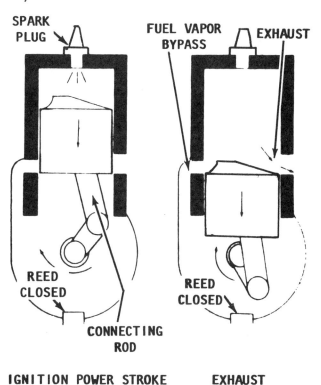

Complete piston cycle of a two-cycle engine, depicting intake, power, and exhaust.

bustion portion of the cylinder and is compressed, as the piston moves upward.

This particular powerhead does not have intake cover plates, because the intake passage is molded into the cylinder wall as described in the previous paragraph. Therefore, if these powerheads and being serviced, disregard the sections covering intake cover plates.

3-2 CHAPTER ORGANIZATION

This chapter is divided into 14 main service sections. Each section covers a particular area of service and outlines complete instructions for the work to be performed. Because of the many countless number of outboard units in the field, it would be impractical and almost impossible to give detailed procedures for removal and installation of each bolt, carburetor, starter, and other "buildup" type units.

Therefore, the sections, for the particular powerhead work to be performed, begin with the preliminary access tasks completed. As an example, disassembly of the powerhead begins with the necessary hood, cowling, and accessories removed.

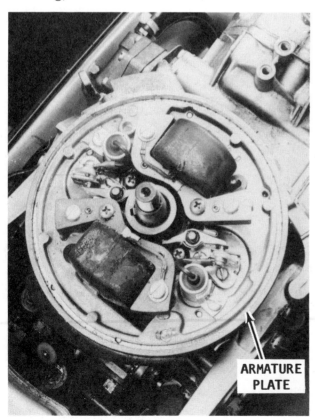

Armature plate prior to powerhead disassembling.

The information is presented in a logical sequence for complete powerhead overhaul. The instructions can be followed generally for almost any size horsepower engine. In rare cases, where the procedures differ depending on the model being serviced, separate steps are included. One example is the three different type of crankshaft installations.

The illustrations accompanying the text are from different size units and the captions clearly identify which model is covered.

Exploded drawings, showing principle parts, for the various size powerheads are included at the end of the chapter.

Special tools may be called out in certain instances. These tools may be purchased from the local Johnson/Evinrude dealer or directly from Customer Services Department, Outboard Marine Corporation (OMC), Waukegan, Illinois, 60085.

The chapter ends with Break-in Procedures, Section 3-16, to be performed after the powerhead has been assembled, all accessories installed, and the powerhead mounted on the exhaust housing.

Torque Values

All torque values must be met when they are specified. Many of the outboard castings and other parts are made of aluminum. The torque values are given to prevent stretching the bolts, but more importantly to protect the threads in the aluminum. It is extremely important to tighten the connecting rods to the proper torque value to ensure proper service. The head bolts are probably the next most important torque value.

Powerhead Components

Service procedures for the carburetors, fuel pumps, starter, and other powerhead components are given in their respective chapters of this manual. See the Table of Contents.

Reed Installation

All reeds on Johnson/Evinrude engines covered in this manual are installed just behind the carburetor behind the intake manifold.

Cleanliness

Make a determined effort to keep parts and the work area as clean as possible.

Parts **MUST** be cleaned and thoroughly inspected before they are assembled, installed, or adjusted. Use proper lubricants, or their equivalent, whenever they are recommended.

Keep rods and rod caps together as a set to ensure they will be installed as a pair and in the proper sequence.

Needle bearings **MUST** remain as a complete set. **NEVER** mix needles from one set with another. If only one needle is damaged, the complete set **MUST** be replaced.

3-3 POWERHEAD DISASSEMBLING

Preliminary Work

Before the powerhead can be disassembled, the battery must be disconnected; fuel lines disconnected; and the carburetor, generator, starter, flywheel, and magneto, all removed. If in doubt as to how these items are to be removed, refer to the appropriate chapter.

After the accessories have been removed, remove the bolts in the front and rear of the powerhead securing the powerhead to the exhaust housing. Lift the powerhead free.

BAD NEWS

If the unit is several years old, or if it has been operated in salt water, or has not had proper maintenance, or shelter, or any number of other factors, then separating the powerhead from the exhaust housing may not be a simple task. An air hammer may be required on the studs to shake the corrosion loose; heat may have to be applied to

Cylinder block water passages corroded preventing proper circulation of coolant water.

the casting to expand it slightly; or other devices employed in order to remove the powerhead. One very serious condition would be the driveshaft "frozen" with the crankshaft. In this case, a circular plug-type hole must be drilled and a torch used to cut the driveshaft. Let's assume the powerhead will come free on the first attempt.

The following procedures pickup the work after these preliminary tasks have been completed.

3-4 HEAD SERVICE

Usually the head is removed and an examination of the cylinders made to determine the extent of overhaul required. However, if the head has not been removed, back out all of the head bolts and lift the head free of the powerhead.

Removing the head from the cylinder block.

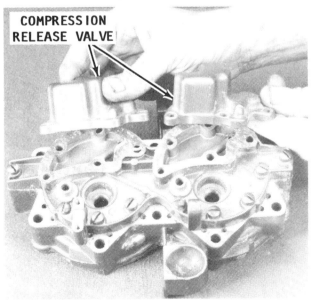

A compression release chamber being removed from the head on an early model 35 hp engine.

Many, but not all, heads have a thermostat installed. In addition to the thermostat, the engine may have a thermostat bypass valve. These two items are easily removed, inspected and cleaned.

Normally, if a thermostat is not functioning properly, it is almost always stuck in the open position. An engine operating at too low a temperature is almost as much a problem as an engine running too hot.

Therefore, during a major overhaul, good shop practice dictates to replace the thermostat and eliminate this area as a possible problem at a later date.

Lay a piece of fine sandpaper or emory paper on a flat surface (such as a piece of glass) with the abrasive side facing up. With the machined face of the head on the sandpaper, move the head in a circular motion to dress the surface. This procedure will also indicate any "high" or "low" spots.

Check the spark plug opening/s to be sure the threads are not damaged. Most marine dealers can insert a heli-coil into a spark plug opening if the threads have been damaged.

On many engines, a sending unit is installed in the head to warn the operator if the engine begins to run too hot. The light on the dash can be checked by turning the ignition switch to the **ON** position, and then ground the wire to the sending unit. The light should come on. If it does not, replace the bulb and repeat the test.

3-5 REED SERVICE

DESCRIPTION

All two-cycle engines have individual ignition and fuel delivery for each cylinder. This means the cylinder is operating independently of the others. The cylinder consists of a top seal, the cylinder, the center seal, and the lower seal. This means each cylinder is completely sealed off from the others.

Therefore, with a two-cylinder powerhead, two sets of reeds are installed, one for each cylinder. These reeds may be installed on a reed plate or with a reed box, depending on the model engine. One carburetor provides fuel to both sets of reeds.

The reed arrangement operates in much the same manner as the reed in a saxophone

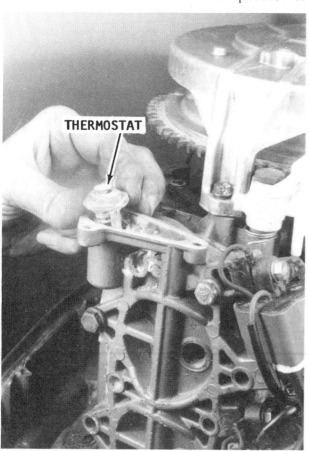

Removing the thermostat from the head.

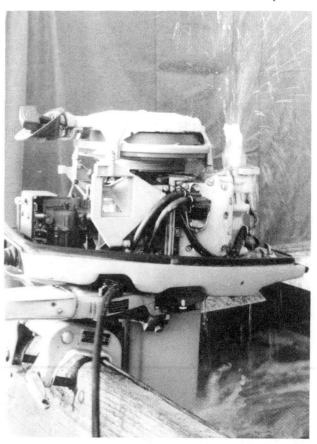

Operating an engine without the thermostat and thermostat cover installed to check the coolant water flow.

or other wind instrument. At rest, the reed is closed and seals the opening to which it is attached. In the case of an outboard engine, this opening is between the crankcase and the carburetor. The reeds are mounted in the intake manifold, just behind the carburetor.

Actual Operation

The piston creates vacuum and pressure as it moves up and down in the cylinder. As the piston moves upward, a vacuum is created in the crankcase pulling the reed open. On the compression stroke, when the piston moves downward, the reed is forced closed.

Reed Designs

A wide range of reeds, reed plates, and reed box installations may be found on an outboard unit, due to the varying designs of the engines. All installations employ the same principle and there is no difference in their operation.

Broken Reed

A broken reed is usually caused by metal fatigue over a long period of time. The failure may also be due to the reed flexing too far because the reed stop has not been adjusted properly or the stop has become distorted. If the reed is broken, the loose piece **MUST** be located and removed, before the engine is returned to service. The piece of reed may have found its way into the crankcase, behind the bypass cover. If the

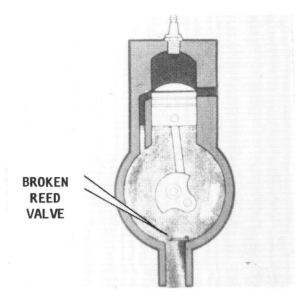

Cross-section of a cylinder to illustrate a broken reed in the crankcase.

broken piece cannot be located, the powerhead must be completely disassembled until it is located and removed.

The accompanying illustration depicts how a broken reed will cause a backflow through the carburetor.

An excellent check for a broken reed on an operating engine is to hold an ordinary business card in front of the carburetor. Under normal operating conditions, a very small amount of fine mist will be noticeable, but if fuel begins to appear rapidly on the card from the carburetor, one of the reeds is broken and causing the backflow through the carburetor and onto the card.

A broken reed will cause the engine to operate roughly and with a "pop" back through the carburetor.

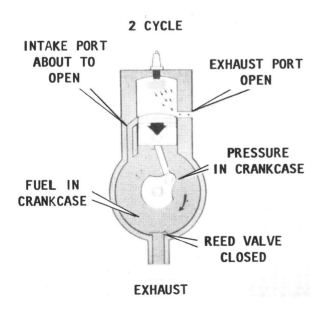

Diagram to illustrate operation of a two-cycle outboard engine.

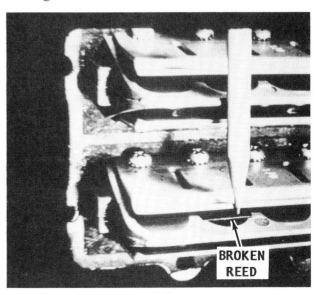

Reed box with a broken reed.

3-8 POWERHEAD

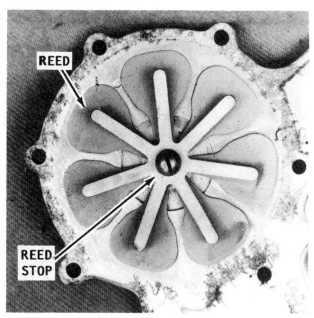

Reed stops centered over the reeds.

Reed Stops

If the reed stops have become distorted, the most effective corrective action is to replace the stop instead of making an attempt at adjustment.

Reed to Base Plate Check

The specified clearance of the reed from the base plate, when the reed is at rest, is 0.010" (0.254 mm) at the tip of the reed.

An alternate method of the checking the reed clearance is to hold the reed up to the sunlight and look through the back side. Some air space should be visible, but not a great amount. If in doubt, check the reed at the tip with a feeler gauge. The maximum clearance should not exceed 0.010" (0.254 mm).

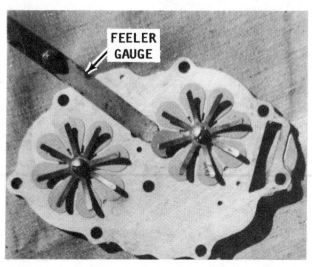

Using a feeler gauge to measure the clearance between the reed tip and the reed plate.

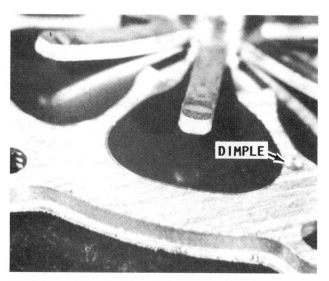

Close view showing the dimple on the reed plate. The reed leaves must straddle the dimple and be centered over the openings for proper operation.

The reeds must **NEVER** be turned over in an attempt to correct a problem. Such action would cause the reed to flex in the opposite direction and the reed would break in a very short time.

REED VALVE ADJUSTMENT

In many instances, the reed is placed on the reed plate in such a manner to cover the openings in the plate. As shown in the accompanying illustration, a small indent is manufactured into the face of the plate. The leaves of the reed should be centered on this identation for proper operation. If the reed is being replaced, both reeds **AND** the reed stops should be replaced as a set.

V-Type Reed Boxes

As the name implies, these reed boxes are shaped in a "V" with a set of reeds and

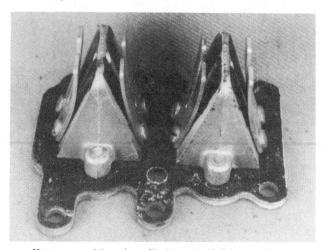

V-type reed box installed on the 9.5 hp engines.

REED SERVICE 3-9

Front view of the reed box showing the Phillips screws that must be removed before the box can be removed from the plate.

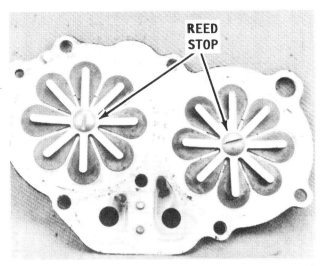

Back side of the same plate shown at the bottom of the previous column. Notice how the reed stops are centered over the reeds.

stops on both arms of the "V". If a problem develops with this type reed box, it is strongly recommended that the complete assembly be replaced -- reeds, box, and stops. The assembly may be purchased as a complete unit and the cost will usually not exceed the time, effort, and problems encountered in an attempt to replace only one part.

CLEANING AND SERVICE

Always handle the reeds with the utmost care. Rough treatment will result in the reeds becoming distorted and will affect their performance.

Wash the reeds in solvent, and blow them dry with compressed air from the **BACK SIDE ONLY**. Do not blow air through the reed from the front side. Such action would cause the reed to open and fly up against the reed stop. Wipe the front of the reed dry with a lint free cloth.

Clean the base plate thoroughly by removing any old gasket material.

Secure the reed blocks together with screws and nuts tightened to the torque value given in the Appendix.

Check for chipped or broken reeds. Observe that the reeds are not preloaded or standing open. Satisfactory reeds will not adhere to the reed block surface, but still there is not more than 0.010" (0.254 mm) clearance between the reed and the block surface.

DO NOT remove the reeds, unless they are to be replaced. **ALWAYS** replace reeds in sets. **NEVER** turn used reeds over to be used a second time.

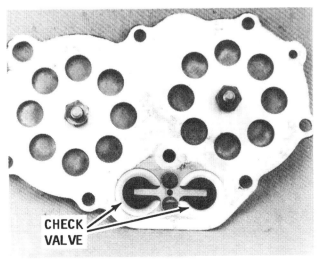

Front view of a reed plate with check valves. This arrangement is only used with a pressure fuel tank.

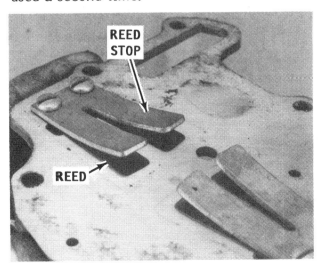

Reed stop installation with flat bars acting as the reed stop. The bars are the same width as the reed.

3-10 POWERHEAD

Check the reed location over the reed block, or plate openings to be sure the reed is centered.

The reed assemblies are then ready for installation.

Small Engines

Disassemble the reed block by first removing the screws securing the reed stops and reeds to the reed block, and then lifting the reed stops and reeds from the block.

Clean the gasket surfaces of the reed block or plate. Check the surfaces for deep grooves, cracks, or any distortion that could cause leakage. Replace the reed block or plate if it is damaged.

After new reeds have been installed, and the reed stop and attaching screws have been tightened to the required torque value, check the new reeds as outlined in the following paragraphs.

Check to be sure the reeds are not preloaded. They should not adhere to the block or plate, and still the clearance between the reed and the block surface, should not be more than 0.010" (0.254 mm). **DO NOT** remove the reeds, unless they are to be replaced. **ALWAYS** replace reeds in sets. **NEVER** turn used reeds over to be used a second time.

Lay the reeds on a flat surface and measure all the reed stops. If there is a great difference between the stops, the entire reed stop assembly should be replaced. Any attempt to bend and get all the stops equal and level would be almost impossible.

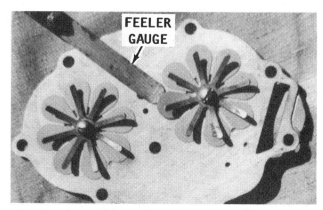

Using a feeler gauge to measure the clearance between the reed tip and the reed plate.

INSTALLATION

Procedures to install the reeds to the powerhead will be found in Section 3-15, Cylinder Block Service, under Reed Box Installation.

3-6 BYPASS COVERS

On some small horsepower engines the powerhead does not contain bypass covers. The bypass covers the area the fuel travels from the crankcase up the side of the powerhead and into the cylinder.

Seldom does a bypass cover cause any problem. On some model engines, a fuel pump may be attached to one of the bypass covers.

Close view showing the dimple on the reed plate. The reed leaves must straddle the dimple and be centered over the openings for proper operation.

Removing the intake cover from a typical powerhead.

BYPASS COVERS 3-11

During a normal overhaul, the bypass covers should be removed, cleaned, and new gaskets installed. Identify the covers to ensure installation in the same location from which they are removed.

INSTALLATION

Procedures to install the bypass covers to the powerhead will be found in Section 3-15, Cylinder Block Service, under Bypass Cover and Exhaust Cover Installation.

3-7 EXHAUST COVER

The exhaust covers are one of the most neglected items on any outboard engine. Seldom are they checked and serviced. Many times an engine may be overhauled and returned to service without the exhaust covers ever having been removed.

One reason the exhaust covers are not removed is because the attaching bolts usually become corroded in place. This means they are very defficult to remove, but the work should be done. Heat applied to the bolt head and around the exhaust cover will help in removal. However, some bolts may still be broken. If the bolt is broken it must be drilled out and the hole tapped with new threads.

The exhaust covers are installed over the exhaust ports to allow the exhaust to leave the powerhead and be transferred to the exhaust housing. If the cover was the only item over the exhaust ports, they would become so hot from the exhaust gases they might cause a fire or a person would be severely burned if they came in contact with the cover.

Therefore, an inner plate is installed to help dissipate the exhaust heat. Two gaskets are installed -- one on either side of the inner plate. Water is channeled to circulate between the exhaust cover and the inner plate. This circulating water cools the exhaust cover and prevents it from becoming a hazard.

On some early model outboards (the late 1950's and early 1960's), the inner plate was constructed of aluminum. Unfortunately, the aluminum would corrode through, especially in a salt water enviroment, and then water could enter the lower cylinder and cause a powerhead failure. The accompanying illustration clearly shows an inner plate corroded through, allowing water to enter the lower cylinder. To correct this corrosion problem, the inner plate is now made of stainless steel material.

A thorough cleaning of the inner plate behind the exhaust covers should be performed during a major engine overhaul. If the integrity of the exhaust cover assembly is in doubt, replace the complete cover including the inner plate.

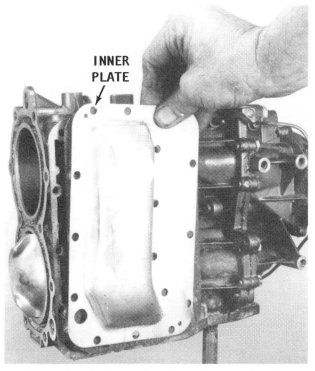

Removing the exhaust cover from the powerhead.

Removing the inner plate from the powerhead.

3-12 POWERHEAD

On powerheads equipped with the heat/electric choke, a baffle is installed on the inside surface of the inner plate. This baffle is heated from the engine exhaust gases. Air passing through the baffle heats the choke and allows the choke to open as engine temperature rises.

CLEANING

Clean any gasket material from the cover and inner plate surfaces. Check to be sure the water passages in the cover and plate are clean to permit adequate passage of cooling water.

Inspect the inlet and outlet hole in the powerhead to be sure they are clean and free of corrosion. The openings in the powerhead may be cleaned with a small size screwdriver.

Clean the area around the exhaust ports and in the webs running up to the exhaust ports. Carbon has a habit of forming in this area.

INSTALLATION

Procedures to install the exhaust covers will be found in Section 3-15, Cylinder Block Service, under Bypass Cover and Exhaust Cover Installation.

The exhaust area of the powerhead, open for inspection and cleaning.

3-8 TOP SEAL

The top seal maintains vacuum and pressure in the crankcase at the top cylinder.

REMOVAL

This seal can only be removed using one of two methods.

The first, is by using a special puller while the powerhead is still assembled. If the puller is used, thread the end of the puller into the seal. After the puller is secured to the seal, remove the seal from the powerhead by tightening the center screw on the puller. **DO NOT** attempt to use any other type of tool to remove this seal or the powerhead flanges will be damaged. If the flanges are damaged, the block must be replaced.

The second method is to remove the seal during powerhead disassembling. After the crankcase cover has been removed, the seal will be loose and can be easily removed by holding onto the bearing and prying the seal out.

INSTALLATION

To install the seal with the powerhead assembled, coat the outside diameter of the

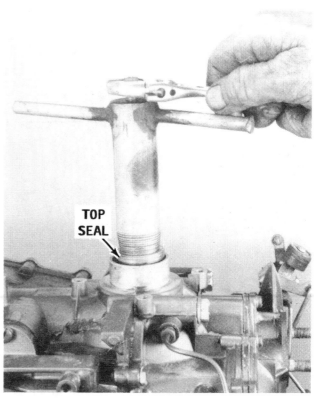

Using the proper tools to remove the top seal.

seal with OMC Seal Compound. Use the special tool to tap the seal **EVENLY** into place around the crankshaft.

If the powerhead has been disassembled, a socket or other similar type tool of equal diameter as the seal may be used to tap it into the bearing. If the powerhead being serviced does not have the seal in the bearing, lay the seal in the recess of the block. When the crankcase cover is installed, the seal will be in place.

3-9 BOTTOM SEAL

The bottom seal has equal importance as the top seal. This seal is installed to maintain vacuum and pressure in the lower half of the crankcase for the lower cylinder.

The bottom seal will vary, depending on the model engine being serviced. The following procedures and accompanying illustrations cover the most common Johnson/Evinrude bottom seal installed.

Seal Mounted on the Driveshaft

When the powerhead is removed, observe around the driveshaft at the lower end, and the seal will be visible. The seal consists of a gasket, plate, an O-ring, lower seal bearing, spring, washer, and a pin. The pin is installed through the driveshaft and holds the seal upward and in place.

As the powerhead is lowered down over the driveshaft during installation, the seal is held in place and will hold the vacuum and pressure created when the engine is operating.

Removal

With the powerhead assembled, it is a simple matter to reach into the exhaust housing and remove the seal and then replace the gasket and O-ring. Check the spring, to be sure it is not distorted, and the washer for damage.

Seal Mounted on the Crankshaft

This bottom seal prevents exhaust fumes from entering the crankcase, and holds pressure and vacuum inside. The seal consists of a quadrant ring, O-ring, retainer washer, spring, another washer, and a snap ring.

Removal

To remove this seal from the lower end of the crankshaft, use a pair of Tru-arc pliers and **CAREFULLY** remove the snap ring. **TAKE CARE** not to lose any of the parts due to the spring pressure against the snap ring. Notice how the quadrant O-ring fits inside the seal. This ring is also removable. Observe how the seal has a raised edge on one side. This raised edge **MUST** face upward when the seal is installed.

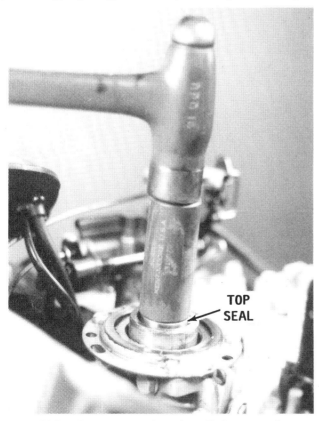

Using the proper tools to install the top seal.

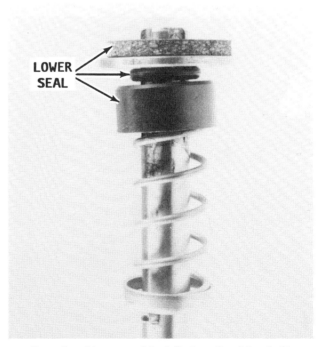

Powerhead lower seal installed on the driveshaft.

3-14 POWERHEAD

Using a pair of truarc pliers to remove the snap ring from the crankshaft.

INSPECTION

Check to be sure the spring has good tension. Check to be sure the washers are not distorted. The quadrant ring should be **DISCARDED** and a new one installed.

Good shop practice dictates the quadrant seal be replaced each time the lower seal is serviced.

Check the groove in the lower end of the crankshaft where the truarc ring fits. If the

Quadrant ring installed in its retainer at the bottom seal.

Removing the quadrant seal and retainer from the crankshaft.

groove is not clean, the ring will snap out and the lower sealing qualities will be lost. If the groove is badly corroded, the crankshaft must be replaced.

Seal Pressed In Lower Bearing

This type lower seal is very similar to the top seal. The seal is pressed inside the lower bearing. Therefore, it is necessary to

Top bearing (left) and bottom bearing (right). Both of these bearings require seals.

Pressing a new seal into a bearing.

remove the crankcase cover and lower bearing before the seal can be removed. The seal is then pressed out of the bearing, using an arbor press, and a new seal installed.

3-10 CENTERING PINS

All Johnson/Evinrude outboard engines have at least one, and in most cases two, centering pins installed through the crankcase cover. These pins index into matching holes in the powerhead block when the crankcase cover is installed. These pins center the crankcase cover on the powerhead block.

The centering pins are tapered. The pins must be carefully checked to determine how they are to be removed from the cover. In most cases the pin is removed by using a center punch and tapping the pin towards the carburetor or intake manifold side of the crankcase.

When removing a centering pin, hold the punch securely onto the pin head, then strike the punch a good hard forceful blow. **DO NOT** keep beating on the end of the pin, because such action would round the pin head until it would not be possible to drive it out of the cover.

Centering pins are the first item to be installed in the cover when replacing the crankcase cover.

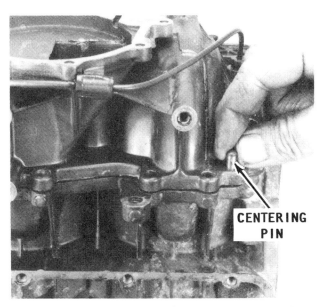
Removing a centering pin from the cylinder block.

3-11 MAIN BEARING BOLTS AND CRANKCASE SIDE BOLTS

The main bearing bolts are installed through the crankcase cover into the powerhead block. Most engines have two bolts installed for the top main bearing, two for the center main bearing, and two for the lower main bearing.

In many cases the upper and lower main bearing bolts are **DIFFERENT** lengths. Therefore, take time to tag and identify the bolts to ensure they will be installed in the same location from which they were removed.

The crankcase side bolts are installed along the edge of the crankcase cover to secure the cover to the cylinder block.

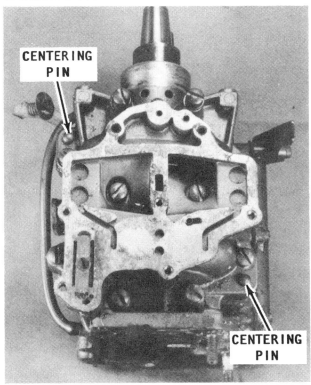
Cylinder block with the two centering pins installed.

Removing the main bearing bolts from the powerhead.

3-16 POWERHEAD

These bolts usually have a 7/16" head and all must be removed before the crankcase cover can be removed. Remove the crankcase side bolts.

Remove the main bearing bolts. Two bolts installed in the center are behind the reeds. Normally these two are not actually bolts, but Allen head screws. All six main bearing bolts must be removed before the crankcase cover can be removed.

INSTALLATION

Main bearing bolt and the crankcase side bolt installation is given in Section 3-15, Cylinder Block Assembling, under Main Bearing and Crankcase Side Bolt Installation.

3-12 CRANKCASE COVER

REMOVAL

After all side bolts and main bearing bolts have been removed, use a soft-headed mallet and tap on the bottom side of the crankshaft. A soft, hollow sound should be heard indicating the cover has broken loose from the crankcase. If this sound is not heard, check to be sure all the side bolts and main bearing bolts have been removed. **NEVER** pry between the cover and the crankcase or the cover will surely be distorted..If the cover is distorted, it will fail to make a proper seal when it is installed.

Once the crankshaft has been tapped, as described, and the proper sound heard, the cover will be jarred loose and may be removed.

CLEANING AND INSPECTING

Wash the cover with solvent, and then dry it thoroughly. Check the mating surface to the cylinder block for damage that may affect the seal.

Inspect the labyrinth seal grooves at the center main bearing area to be sure they are clean and not damaged in any manner.

INSTALLATION

Installation procedures for the crankcase cover are given in Section 3-15, Cylinder Block Service, under Crankcase Cover Installation.

3-13 CONNECTING RODS AND PISTONS

The connecting rods and their rod caps are a **MATCHED set.** They absolutely **MUST** be identified, kept, and installed as a set. Under no circumstances should the connecting rod and caps be interchanged. Therefore, on a multiple piston engine, **TAKE TIME AND CARE** to tag each rod and rod cap; to keep them together as a set while they are on the bench; and to install them into the same cylinder from which they were removed as a set.

Cylinder block after the crankcase cover has been removed.

Crankcase cover with the labyrinth seal area clearly visible.

ROD AND PISTONS 3-17

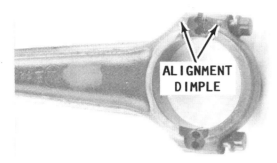

Rod and rod cap with the two alignment dimples shown.

Wrist pin end of the rod with the pressed-in bearing shown.

The connecting rod and its cap on 15 to 40 hp engines are manufactured as a set -- as a single unit. After the complete rod and cap have been made, two holes are drilled through the side of the cap and rod, and the cap is then fractured from the rod. Therefore, the cap must always be installed with its original rod. The cap half of the break can **ONLY** be matched with the other half of the break on the **ORIGINAL** rod.

The rods and caps on the smaller horsepower engines are made of aluminum with babbitt inserts. These rods and caps are manufactured as two separate items.

Inspect the rod and the rod cap before removing the cap from the crankshaft. Under normal conditions, a line or a dot is visible on the top side of the rod and the cap. This identification is an assist to assemble the parts together and in the proper location.

Observe into the block and notice how the rods have a "trough". Also notice the hole in the rod near where the wrist pin passes through the piston. On many rods there is also a hole in the rod at the crank end. These two holes **MUST ALWAYS** face upward during installation.

REMOVAL

To remove the rod bolts from the cap, it is recommended to loosen each bolt just a little at-a-time and alternately. This procedure will prevent one bolt from being completely removed while the other is still tightened to its recommended torque value. Such action may very likely warp the cap.

Remove the bolts as described in the previous paragraph, and then **CAREFULLY** remove the rod cap to prevent loosing the needle bearings installed under the cap, if used.

Remove the needle bearings and cages, if used, from around the crankshaft. Count the needle bearings and insert them into a separate container -- one container for each rod, with the container clearly identified to ensure they will be installed with the proper rod at the crankshaft journal from which they were removed.

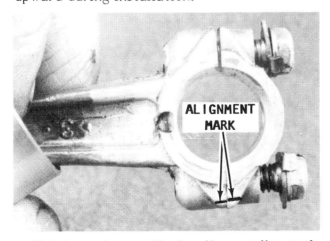

Rod and rod cap with the alignment line marks shown.

Removing the bolts from the rod cap.

Tap the piston out of the cylinder from the crankshaft side. Immediately attach the proper rod cap to the rod and hold it in place with the rod bolts. The few minutes involved in securing the cap with the rod will ensure the matched cap remains with its mating rod during the cleaning and assembling work.

Identify the rod to ensure it will be installed into the cylinder from which it was removed.

Remove and identify the other rod caps, needle bearings and cages, and rods with pistons, in the same manner.

DISASSEMBLY

Before separating the piston from the rod, notice the location of the piston in relation to the rod. Observe the hole in the rod trough on one side of the rod near the wrist pin opening and another at the lower end. These holes must face toward the **TOP** of the engine during installation.

Observe the slanted edge and the sharp edge of the dome-type piston. The slanted edge **MUST** face toward the exhaust side of the cylinder and the sharp edge toward the intake side during installation.

When the rod is installed to the piston, the relationship of the rod can only be one way. The rod holes must face upward and the piston must face as described in the previous paragraph.

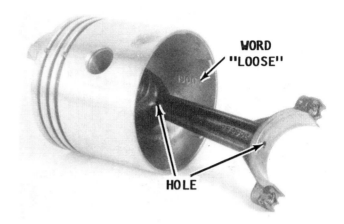

Identifying word "LOOSE" on the inside of the piston skirt and the hole in the rod at the wrist pin end. The wrist pin must be driven from the loose side of the piston out the tight side, as described in the text.

Observe into the piston skirt. On most model pistons, notice the **"L"** stamped on the boss through which the wrist pin passes. The letter mark identifies the "loose" side of the piston and indicates side of the piston from which the wrist pin must be driven out without damaging the piston. Some pistons may have the full word **"LOOSE"** stamped on the inside of the piston skirt.

If the piston does not have the **"L"** or the word **LOOSE** stamped, the wrist pin may be driven out in either direction.

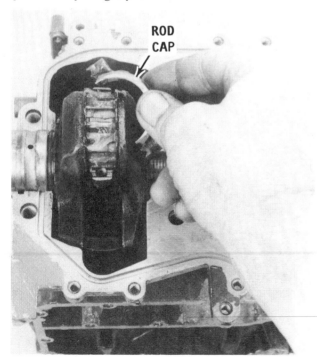

Removing the rod cap from the rod.

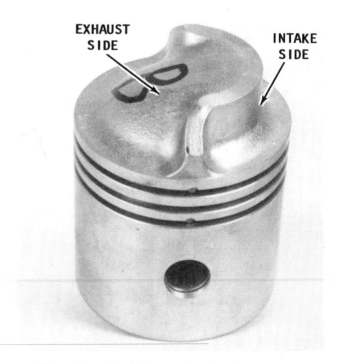

Close view of a piston with the slanted edge and sharp edges identified. The piston can only be installed one way for proper operation.

It may be necessary to heat the piston in a container of boiling water in order to press the wrist pin free.

Remove the retaining clips from each end of the wrist pin. Some clips are spring wire type and may be worked free of the piston using a screwdriver. Other model pistons have a truarc snap ring. This type of ring can only be successfully removed using a pair of truarc pliers.

Place the piston in an arbor press using the **PROPER** size cradle for the piston being serviced, and with the **LOOSE** side of the piston facing **UPWARD**.

The wrist pin must be driven out **FROM** the loose side. This may not seem reasonable, but there is a very simple explanation. By placing the piston in the arbor press cradle with the tight side down, and the arbor ram pushing from the loose side, the piston has good support and will not be distorted. If the piston is placed in the arbor press with the loose side down, the piston would be distorted and unfit for further service.

Many rods have a wrist pin bearing. Some are caged bearings and other are not.

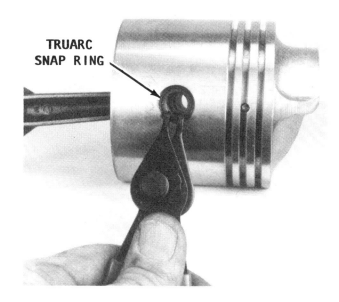

Removing the wrist pin truarc snap ring from the piston.

Snap ring used to retain the wrist pin in the piston.

Heating a piston in hot water to expand the metal slightly.

Removing a snap ring with a pointed tool.

3-20 POWERHEAD

TAKE CARE not to lose any of the bearings when the wrist pin is driven free of the piston.

Alternate Removal Method

If the piston does not have the "L" or the word **LOOSE** stamped, the wrist pin may be driven out in either direction.

If an arbor press or cradle is not available, proceed as follows: Heat the piston in a container of very hot water for about ten minutes. Heating the piston will cause the metal to expand ever so slightly, but ease the task of driving the pin out. Assume a sitting position in a chair, on a box, whatever. Next, lay a couple towels over your legs. Hold your legs tightly together to form a cradle for the piston above your knees. Set the piston between your legs with the **LOOSE** side of the piston facing upward. Now, drive the wrist pin free using a drift pin with a shoulder. The drift pin will fit into the hole through the wrist pin and the shoulder will ride on the edge of the wrist pin. Use sharp hard blows with a hammer. Your legs will absorb the shock without damaging the piston. If this method is used on a regular basis during the busy season, your legs will develop black-and-blue areas, but no problem, the marks will disappear in a few days.

Removing the wrist pin using a drift pin. The piston can be supported between your legs as described in the text.

Rod, rod cap, wrist pin, and wrist pin bearing, after removal.

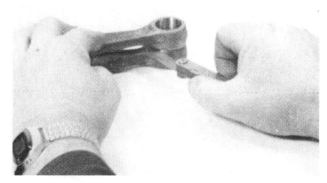

Testing two rods at the wrist pin end for warpage.

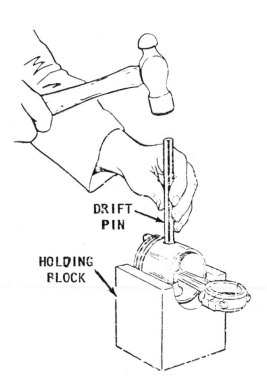

Removing the wrist pin using a holding block.

Testing two rods at the rod cap end for warpage.

Needle bearings and cages unfit for further service.

ROD INSPECTION AND SERVICE

If the rod has needle bearings, the needles should be replaced anytime a major overhaul is performed. It is not necessary to replace the cages, but a complete **NEW** set of needles should be purchased and installed.

Place each connecting rod on a surface plate and check the alignment. If light can be seen under any portion of the machined surfaces, or if the rod has a slight wobble on the plate, or if a 0.002" feeler gauge can be inserted between the machined surface and the surface plate, the rod is bent and unfit for further service.

Inspect the connecting rod bearings for rust or signs of bearing failure. **NEVER** intermix new and used bearings. If even one bearing in a set needs to be replaced, all bearings at that location **MUST** be replaced.

Inspect the bearing surface of the rod and the rod cap for rust and pitting.

Inspect the bearing surface of the rod and the rod cap for water marks. Water marks are caused by the bearing surface being subjected to water contamination, which causes "etching". The etching resembles the size of the bearing as shown in the accompanying illustration.

Inspect the bearing surface of the rod and rod cap for signs of spalling. Spalling is

The wrist pin bearing should be carefully checked to be sure the needles turn freely and there is no sign of corrosion.

Badly rusted and corroded crankshaft from a submerged engine. This crankshaft is no longer fit for service.

the loss of bearing surface, and resembles flaking or chipping. The spalling condition will be most evident on the thrust portion of the connecting rod in line with the I-beam. Bearing surface damage is usually caused by improper lubrication.

Check the bearing surface of the rod and rod cap for signs of chatter marks. This condition is identified by a rough bearing surface resembling a tiny washboard. The condition is caused by a combination of low-speed low-load operation in cold water, and is aggravated by inadequate lubrication and improper fuel. Under these conditions, the crankshaft journal is hammered by the connecting rod. As ignition occurs in the

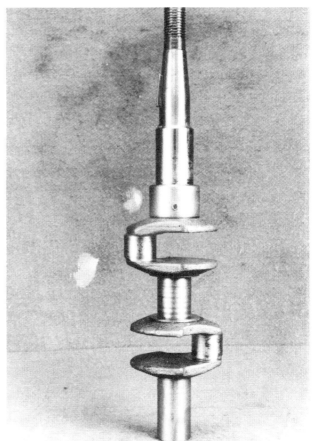

A crankshaft cleaned and ready for installation.

3-22 POWERHEAD

cylinder, the piston pushes the connecting rod with tremendous force, and this force is transferred to the connecting rod journal.

Since there is little or no load on the crankshaft, it bounces away from the connecting rod. The crankshaft then remains immobile for a split second, until the piston travel causes the connecting rod to catch up to the waiting crankshaft journal, then hammers it.

In some instances, the connecting rod crankpin bore becomes highly polished.

While the engine is running, a "whirr" and/or "chirp" sound may be heard when the engine is accelerated rapidly from idle speed to about 1500 rpm, then quickly returned to idle. If chatter marks are discovered, the crankshaft and the connecting rods should be replaced.

Inspect the bearing surface of the rod and rod cap for signs of uneven wear and possible overheating. Uneven wear is usually caused by a bent connecting rod. Overheating is identified as a bluish bearing surface color and is caused by inadequate lubrication or operating the engine at excessive high rpm.

Inspect the needle bearings, if installed. A bluish color indicates the bearing became very hot and the complete set for the rod **MUST** be replaced, no question.

Service the connecting rod bearing surfaces according to the following procedures and precautions:

a- Align the etched marks on the knob side of the connecting rod with the etched marks on the connecting rod cap.

b- Tighten the connecting rod cap attaching bolts securely.

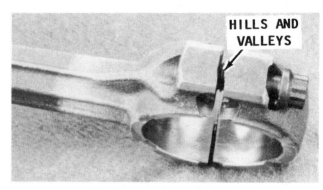

Rod cap separated slightly from its matching rod. Notice the matching hills and valleys.

c- Use **ONLY** crocus cloth to clean bearing surface at the crankshaft end of the connecting rod. **NEVER** use any other type of abrasive cloth.

d- Insert the crocus cloth in a slotted 3/8" diameter shaft. Chuck the shaft in a drill press and operate the press at high speed and at the same time, keep the connecting rod at a 90° angle to the slotted shaft.

e- Clean the connecting rod **ONLY** enough to remove marks. **DO NOT** continue once the marks have disappeared.

f- Clean the piston pin end of the connecting rod using the method described in Steps d and e, but using 320 grit Carborundum cloth instead of crocus cloth.

g- Thoroughly wash the connecting rods to remove abrasive grit. After washing, check the bearing surfaces a second time.

h- If the connecting rod cannot be cleaned properly, it should be replaced.

i- Lubricate the bearing surfaces of the connecting rods with light-weight oil to prevent corrosion.

PISTON AND RING INSPECTION AND SERVICE

Inspect each piston for evidence of scoring, cracks, metal damage, cracked piston pin boss, or worn pin boss. Be especially

*Installing the rod cap onto the rod in preparation for cleaning the inside surface. The cap **MUST** always be kept with its matching rod.*

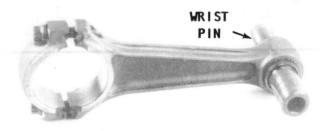

Testing the wrist pin end of the rod prior to installation.

critical during inspection if the engine has been submerged.

Carefully check each wrist pin to be sure it is not the least bit bent. If a wrist pin is bent, the pin and piston **MUST** be replaced as a set, because the pin will have damaged the boss when it was removed.

Check the wrist pin bearings. If the bearing is the pressed-in type, use your finger and determine the bearing is in good condition with no indication of binding or "rough" spots. If the wrist pin bearing is the removable type, the needle should be replaced.

Grasp each end of the ring with either a ring expander or your thumbnails, open the ring and remove it from the piston. Many times, the ring may be difficult to remove because it is "frozen" in the piston ring groove. In such a case, use a screwdriver and pry the ring free. The ring may break, but if it is difficult to remove, it **MUST** be replaced.

OBSERVE the pin in each ring groove of the piston. The ends of the ring **MUST** straddle this pin. The pin prevents the ring from rotating while the engine is operating. This fact is the direct opposite of a four-cycle engine where the ring must rotate. In a two-cycle engine, if the ring is permitted to rotate, at one point, the opening between the ring ends would align with either the intake or exhaust port in the cylinder. At

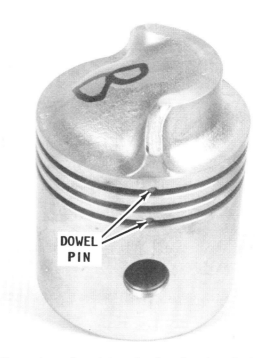

Close view of a piston showing the ring pin in the groove.

that time, the ring would expand very slightly, catch on the edge of the port, and **BREAK**.

Therefore, when checking the condition of the piston, **ALWAYS** check the pin in each groove to be sure it is tight. If one pin is the least bit loose, the piston **MUST** be replaced, without question. Never attempt to replace the pin, it is **NEVER** successful.

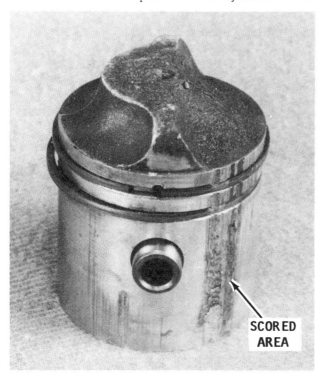

Piston badly scored and no longer fit for service.

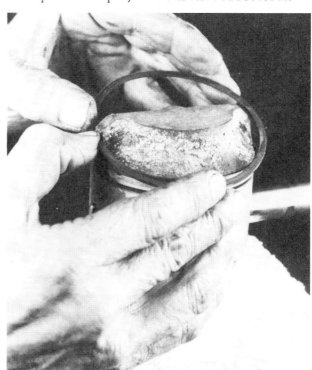

Removing the rings from the piston.

Check the piston ring grooves for wear, burns, distortion or loose locating pins. During an overhaul, the rings should be replaced to ensure lasting repair and proper engine performance after the work has been completed.

Clean the piston dome, ring grooves and the piston skirt. Clean the piston skirt with a crocus cloth.

Clean carbon deposits from the top of the piston using a soft wire brush, carbon removal solution, or by sand blasting. If a wire brush is used, **TAKE CARE** not to burr or round machined edges.

Wear a pair of good gloves for protection against sharp edges, and clean the piston ring grooves using the recessed end of the proper broken ring as a tool. **NEVER** use a rectangular ring to clean the groove for a tapered ring, or use a tapered ring to clean the groove for a rectangular ring.

NEVER use an automotive-type ring groove cleaner to clean piston ring grooves, because this type of tool could loosen the piston ring locating pins. **TAKE CARE** not to burr or round the machined edges. Inspect the piston ring locating pins to be sure they are tight. There is one locating pin in each ring groove. If one locating pin is loose, the piston must be replaced. Never attempt to replace the pin, it is **NEVER** successful.

Oversize Pistons and Rings

Scored cylinder blocks can be saved for further service by reboring and installing

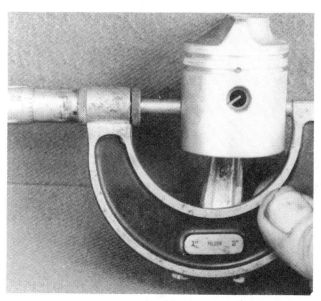

Using a micrometer to check the roundness of the piston.

oversize pistons and piston rings. **ONE MORE WORD:** Oversize pistons and rings are not available for all engines. At the time of this printing, the sizes listed in the Appendix were available. Check with the parts department at your local dealer for the model engine you are servicing, and to be sure the factory has not deleted a size from their stock.

ASSEMBLING

CRITICAL WORDS

Two conditions absolutely **MUST** exist when the piston and rod assembly are installed into the cylinder block.

The slanted side of the piston must face toward the exhaust side of the cylinder.

The hole in the rod near the wrist pin opening and at the lower end of the rod must face **UPWARD**.

*Cleaning the piston ring grooves. An automotive type ring groove cleaner should **NEVER** be used.*

*An automotive ring compressor should **NEVER** be used to install the rings for a two-cycle engine.*

Therefore, the rod and piston **MUST** be assembled correctly in order for the assembly to be properly installed into the cylinder. Soak the piston in a container of very hot water for about ten minutes. Heating the piston will cause it to expand ever so slightly, but enough to allow the wrist pin to be pressed through without difficulty.

Before pressing the wrist pin into place, hold the piston and rod near the cylinder block and check to be sure both will be facing in the right direction when they are installed.

Pack the wrist pin needle bearing cage with needle bearing grease, or a good grade of petroleum jelly. Load the bearing cage with needles and insert it into the end of the rod.

Slide the rod into the piston boss and check a second time to be sure the slanted side of the piston is facing toward the exhaust side of the cylinder and the hole in the rod is facing upward.

Place the piston and rod in the arbor press with the **LOOSE** or stamped **"L"** side of the piston facing **UPWARD**. Press the wrist pin through the piston and rod. Continue to press the wrist pin through until the groove in the wrist pin for the lock ring is visible on both ends of the pin. Remove the assembly from the arbor press. Install the retaining ring onto each end of the wrist pin. Some models have a wire ring, and others have a truarc ring. Use a pair of truarc pliers to install the truarc ring.

Fill the piston skirt with a rag, towel, shop cloths, or other suitable material. The

Wrist pin entering the piston from the "LOOSE" side.

rag will prevent the rod from coming in contact with the piston skirt while it is laying on the bench. If the rod is allowed to strike the piston skirt, the skirt may become distorted.

Assemble the other pistons, rods, and wrist pins in the same manner. Fill the skirt with rags as protection until the assembly is installed.

Alternate Assembling Method

If an arbor press is not available, the piston may be assembled to the rod in much the same manner as described for disassembling.

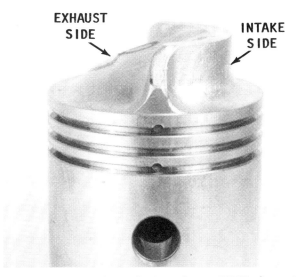

*The slanted side of the piston **MUST** face the exhaust port and the sharp edge face the intake port.*

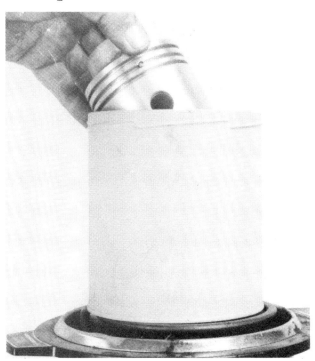

Heating the piston in hot water to expand the metal slightly as an assist to installing the wrist pin.

First, soak the piston in a container of very hot water for about ten minutes.

Before pressing the wrist pin into place, hold the piston and rod near the cylinder block and check to be sure both will be facing in the right direction when they are installed.

Pack the wrist pin needle bearing cage with needle bearing grease, or a good grade of petroleum jelly. Load the bearing cage with needles and insert it into the end of the rod.

Slide the rod into the piston boss and check a second time to be sure the slanted side of the piston is facing toward the exhaust side of the cylinder and the hole in the rod is facing upward.

Now, assume a sitting position and lay a couple towels over your lap. Hold your legs tightly together to form a cradle for the piston above your knees. Set the piston between your legs with the **LOOSE** side of the piston facing upward. Now, drive the wrist pin through the piston using a drift pin with a shoulder. The drift pin will fit into the hole through the wrist pin and the shoulder will ride on the end of the wrist pin. Use sharp hard blows with a hammer. Your legs will absorb the shock without damaging the piston. If this method is used on a regular basis during the busy season, your legs will develop black-and-blue areas, but no problem, the marks will disappear in a few days.

Continue to drive the wrist pin through the piston until the groove in the wrist pin for the lockring is visible at both ends. Install the retaining spring wire or truarc ring onto each end of the wrist pin.

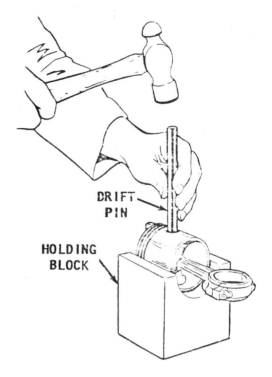

Install the wrist pin using a holding block.

Fill the piston skirt with a rag, towel, shop cloths, or other suitable material. The rag will prevent the rod from coming in contact with the piston skirt while it is laying on the bench. If the rod is allowed to strike the piston skirt, the skirt may become distorted.

Assemble the other pistons, rods, and wrist pins in the same manner. Fill the skirt with rags as protection until the assembly is installed.

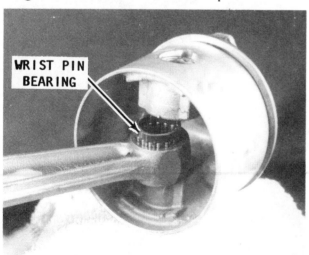

Installing the rod and wrist pin bearing into the piston.

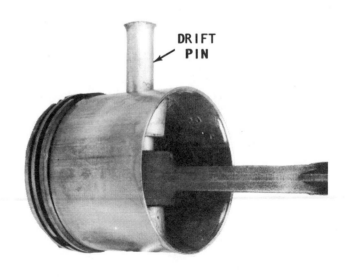

Installing the wrist pin without an arbor press. The piston can be held in your lap as described in the text.

CRANKSHAFT 3-27

Filling the inside of the piston to protect the skirt from being struck by the rod before the piston is installed in the cylinder block.

INSTALLATION

Piston and rod assembly installation procedures will be found in Section 3-15, Cylinder Block Service under Piston Installation

3-14 CRANKSHAFT

REMOVAL

Lift the crankshaft assembly from the block. On some models, especially the larger horsepower engines, it may be necessary to use a soft-headed mallet and tap on the bottom side of the crankshaft to jar it loose. As the crankshaft is lifted, **TAKE CARE** to work the center main bearing loose. This center bearing is a split bearing held together with a snap wire ring. On some models, the bottom half of the bearing may be stuck in the cylinder block. Therefore, the crankshaft and the center main bearing must be worked free of the block together.

If servicing a 15 hp to 40 hp engine, observe how the center main bearing, and the top and bottom main bearings all have a hole in the outside circumference. Notice

Crankshaft with the upper, center, and lower main bearings ready to be removed.

Crankcase cover with the labyrinth seal area clearly visible.

the locating pins in the cylinder block. The purpose of this arrangement is to prevent the bearing shell from rotating. During assembling, the holes in the bearings **MUST** index with the pins in the block. Also notice the grooves in the block on one side of the center main bearing. Observe the grooves in the crankcase cover. This arrangement of grooves forms what is commonly known as a "labyrinth" seal. The grooves fill with oil and/or fuel creating a seal between the cylinders.

On the smaller horsepower engines, babbitt bearings are used for the center main with needle bearings installed for the upper and lower main bearings.

CLEANING AND INSPECTION

Inspect the splines for signs of abnormal wear. Check the crankshaft for straightness. Inspect the crankshaft oil seal surfaces to be sure they are not grooved, pitted or scratched. Replace the crankshaft if it is severely damaged or worn. Check all crankshaft bearing surfaces for rust, water

Badly rusted and corroded crankshaft from a submerged engine. This crankshaft is no longer fit for service.

marks, chatter marks, uneven wear or overheating. Clean the crankshaft surfaces with crocus cloth.

Clean the crankshaft and crankshaft bearing with solvent. Dry the parts, but **NOT** the bearing, with compressed air. Check the crankshaft surfaces a second time. Replace the crankshaft if the surfaces cannot be cleaned properly for satisfactory service. If the crankshaft is to be installed for service, lubricate the surfaces with light oil.

The top and lower bearing may be easily removed from the crankshaft. The center main bearing has a spring steel wire securing the two halves together. Remove the wire, and then the outer sleeve, then the needle bearings. **TAKE CARE** not to lose any of the needles. The outer shell is a fractured break type unit. Therefore, the two halves of the shell **MUST** absolutely be kept as a set.

Check the crankshaft bearing surfaces to be sure they are not pitted or show any signs of rust or corrosion. If the bearing surfaces are pitted or rusted, the crankshaft and bearings must be replaced.

During an engine overhaul to this degree, it is a good practice to remove the seal

Crankshaft with a badly corroded "throw". This crankshaft is unfit for further service.

from the top main bearing. If the same type of seal is used in the bottom main bearing, remove that seal also.

ASSEMBLING

Insert the proper number of needle bearings into the center main bearing cage. Install the outer sleeve over the bearing cage. Check to be sure the two halves of the outer sleeve are matched. Again, these two halves are manufactured as a single unit and then broken. Therefore, the hills and valleys of the break absolutely **MUST** match during installation.

Snap the retaining ring into place around the bearing. Slide the upper bearing onto the crankshaft journal at the upper end and the lower bearing onto the lower end. Rotate the installed bearings to be sure there is no evidence of binding or rough spots. The crankshaft is now ready for installation.

INSTALLATION

Installation procedures are given in Section 3-15, Cylinder Block Service, under Crankshaft Installation.

3-15 CYLINDER BLOCK SERVICE

Inspect the cylinder block and cylinder bores for cracks or other damage. Remove

Crankshaft with the upper, center, and lower main bearings installed. Check to be sure the snap ring on the center main bearing is installed.

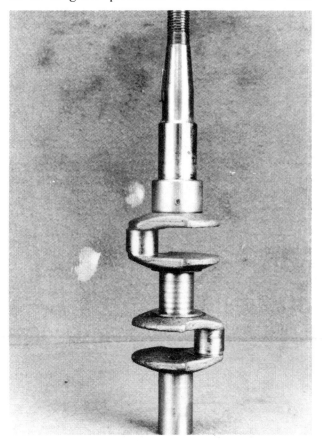

A crankshaft cleaned and ready for installation.

carbon with a fine wire brush on a shaft attached to an electric drill or use a carbon remover solution.

Use an inside micrometer or telescopic gauge and micrometer to check the cylinders for wear. Check the bore for out-of-round and/or oversize bore. If the bore is tapered, out-of-round or worn more than 0.003" - 0.004" (0.076 mm - 0.102 mm) the cylinders should be rebored and oversize pistons and rings installed.

GOOD WORDS:

Oversize piston weight is approximately the same as a standard size piston. Therefore, it is **NOT** necessary to rebore all cylinders in a block just because one cylinder requires reboring. The APBA (American Power Boat Association) accepts and permits the use of 0.015" (0.381 mm) oversize pistons.

Hone the cylinder walls lightly to seat the new piston rings, as outlined in the Honing Procedures Section in this chapter. If the cylinders have been scored, but are not out-of-round or the sleeve is rough, clean the surface of the cylinder with a cylinder hone as described in Honing Procedures, next section.

SPECIAL WORD

Cylinder sleeves may be installed on some models, but the cost is very high.

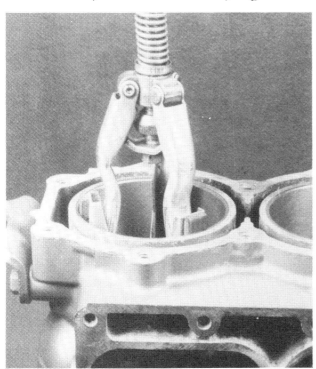

Resurfacing a cylinder wall using a honing tool.

HONING PROCEDURES

To ensure satisfactory engine performance and long life following the overhaul work, the honing work should be performed with patience, skill, and in the following sequence:

a- Follow the hone manufacturer's recommendations for use of the hone and for cleaning and lubricating during the honing operation.

b- Pump a continuous flow of honing oil into the work area. If pumping is not practical, use an oil can. Apply the oil generously and frequently on both the stones and work surface.

c- Begin the stroking at the smallest diameter. Maintain a firm stone pressure against the cylinder wall to assure fast stock removal and accurate results.

d- Expand the stones as necessary to compensate for stock removal and stone wear. The best cross-hatch pattern is obtained using a stroke rate of 30 complete cycles per minute. Again, use the honing oil generously.

e- Hone the cylinder walls **ONLY** enough to de-glaze the walls.

f- After the honing operation has been completed, clean the cylinder bores with hot water and detergent. Scrub the walls with a stiff bristle brush and rinse thoroughly with hot water. The cylinders **MUST** be cleaned well as a prevention against any abrasive material remaining in the cylinder bore. Such material will cause rapid wear of new piston rings, the cylinder bore, and the bearings.

g- After cleaning, swab the bores several times with engine oil and a clean cloth, and then wipe them dry with a clean cloth. **NEVER** use kerosene or gasoline to clean the cylinders.

h- Clean the remainder of the cylinder block to remove any excess material spread during the honing operation.

WORDS OF ADVICE

If new rings are to be installed, each ring from the package **MUST** be checked in the cylinder. Errors happen. Men and machines can make mistakes. The wrong size ring can be included in a package with the proper part number.

Therefore, check **EACH** ring, one at-a-time as follows: Turn the ring sideways and lower it a couple inches into the cylinder bore. Now, turn the ring horizontal in the cylinder. It is now in its normal operating position, but without the piston. Next, use a feeler gauge and measure the distance (the gap) between the ends of the ring. The maximum and minimum allowable ring gap is listed in the Specifications in the Appendix.

Turn the piston upside down and slide it in and out of the cylinder. The piston should slide without any evidence of binding.

ASSEMBLING

SPECIAL WORD

The cylinder block assembling work should proceed quickly and without interruptions. If the work is partially completed and then left for any period of time, sealant may become hard, parts may be moved and their identity for a particular cylinder lost, or an important step may be bypassed, overlooked, or forgotten.

The following procedures pickup the work of assembling the cylinder block **AFTER** the various parts have been serviced and assembled. Procedures for each area are found in this chapter under separate headings.

PISTON AND ROD ASSEMBLY INSTALLATION

Several different methods are possible to install the piston and rod assembly into the cylinder. The following procedures are outlined for the do-it-yourselfer, working at home without the advantage of special tools.

First, purchase a special hose clamp with a strip of metal inside the clamp, as shown

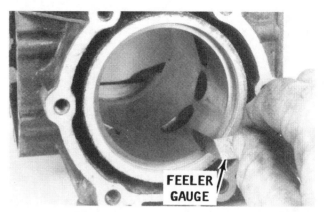

Using a feeler gauge to check the ring end gap with the ring in the cylinder.

in the accompanying illustration. This piece of metal on the inside allows the outside portion of the clamp to slide on the inner strip without causing the ring to rotate.

Actually, to our knowledge a Mercruiser dealer is the only place such a clamp may be purchased. At the Mercruiser marine dealer, ask for an exhaust bellows hose clamp. The design of this hose clamp prevents the clamp and the piston ring from turning as the clamp is tightened. **DO NOT** attempt to use an ordinary hose clamp from an automotive parts house because such a clamp will cause the piston ring to rotate as the clamp is tightened. The ring **MUST NOT** rotate, because the ring ends must remain on either side of the dowel pin in the ring groove.

Next, coat the inside surface of the cylinder with a film of light-weight oil. Coat the exterior surface of the piston with the oil.

TAKE TIME

Take just a minute to notice how the piston rings are manufactured. Each end of

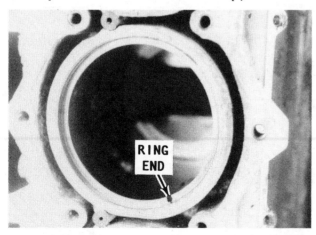

Checking the ring gap clearance by inserting the ring in the cylinder, as described in the text.

Proper hose clamp to install the rings if a ring compressor is not available.

CYLINDER BLOCK

the ring has a small cutout on the inside circumference. Now, visualize the ring installed in the piston groove. The ring ends must straddle the pin installed in each piston groove. As the ring is tightened around the piston, the ends will begin to come together. When the piston is installed into the cylinder bore, the two ends of the ring will come together and the cutout edge will be up against the pin. For this reason, **CARE** must be exercised when installing the rings onto the piston and when the piston is installed into the cylinder.

Install only the bottom ring into the bottom piston groove. Do not expand the ring any further than necessary, to prevent it from breaking.

Install the ring into the piston groove with the ends of the ring straddling the pin in the groove. The ring ends **MUST** straddle the pin to prevent the ring from rotating during engine operation. In a two-cycle engine, if the ring is permitted to rotate, at one point the opening between the ring ends would align with either the intake or exhaust port in the cylinder, the ring would expand very slightly, catch on the edge of the port, and **BREAK**.

CAREFULLY insert the rod and the piston skirt down into the cylinder.

GOOD WORDS

The following four areas must be checked at this point in the assembling work.

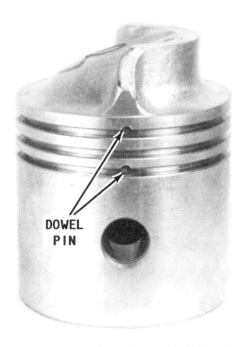

Piston ring groove pins. The ends of the ring must straddle the pin.

Installing the hose clamp over the ring prior to moving the piston further into the cylinder.

a- The piston and rod are being installed into the same cylinder from which they were removed.

b- The hole in the rod is facing **UPWARD**.

c- The slanted side of the piston is **TOWARD** the exhaust side of the cylinder.

d- The ends of the bottom ring straddle the pin in the piston groove.

Push the piston into the cylinder until the bottom ring, just installed, is about an inch from the surface of the cylinder block.

Tapping the piston into the cylinder with a soft-headed mallet.

Using a ring expander to install the ring onto the piston during piston installation.

When installing the piston into a small horsepower powerhead, it is possible to compress the ring with the fingers of each hand, and then to push the piston into the cylinder with your thumbs.

Install the hose clamp over the piston and bottom ring. Tighten the hose clamp with one hand and at the same time rotate the clamp back-and-forth slightly with the other hand. This "rocking" motion of the clamp as it is tightened will convince you the ring ends are properly positioned on either side of the pin. Continue to tighten the clamp, and "rocking" the clamp until

Checking the flexibility of the rings through intake port.

the clamp is against the piston skirt. At this point, the ring ends will be together and the cutout on each ring end will be against the pin.

Tap the piston with the end of a wooden tool handle until the ring enters the cylinder. Remove the hose clamp.

Install the remaining rings in the same manner, one at-a-time, making sure the ends of each ring straddle the pin in the piston groove.

Notice how the ring pins are staggered from one groove to the next, by $180°$.

After the last ring has been installed and the clamp removed, tap the piston into the bore until the crown is about even with the cylinder block surface.

Install the other pistons in exactly the same manner.

Turn the cylinder block upside down with the top of the block to your **LEFT**. Remove the bolts and rod caps from each rod. Set each rod cap in a definite position to ensure each will be installed onto the rod from which it was removed.

Both pistons installed in the powerhead. Notice the cylinder identification on each piston. The slanted side of the each piston is facing toward the exhaust port.

Checking the flexibility of the rings through the exhaust port.

CYLINDER BLOCK 3-33

Crankshaft with the upper, lower, and center main bearings installed. Notice the hole in each bearing. Matching pins in the cylinder block must index into these holes during crankshaft installation.

CRANKSHAFT INSTALLATION LARGE HORSEPOWER ENGINES 15 HP TO 40 HP

The following procedures outline steps to install a crankshaft with needle upper, center, and lower main bearings. The upper and lower mains are complete bearings and cannot be disassembled. The center bearing is caged. Installation procedures for small horsepower crankshafts with babbitt upper, lower, and center main bearings and with babbitt rod bearings are given in the following sections.

Observe the pin installed in each main bearing recess. Notice the hole in each main bearing outer shell. During installation, the hole in each bearing shell **MUST** index over the pin in the cylinder block.

Hold the crankshaft over the cylinder block with the upper end to your **LEFT**. Now, lower the crankshaft into the block, and at the same time, align the hole in each bearing to enable the pin in the block to

Crankshaft installed into the cylinder block.

index with the hole. Rotate each bearing slightly until all pins are properly indexed with the matching bearing hole. Once all pins are indexed, the crankshaft will be properly seated.

Apply needle bearing grease to each bearing cage. Coat the rod half of the bearing area with needle bearing grease. Needle bearing grease **MUST** be used because other types of grease will not thin out and dissipate. The grease must disipate to allow the gasoline and oil mixture to enter and lubricate the bearing. If needle bearing grease is not available, use a good grade of petroleum jelly (Vasoline).

Insert the proper number of needle bearings into each cage. Set the bearing cage into the bottom half of the rod. With your fingers on each side of the rod, pull up on the rod and bring the rod up to the bottom

Bearing locating pins in the cylinder block. Each pin must index into a hole in the bearing shown in the illustration at the top of this column.

Cage and needle bearings installed into the lower portion of the rod.

3-34 POWERHEAD

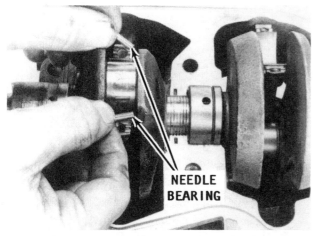

Installing a needle bearing on each side of the crankshaft.

side of the crankshaft. Put one needle bearing on each side of the crankshaft. Using needle bearing grease load the other cage and install the needle bearings into the cage. Lower the cage onto the crankshaft journal.

Install the proper rod cap to the rod with the identifying mark or dimple properly aligned to ensure the cap is being installed in the same position from which it was removed. Tighten the rod bolts fingertight, and then just a bit more.

Use a "scratchall", pick, or similar tool and move it back-and-forth on the outside surface of the rod and cap. Make the movement across the mating line of the rod and cap. The tool should not catch on the rod or on the cap. The rod cap must seat squarely with the rod. If not, tap the cap until the "scratchall" will move back-and-forth on the rod and cap across the mating

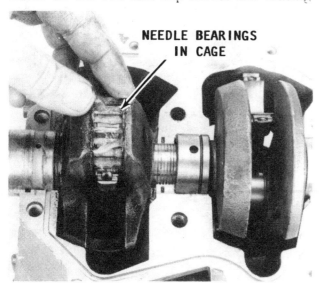

Lowering the cage and needle bearings over the top of the crankshaft.

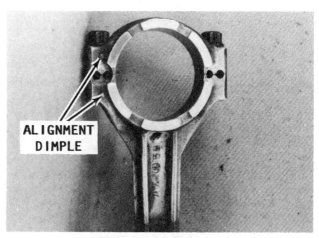

Rod and cap showing the alignment dimples.

Installing the rod cap over the needle bearings and cage.

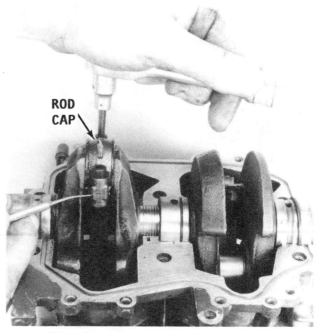

Installing the rod cap bolts, and at the same time, checking the cap alignment with the rod using a pick.

CYLINDER BLOCK 3-35

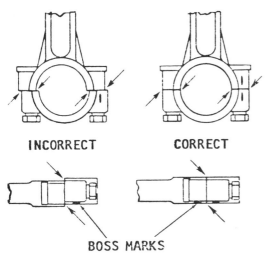

Correct and incorrect rod cap alignment.

line without any feeling of catching. Any step on the outside will mean a step on the inside of the rod and cap. Just a whisker of a lip, will cause one of the needle bearings to catch and fail to rotate. The needle will quickly flatten, and the rod will begin to "knock". Needle bearings **MUST** rotate or the function of the bearing is lost.

Tighten the rod cap bolts alternately and evenly in three rounds to the torque value given in the Torque Table in the Appendix. Tighten the bolts to 1/2 the torque value on the first round, to 3/4 the torque value on the second round, and to the full torque value on the third and final round. On each round, check with the pick to be sure the cap remains seated squarely.

Install the other rod cap/s in the same manner.

After the rods have been connected to the crankshaft, rotate the crankshaft until the rings on one cylinder are visible through the exhaust port. Use a screwdriver and push on each ring to be sure it has spring tension. It will be necessary to move the piston slightly, because all of the rings will not be visible at one time. If there is no spring tension, the ring was broken during installation. The piston must be removed and a new ring installed. Repeat the tension test at the intake port. Check the other cylinder/s in the same manner.

CRANKSHAFT INSTALLATION
SMALL HORSEPOWER ENGINES
1.5 HP W/TOP NEEDLE MAIN BRG. AND BABBITT CTR. AND BOTTOM
9.5 HP W/TOP AND BOTTOM NEEDLE BRG. AND CTR. BABBITT BRG.
5.0 HP, 5.5 HP, AND 6.0 HP W/ALL BABBITT MAIN BEARINGS

This section provides detailed instructions to install a small horsepower crankshaft with any of the bearing combinations listed in the heading. Some of the engines covered in these paragraphs have rod liners, others do not as follows:

With rod bearing liner -- 1.5 hp, 5.0 hp, 5.5 hp, 6 hp, and 9.5 hp from 1956 thru 1960.

With no rod bearing liner -- 5.5 hp and 10 hp 1961 thru 1964.

The procedures pickup the work after the piston/s have been installed, as described earlier in this section.

Lower the crankshaft into place in the cylinder block with the long threaded shank end at the top of the cylinder block. (It is a known fact, in more than just a few shops around the country, because of haste, the crankshaft installation work has proceeded with the short end at the top.)

Checking the flexibility of the rings through the exhaust port.

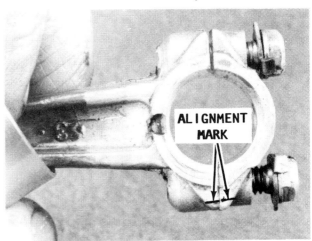

Rod and cap with the alignment marks visible.

If servicing the 9.5 hp or a 10 hp engine, the hole in the upper and lower main bearing **MUST** index into the pin in the cylinder block.

Some engines may have a lining arrangement as listed in the heading of this section. The lining is made in two parts. Install the liner half into the rod, then install the bearings as described in the next paragraph. The matching liner is to be installed into the rod cap.

Coat the rod half, of the bearing area, with needle bearing grease. Needle bearing grease **MUST** be used because other types of grease will not thin out and dissipate. The grease must dissipate to allow the gasoline and oil mixture to enter and lubricate the bearing. If needle bearing grease is not available, use a good grade of petroleum jelly (Vasoline).

Load the rod half of the rod bearing with needle bearings. Next, bring the rod up to the crankshaft rod journal. Coat the crankshaft journal with needle bearing grease. Place the needle bearings around the crankshaft jounal.

Position the rod cap, with the liners (if used) over the needle bearings. Install the rod cap bolts and lockwashers. Bring the bolts up fingertight, and then just a bit more.

If the liners are used, the cap and rod automatically align properly. If liners are not used, a dowel pin is installed in the rod cap. This pin will index into a hole in the rod for proper alignment.

Tighten the rod cap bolts alternately and evenly in three rounds to the torque value given in the Specifications in the Appendix. Tighten the bolts to 1/2 the torque value on the first round, to 3/4 the torque value on

Rod cap with liner ready for installation around the crankshaft.

the second round, and to the full torque value on the third and final round. On each round, check with the pick to be sure the cap remains seated squarely.

After the rod cap bolts have been tightened to the required torque value and the installation appears satisfactory, bend the bolt locking tabs upward to prevent the bolts from loosening.

Install the other rod cap/s in the same manner.

After the rods have been connected to the crankshaft, rotate the crankshaft until the rings on one cylinder are visible through the exhaust port. Use a screwdriver and push on each ring to be sure it has spring tension. It will be necessary to move the piston slightly, because all of the rings will not be visible at one time. If there is no spring tension, the ring was broken during installation. The piston must be removed and a new ring installed. Repeat the tension test at the intake port. Check the other cylinder/s in the same manner.

Needle bearings installed in the rod cap liner and around the crankshaft.

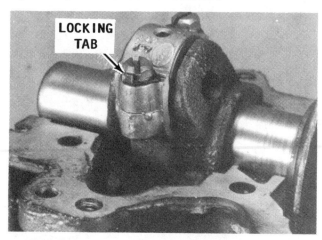

The locking tabs must be bent upward after the rod cap bolts have been tightened to the proper torque value.

CYLINDER BLOCK 3-37

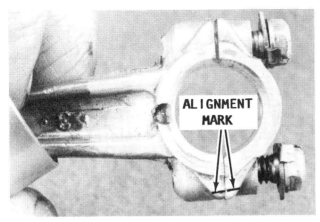

Rod and cap with the alignment marks visible.

CRANKSHAFT INSTALLATION SMALL HORSEPOWER ENGINES BABBIT MAIN BEARINGS AND BABBITT ROD BEARINGS ALL 3.0 HP, 4.0 HP, AND 7.5 HP ENGINES

This section provides detailed instructions to install a small horsepower crankshaft with babbitt upper, lower, and center main bearings, and with babbitt rod bearings on the engines listed in the heading.

The procedures pickup the work after the piston/s have been installed, as described earlier in this section.

Lower the crankshaft into place in the cylinder block with the long threaded shank end at the top of the cylinder block. (It is a known fact, in more than just a few shops around the country, because of haste, the crankshaft installation work has proceeded with the short end at the top.)

Pull the rod up to the crankshaft journal. Position the rod cap over the crankshaft journal. Install the rod cap bolts and lockwashers. Bring the bolts up fingertight, and then just a bit more.

Tighten the rod cap bolts alternately and evenly in three rounds to the torque value

Tightening the rod cap bolts to the proper torque value.

Using two hammers to fit the rod cap to the crankshaft.

given in the Torque Table in the Appendix. Tighten the bolts to 1/2 the torque value on the first round, to 3/4 the torque value on the second round, and to the full torque value on the third and final round. On each round, check with the pick to be sure the cap remains seated squarely.

Repeat the procedure for the other rod and cap.

After the other rod cap has been installed and the bolts tightened to the proper torque value, hold one hammer on one side of the rod and cap, and at the same time tap the other side of the rod and cap with the other hammer. Tap lightly on the top of the cap. Reverse the hammer positions and tap the opposite sides of the rod and cap. This procedure will "fit" the rod and cap to the crankshaft journal.

Checking movement of the pistons and crankshaft with the flywheel temporarily installed.

Repeat the "fitting" procedure for the other rod and cap.

Once the installation procedure appears satisfactory and all work has been completed, bend the bolt locking tabs upward to prevent the bolts from loosening.

CRANKCASE COVER INSTALLATION

First, check to be sure the mating surfaces of the crankcase cover and the cylinder block are clean. Pay particular attention to the labyrinth seal grooves in the center main bearing area. The mating surfaces and the seal grooves **MUST** be free of any old sealing compound or other foreign material.

CRITICAL WORDS

The remainder of the cylinder block installation work should be performed **WITHOUT** interruption. Do not begin the work if a break in the sequence is expected, coffee, lunch, whatever. The sealer will begin to set almost immediately, therefore, the crankcase cover installation, main bearing bolt installation and tightening, and the side bolt installation and tightening must move along rapidly.

Apply just a small amount of 1000 Sealer into the groove in the cylinder block to hold the seal in place. Install a new "spaghetti" seal into the groove, if used.

Installing sealer to the cylinder block when the "spaghetti" seal is not used.

Installing the tapered pin through the crankcase cover into the cylinder block.

Installing sealer to the "spaghetti" seal in the crankcase cover.

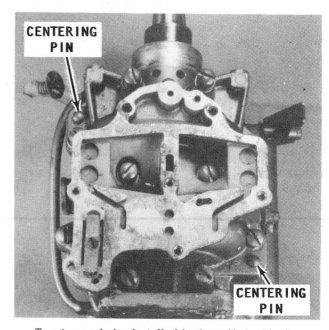

Two tapered pins installed in the cylinder block.

CYLINDER BLOCK 3-39

SPECIAL NOTE

Some model powerheads do not use the "spaghetti" seal. If the engine being serviced does not use the seal, apply a coating of 1000 Sealer to the mating surface of the cylinder block and the crankcase cover. The remaining installation instructions apply to all powerheads.

After the seal on both sides of the cylinder block have been installed, if used, apply a light coating of 1000 Sealer to the outside edge of the "spaghetti" seal.

Next, lower the crankcase cover into place on the cylinder block. Install the two guide centering pins through the cover and into the block. The centering pins are tapered. Therefore, check the crankcase and notice which side has the large hole and which has the small hole. The pin must be inserted into the large hole first. If the pin is installed into the small hole first, the crankcase cover or the cylinder block will break.

MAIN BEARING BOLT AND CRANKCASE SIDE BOLT INSTALLATION

Apply a coating of 1000 Sealer to the threads of the main bearing bolts. Install and tighten the main bearing bolts finger-tight and then just a bit more.

Tighten the main bearing bolts alternately and evenly in three rounds to the torque value given in the Torque Table in the Appendix. Be sure to check the Specifications in the Appendix for the engine being serviced.

Tighten the bolts to 1/2 the total torque value on the first round, to 3/4 the total torque value on the second round, and to the full torque value on the third and final round.

As an example: If the total torque value specified is 200 ft-lbs, the bolts should be tightened to 100 ft-lbs on the first go-around; to 150 ft-lbs on the second round; and to the full 200 ft-lbs on the third round.

Install and tighten the crankcase side bolts to the torque value given in the Appendix.

Install the Woodruff key in the crankshaft. Slide the flywheel onto the crankshaft. Rotate the flywheel through several revolutions and check to be sure all moving parts indicate smooth operation without evidence of binding or "rough" spots.

Remove the flywheel and the Woodruff key.

BOTTOM SEAL INSTALLATION
15 HP TO 40 HP ENGINES

This type of seal is attached to the crankshaft on the 15 and 40 hp engines. On the smaller engines, a seal is used on the crankshaft with a spring, O-ring, and gasket. These items push up against the bottom of the powerhead to affect the seal.

Install the quadrant O-seal into the quadrant seal holder with the lip of the seal facing toward the **BOTTOM** of the cylinder block.

Apply a small amount of light-weight oil onto the quadrant O-ring, and then slide the ring onto the crankshaft. Slide the large washer, spring, and small washer, onto the crankshaft, and secure them in place with the truarc snap ring.

Installing the main bearing bolts through the crankcase cover into the cylinder block.

Installing the lower main seal assembly to the crankshaft.

3-40 POWERHEAD

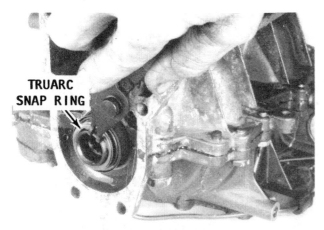

Using a pair of truarc pliers to install the truarc snap ring onto the crankshaft to secure the bottom seal in place.

EXHAUST COVER AND BYPASS COVER INSTALLATION

Coat both sides of a **NEW** gasket with 1000 Sealer, and then place the gasket in position on the exhaust side of the cylinder block. Install the inner plate. Coat both sides of another **NEW** gasket with sealer, and then install the gasket and exhaust cover.

Secure the exhaust cover in place with the attaching hardware.

Coat both sides of a **NEW** gasket with sealer, and then place it in position on the cylinder block. Install the bypass covers and secure them in place with the attaching hardware. If a fuel pump is used, be sure the same bypass cover is installed in the position from which it was removed.

Installing the intake bypass covers. One is already in place.

REED BOX INSTALLATION

Install the reed box and intake manifold onto the cylinder block. A gasket is usually installed on both sides of the reed box. The reeds and reed stops face inward toward the cylinder.

On 15 hp to 40 hp engines, a screw is installed in the center of the reed box into the cylinder block. This center screw is installed first, then the intake manifold is installed and secured in place.

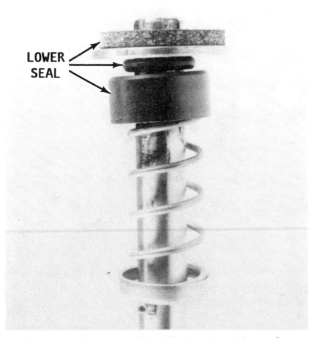

Powerhead lower seal installed on the driveshaft.

Installing the exhaust manifold.

Installing the gasket and reed plate to the powerhead.

HEAD INSTALLATION

Place a **NEW** head gasket in place on the cylinder block. **NEVER** use automotive type head gasket sealer. The chemicals in the sealer will cause electrolytic action and eat the aluminum faster than you can get to the bank for money to buy a new cylinder block.

Install the head bolts and tighten them fingertight, then just a bit more. Tighten the bolts alternately and evenly in three rounds to the torque value specified in the Appendix. On the first round tighten the bolts to 1/2 the total torque value, on the second round to 3/4 the total torque value, and to the full torque value on the third and final round.

POWERHEAD INSTALLATION

Install the assembled powerhead to the exhaust housing and tighten the attaching bolts alternately and evenly in three rounds to the torque value specified in the Appendix. Tighten the bolts to 1/2 the torque value on the first round, to 3/4 the total torque value on the second round, and to the full torque value on the third and final round.

Installing the front gasket to the reed plate.

Installing the intake manifold over the reed plate.

Install all powerhead accessories including the flywheel, carburetor, magneto, starter, etc. If any doubts or difficulties are encountered, follow the procedures outlined in the chapters covering the particular component.

The complete outboard unit is now ready to be started and "broke-in" according to the procedures outlined in the next section.

3-16 BREAK-IN PROCEDURES

Mount the engine in a test tank or body of water.

CAUTION: Water must circulate through the lower unit to the engine any time the engine is run to prevent damage to the water pump in the lower unit. Just five seconds without water will damage the water pump.

Tightening the head bolts to the proper torque value.

As soon as the engine starts, **CHECK** to be sure the water pump is operating. If the water pump is operating, a water mist will be discharged from the exhaust relief holes at the rear of the drive shaft housing.

During the first 10 hours of operation, **DO NOT** operate the engine at full throttle (except for **VERY** short periods). Perform the break-in as follows:

a- Operate at 1/2 throttle, approximately 2500 to 3500 rpm, for 2 hours.

b- Operate at any speed after 2 hours **BUT NOT** at sustained full throttle until another 8 hours of operation.

c- Mix gasoline and oil during the break-in period, total of 10 hours, at a ratio of 50:1.

d- While the engine is operating during the initial period, check the fuel, exhaust, and water systems for leaks.

e- Refer to Chapter 5 for synchronizing procedures.

After the test period, disconnect the fuel line. Remove the engine from the test tank. Install the engine hood.

Testing a 35 horsepower unit with a pressure fuel tank.

Operating an engine in a test tank following an overhaul.

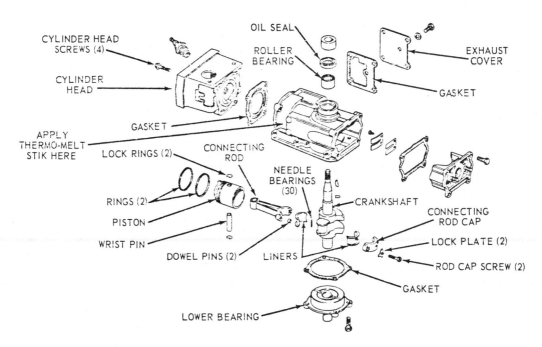

Exploded drawing of a 1.5 hp powerhead -- 1968-70, with principle parts identified.

CYLINDER BLOCK 3-43

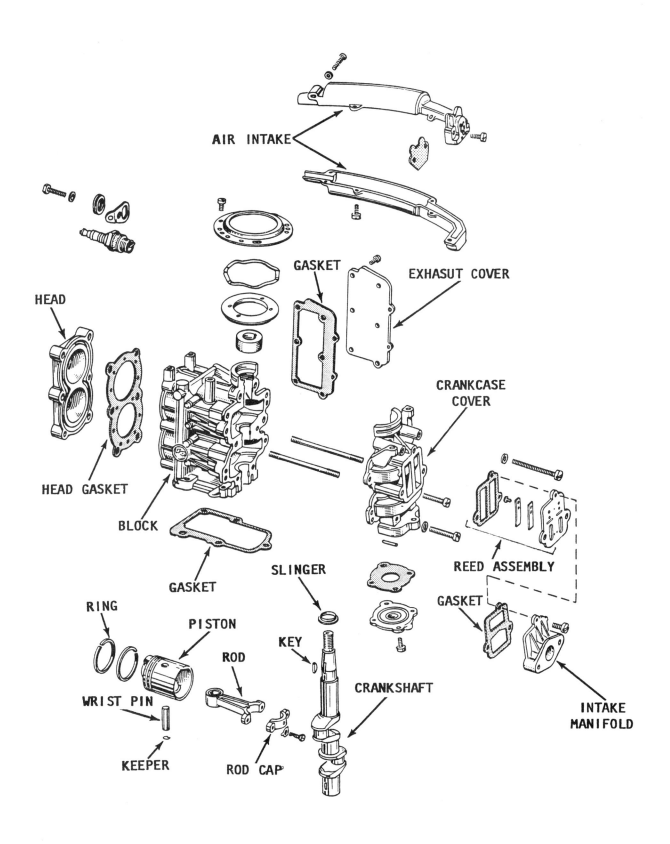

Exploded drawing of a 3.0 hp powerhead -- 1956 and the 4.0 hp powerhead -- 1969-70, with principle parts identified. The 4.0 hp unit uses a bearing and seal on the crankshaft instead of the oil slinger.

3-44 POWERHEAD

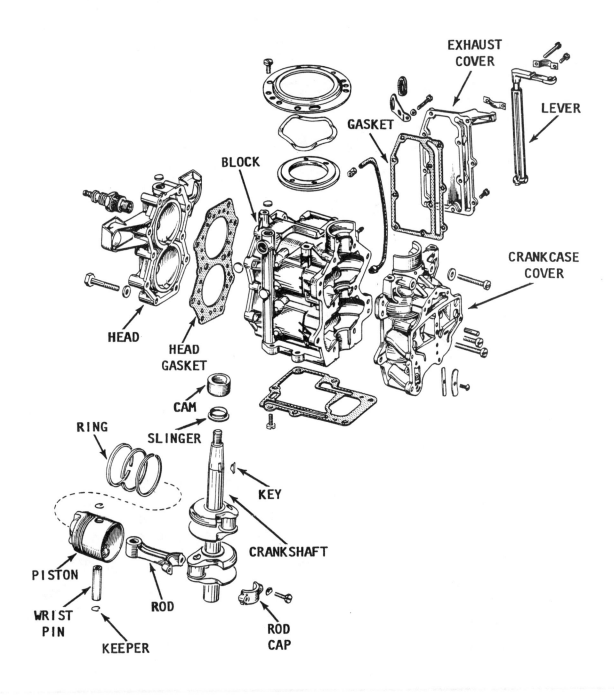

Exploded drawing of a 5.5 hp powerhead -- 1956-60, principle parts identified.

CYLINDER BLOCK 3-45

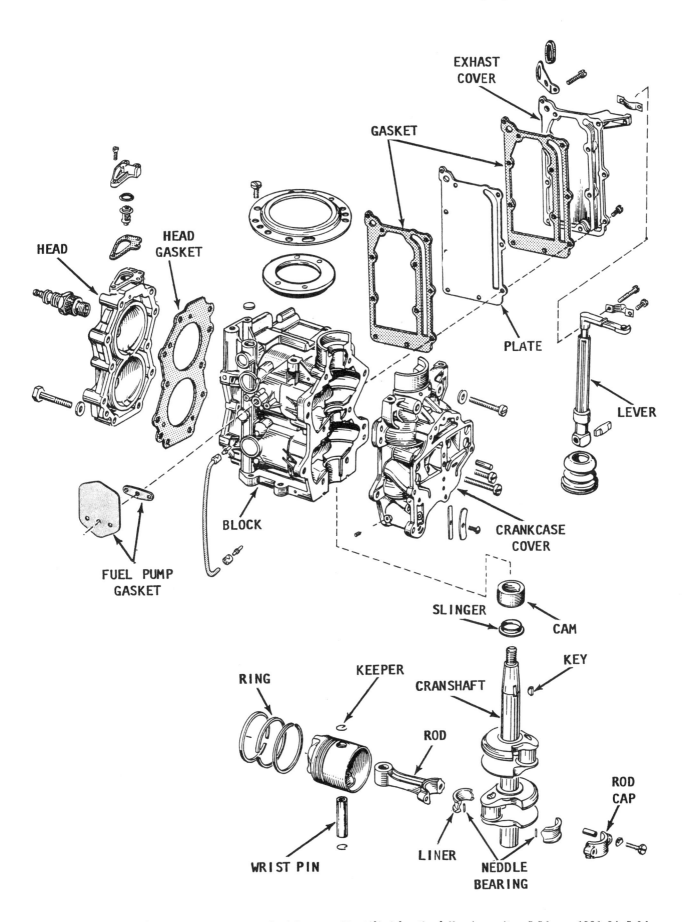

Exploded drawing of a powerhead, with principle parts identified for the following units: 5.5 hp -- 1961-64; 5.0 hp -- 1965-68; 6.0 hp -- 1965-70. The 6.0 hp unit uses an oil slinger instead of the seal.

3-46   POWERHEAD

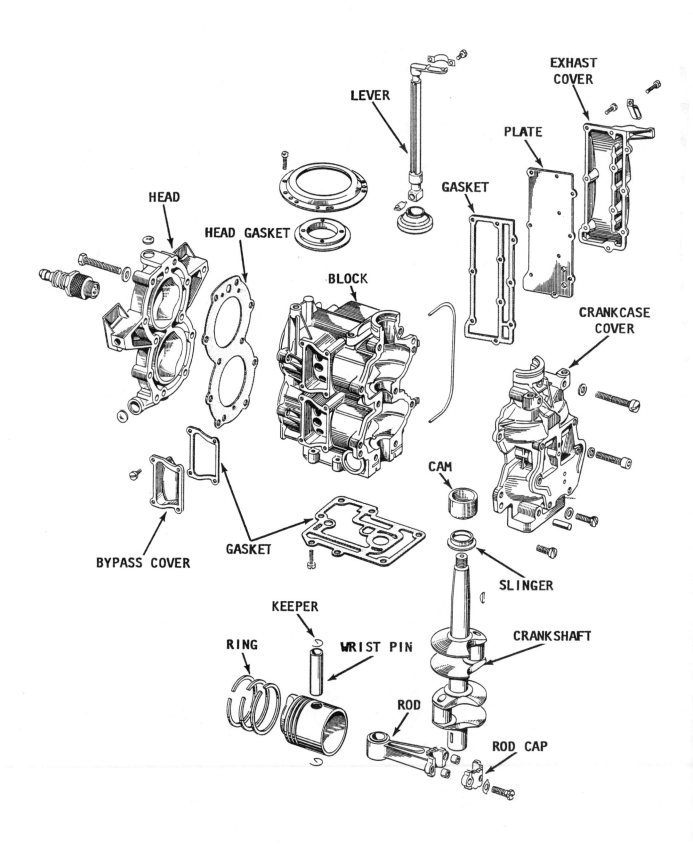

Exploded drawing of a 7.5 hp powerhead -- 1956-58, with principle parts identified.

CYLINDER BLOCK 3-47

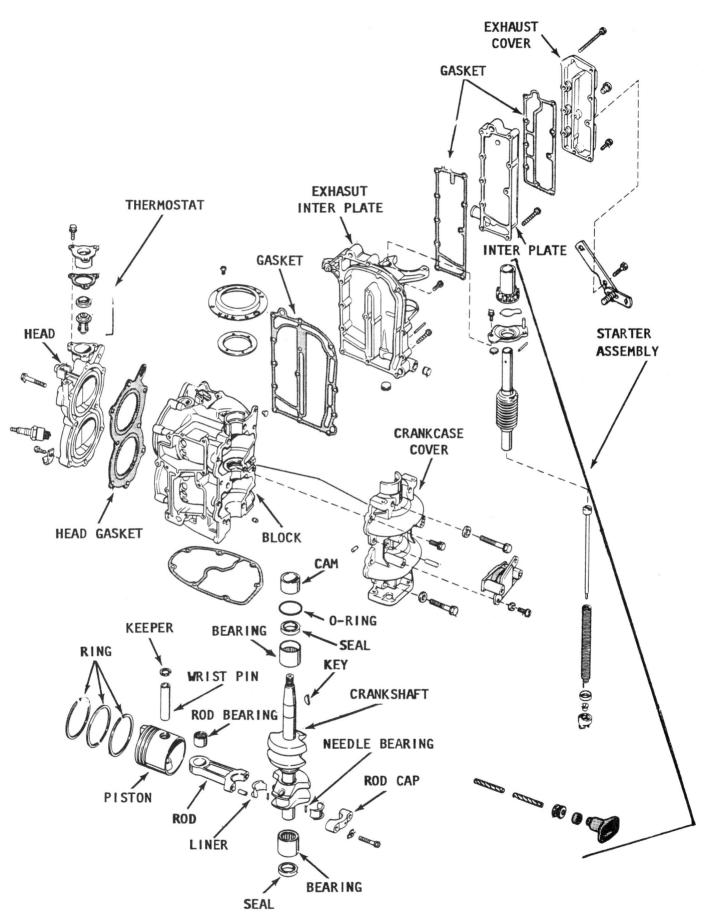

Exploded drawing of a 9.5 hp powerhead -- 1964-70, with principle parts identified.

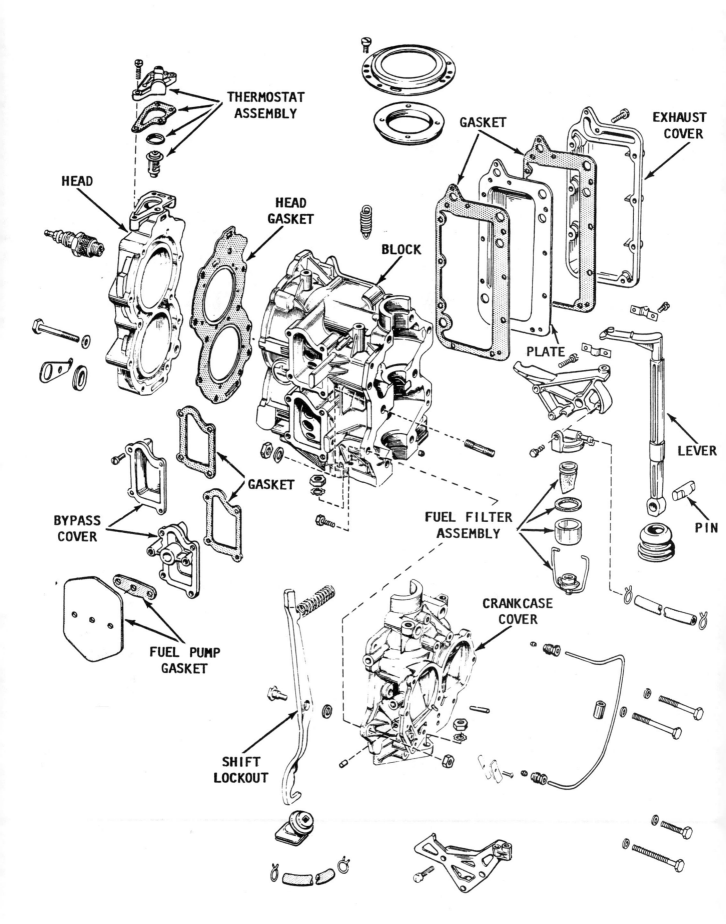

Exploded drawing of a 10 hp powerhead -- 1956-63, with principle parts identified.

CYLINDER BLOCK   3-49

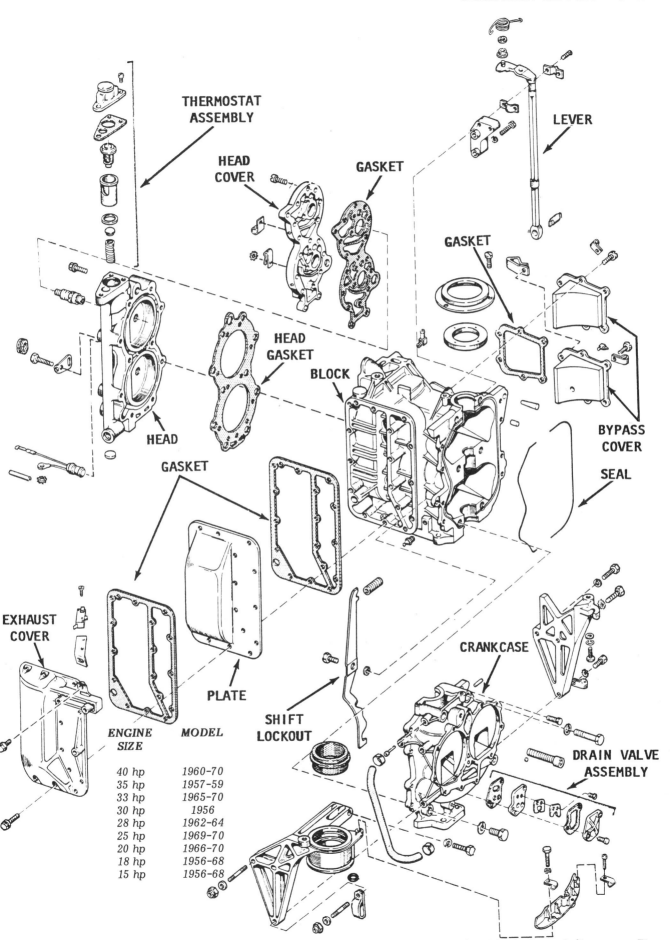

Exploded drawing of a powerhead, with principle parts idetntified for the units listed in the lower left corner. The crankshaft and associated parts are shown on the following page.

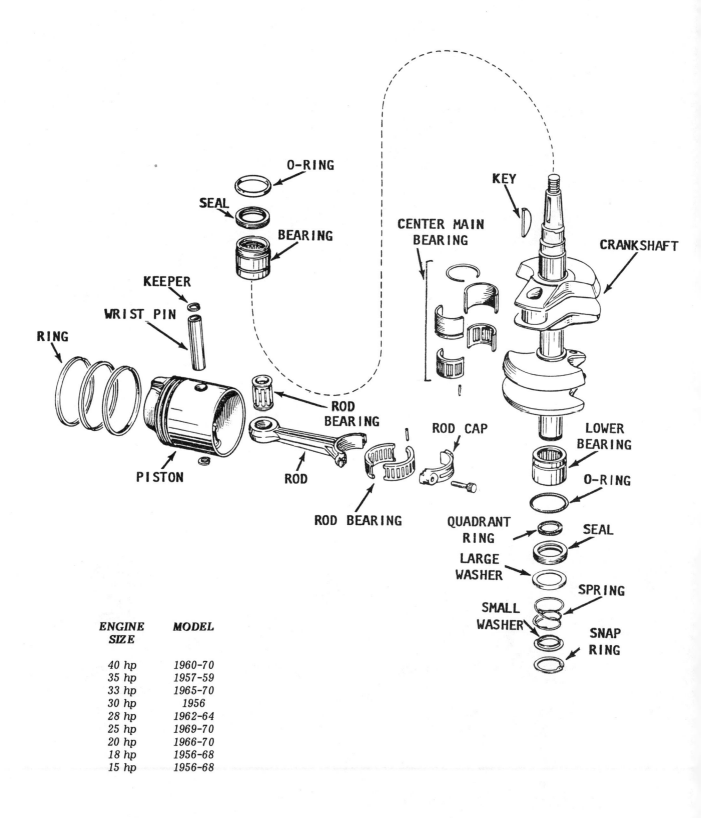

Exploded drawing of the crankshaft and associated parts for the powerhead shown on the previous page.

4
FUEL

4-1 INTRODUCTION

The carburetion and ignition principles of two-cycle engine operation **MUST** be understood in order to perform a proper tune-up on an outboard motor.

If you have any doubts concerning your understanding of two-cycle engine operation, it would be best to study the operation theory section in the first portion of Chapter 3, before tackling any work on the fuel system.

4-2 GENERAL CARBURETION INFORMATION

The carburetor is merely a metering device for mixing fuel and air in the proper proportions for efficient engine operation. At idle speed, an outboard engine requires a mixture of about 8 parts air to 1 part fuel. At high speed or under heavy duty service, the mixture may change to as much as 12 parts air to 1 part fuel.

Float Systems

A small chamber in the carburetor serves as a fuel reservoir. A float valve admits fuel into the reservoir to replace the fuel consumed by the engine.

Fuel level in each chamber is extremely critical and must be maintained accurately. Accuracy is obtained through proper adjustment of the float. This adjustment will provide a balanced metering of fuel to each cylinder at all speeds.

Following the fuel through its course, from the fuel tank to the combustion chamber of the cylinder, will provide an appreciation of exactly what is taking place. In order to start the engine, the fuel must be moved from the tank to the carburetor by a squeeze bulb installed in the fuel line.

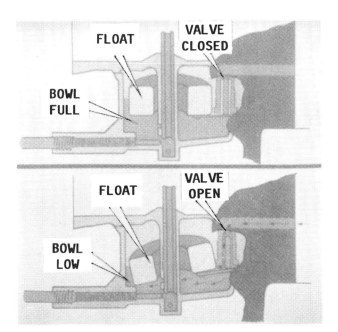

Fuel flow principle of a modern carburetor.

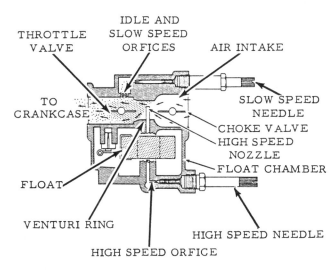

Fuel flow through the venturi, showing principle and related parts controlling intake and outflow.

4-2 FUEL

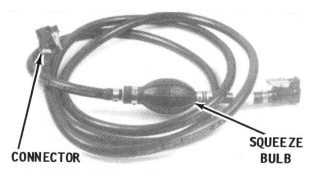

Typical fuel line with squeeze bulb and quick-disconnect fitting at each end. These items may be purchased as an assembled unit.

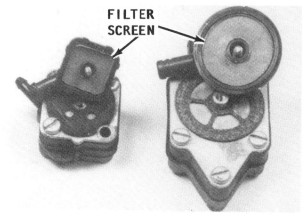

Two different type fuel pumps with the covers removed to show the filter screen. These pumps cannot be rebuilt. The only service possible is to clean the screens with solvent and then blow them dry.

On models produced from 1956 through 1959, the fuel tank is equipped with a hand pressure pump. Operation of this pump forces fuel to the carburetor for starting. After engine start, the pressure from one cylinder is fed through a hose to the tank. Therefore, the pressurized fuel tank maintains a steady fuel supply to the engine while the engine is operating. This action is necessary because the fuel pump does not have sufficient pressure to draw fuel from the tank during cranking before the engine starts.

Since 1960, most fuel systems are equipped with a manually-operated squeeze bulb in the line to transfer fuel from the tank to the engine until the engine starts.

After the engine starts, the fuel passes through the fuel pump to the carburetor. All systems have some type of filter installed somewhere in the line between the tank and the carburetor, except the old-style pressure fuel tank systems. With the pressure fuel tank arrangement, the filter is actually an intergral part of the fuel pickup in the tank. Many units have a filter as an integral part of the carburetor.

At the carburetor, the fuel passes through the inlet passage to the needle and seat, and then into the float chamber (reservoir). A float in the chamber rides up and down on the surface of the fuel. After fuel enters the chamber and the level rises to a predetermined point, a tang on the float closes the inlet needle and the flow of entering the chamber is cutoff. When fuel leaves the chamber as the engine operates, the fuel level drops and the float tang allows the inlet needle to move off its seat and fuel once again enters the chamber. In this manner a constant reservoir of fuel is maintained in the chamber to satisfy the demands of the engine at all speeds.

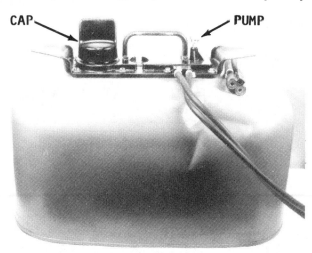

Obsolete pressure-type fuel tank used extensively until about 1958. The system was replaced by the squeeze bulb arrangement shown at the top of this page.

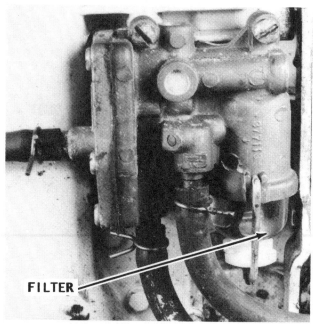

Fuel pump installed on the engine. This type pump may be rebuilt, as explained in the text.

A fuel chamber vent hole is located near the top of the carburetor body to permit atmospheric pressure to act against the fuel in each chamber. This pressure assures an adequate fuel supply to the various operating systems of the engine.

Air/Fuel Mixture

A suction effect is created each time the piston moves upward in the cylinder. This suction draws air through the throat of the carburetor. A restriction in the throat, called a venturi, controls air velocity and has the effect of reducing air pressure at this point.

The difference in air pressures at the throat and in the fuel chamber, causes the fuel to be pushed out metering jets extending down into the fuel chamber. When the fuel leaves the jets, it mixes with the air passing through the venturi. This fuel/air mixture should then be in the proper proportion for burning in the cylinder/s for maximum engine performance.

In order to obtain the proper air/fuel mixture for all engine speeds, high- and low-speed needle valves are installed. On late-model engines, the high-speed needle valve was replaced with a high-speed orifice. There is no adjustment with the orifice type. These needle valves are used to compensate for changing atmospheric conditions. Probably over 50% of the engines covered in this manual have an adjustable high- and low-speed needle valve.

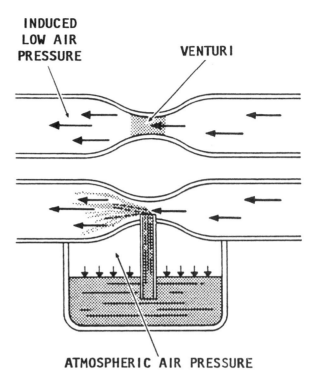

Air flow principle of a modern carburetor.

Engine operation at sea level compared with performance at high altitudes is quite noticeable. A throttle valve controls the volume of air/fuel mixture drawn into the engine. A cold engine requires a richer fuel mixture to start and during the brief period it is warming to normal operating temperature. A choke valve is placed ahead of the metering jets and venturi to provide the extra amount of air required for start and while the engine is cold.

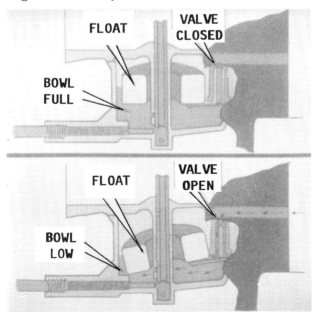

Fuel flow controlled by the float in a typical carburetor. The inlet valve is opened and closed by movement of the float in the carburetor bowl.

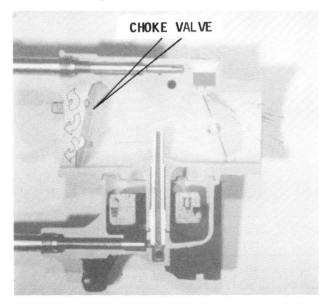

Choke valve location in the carburetor venturi. The choke valve on most Johnson/Evinrude carburetors is located in front of the venturi.

4-4 FUEL

When this choke valve is closed, a very rich fuel mixture is drawn into the engine.

The throat of the carburetor is usually referred to as the "barrel." Carburetors installed on engines covered in this manual all have a single metering jet with a single throttle and choke plate. Single barrel carburetors are fed by one float and chamber.

4-3 FUEL SYSTEM

The fuel system includes the fuel tank, fuel pump, fuel filters, carburetor, connecting lines, with a squeeze bulb, and the associated parts to connect it all together. Regular maintenance of the fuel system to obtain maximum performance, is limited to changing the fuel filter at regular intervals and using **FRESH** fuel. Even with the high price of fuel, removing gasoline that has been standing unused over a long period of time is still the easiest and least expensive preventive maintenance possible.

In most cases this old gas, even with some oil mixed with it, can be used without harmful effects in an automobile using regular gasoline.

If a sudden increase in gas consumption is noticed, or if the engine does not perform

Damaged piston, possibly caused by insufficient oil mixed with the fuel; using too-low an octane fuel; or using fuel that had "soured" (stood too long without a preservative additive).

properly, a carburetor overhaul, including boil-out, or replacement of the fuel pump may be required.

4-4 TROUBLESHOOTING

The following paragraphs provide an orderly sequence of tests to pinpoint problems in the system. If an engine has not been used for some time and fuel has remained in the carburetor, it is possible that varnish may have formed. Such a condition could be the cause of hard starting, or complete failure of the engine to operate.

Fuel Problems

Many times fuel system troubles are caused by a plugged fuel filter, a defective

Commercial additives, such as Sta-bil, may be used to keep the fuel in the tank fresh. Under favorable conditions, such additives will prevent the fuel from "souring" for up to twelve months.

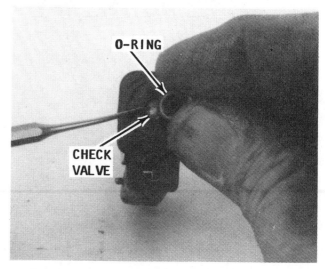

Fuel connector with the O-ring visible. These O-rings have a relative short life and must be replaced at regular intervals, as detailed in the text.

fuel pump, or by a leak in the line from the fuel tank to the fuel pump. Aged fuel left in the carburetor and the formation of varnish could cause the needle to stick in its seat and prevent fuel flow into the bowl. A defective choke may also cause problems. **WOULD YOU BELIEVE,** a majority of starting troubles, which are traced to the fuel system, are the result of an empty fuel tank or aged fuel.

Fuel will begin to sour in three to four months and will cause engine starting problems. Therefore, leaving the motor setting idle with fuel in the carburetor, lines, or tank during the off-season, usually results in very serious problems. A fuel additive such as Sta-Bil may be used to prevent gum from forming during storage or prolonged idle periods.

For many years there has been the widespread belief that simply shutting off the fuel at the tank and then running the engine until it stops is the proper procedure before storing the engine for any length of time. Right? **WRONG.**

First, it is **NOT** possible to remove all fuel in the carburetor by operating the engine until it stops. Considerable fuel is trapped in the float chamber and other passages and in the line leading to the carburetor. The **ONLY** guaranteed method of removing **ALL** fuel is to take the time to remove the carburetor, and drain the fuel.

Secondly, if the engine is operated with the fuel supply shut off until it stops the fuel and oil mixture inside the engine is removed, leaving bearings, pistons, rings, and other parts without any protective lubricant.

Proper procedure involves: Shutting off the fuel supply at the tank; disconnecting the fuel line at the tank; operating the engine until it begins to run **ROUGH;** then stopping the engine, which will leave some fuel/oil mixture inside; and finally removing and draining the carburetor. By disconnecting the fuel supply, all **SMALL** passages are cleared of fuel even though some fuel is left in the carburetor. A light oil should be put in combustion chamber as instructed in the Owners Manual. On some model carburetors, the high-speed orifice plug can be removed to drain the fuel from the carburetor.

Choke Problems

When the engine is hot, the fuel system can cause starting problems. After a hot engine is shut down, the temperature inside the fuel bowl may rise to 200°F and cause the fuel to actually boil. All carburetors are vented to allow this pressure to escape to the atmosphere. However, some of the fuel may percolate over the high-speed nozzle.

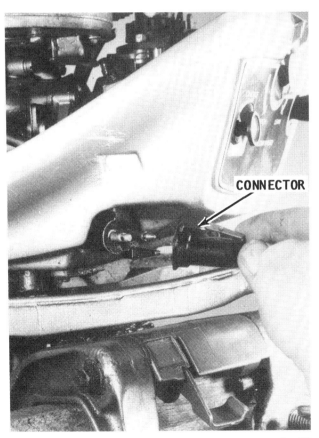

Female fuel line connector ready to be mated with the male portion of the connector.

Fouled spark plug, possibly caused by operator's habit of over-choking or a malfunction holding the choke closed. Either of these conditions delivered a too-rich fuel mixture to the cylinder.

4-6 FUEL

If the choke should stick in the open position, the engine will be hard to start. If the choke should stick in the closed position, the engine will flood making it **VERY** difficult to start.

In order for this raw fuel to vaporize enough to burn, considerable air must be added to lean out the mixture. Therefore, the only remedy is to remove the spark plug/s; ground the leads; crank the engine through about 10 revolutions; clean the plugs; install the plugs again; and start the engine.

If the needle valve and seat assembly is leaking, an excessive amount of fuel may enter the intake manifold in the following manner: After the engine is shut down, the pressure left in the fuel line will force fuel past the leaking needle valve. This extra fuel will raise the level in the fuel bowl and cause fuel to overflow into the intake manifold.

A continuous overflow of fuel into the intake manifold may be due to a sticking inlet needle or to a defective float which would cause an extra high level of fuel in the bowl and overflow into the intake manifold.

FUEL PUMP TESTS

CAUTION: Gasoline will be flowing in the engine area during this test. Therefore, guard against fire by grounding the high-tension wire to prevent it from sparking.

Grounding the spark plug wires during fuel tests to prevent an accidental spark from igniting fuel or fuel vapors. Exercise care when working on the fuel system.

Testing System with Squeeze Bulb

An adequate safety method, is to ground each spark plug lead. Disconnect the fuel line at the carburetor. Place a suitable container over the end of the fuel line to catch the fuel discharged. Insert a small screwdriver into the end of the line to open the check valve, and then squeeze the primer bulb and observe if there is satisfactory fuel flow from the line.

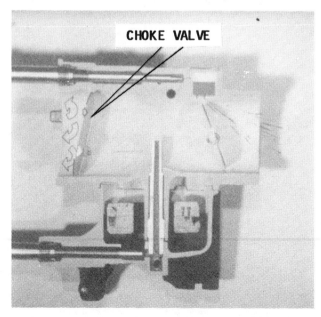

The choke plays a most important role during engine start and in controlling the amount of air entering the carburetor, under various load conditions.

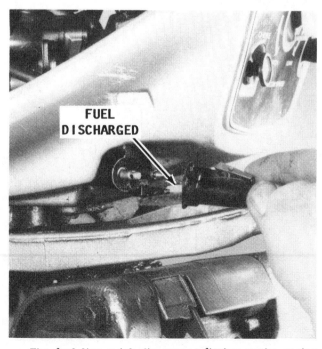

The fuel line quick-disconnect fitting at the engine separated in preparation to making a fuel flow test.

If there is no fuel discharged from the line, the check valve in the squeeze bulb may be defective, or there may be a break or obstruction in the fuel line.

If there is a good fuel flow, then crank the engine. If the fuel pump is operating properly, a healthy stream of fuel should pulse out of the line.

Continue cranking the engine and catching the fuel for about 15 pulses to determine if the amount of fuel decreases with each pulse or maintains a constant amount. A decrease in the discharge indicates a restriction in the line. If the fuel line is plugged, the fuel stream may stop. If there is fuel in the fuel tank but no fuel flows out the fuel line while the engine is being cranked, the problem may be in one of several areas:

1- Plugged fuel line from the fuel pump to the carburetor.

2- Defective O-ring in fuel line connector into the fuel tank.

3- Defective O-ring in fuel line connector into the engine.

4- Defective fuel pump.

5- The line from the fuel tank to the fuel pump may be plugged; the line may be leaking air; or the squeeze bulb may be defective.

6- Defective fuel tank.

7- If the engine does not start even though there is adequate fuel flow from the fuel line, the fuel inlet needle valve and the seat may be gummed together and prevent adequate fuel flow.

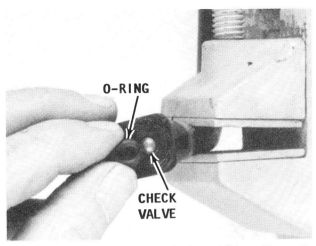

Fuel connector with the O-ring visible. These O-rings have a relative short life and may be the source of fuel problems. The O-rings should be replaced on a regular basis.

FUEL LINE TEST

On most installations, the fuel line is provided with quick-disconnect fittings at the tank and at the engine. If there is reason to believe the problem is at the quick-disconnects, the hose ends can be replaced as an assembly, or new O-rings may be installed. A supply of new O-rings

Fuel line connection at the carburetor on a 9.5 horsepower engine.

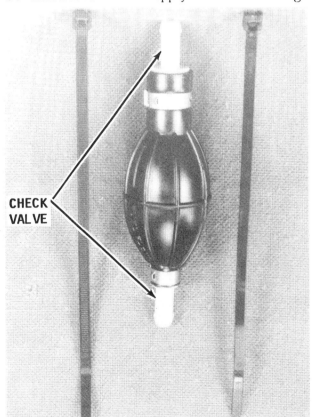

Parts in a squeeze bulb replacement kit include the squeeze bulb, two check valves, and two tie straps to secure the bulb in the line.

4-8 FUEL

A typical squeeze bulb with the directional arrow clearly visible. The squeeze bulb must be installed with the arrow pointing in the direction of fuel flow, toward the engine.

should be carried on board for use in isolated areas where a marine store is not available. For a small additional expense, the entire fuel line can be replaced and eliminate this entire area as a problem source for many future seasons.

The primer squeeze bulb can be replaced in a short time. A squeeze bulb assembly, complete with the check valves installed, may be obtained from the local OMC dealer.

An arrow is clearly visible on the squeeze bulb to indicate the direction of fuel flow. The squeeze bulb **MUST** be installed correctly in the line because the check valves in each end of the bulb will allow fuel to flow in **ONLY** one direction. Therefore, if the squeeze bulb should be installed backwards, in a moment of haste to get the job done, fuel will not reach the carburetor.

To replace the bulb, first unsnap the clamps on the hose at each end of the bulb. Next, pull the hose out of the check valves at each end of the bulb. New clamps are included with a new squeeze bulb.

If the fuel line has been exposed to considerable sunlight, it may have become hardened, causing difficulty in working it over the check valve. To remedy this situation, simply immerse the ends of the hose in boiling water for a few minutes to soften the rubber. The hose will then slip onto the check valve without further problems. After the lines on both sides have been installed, snap the camps in place to secure the line. Check a second time to be sure the arrow is pointing in the fuel flow direction, **TOWARDS** the engine.

TESTING SYSTEM WITH PRESSURE TANK

Some early model small engines produced by OMC used a type of fuel system utilizing two hoses connected to the engine. To prime this system, a primer pump on the fuel tank, is worked to manually pump fuel to the engine. Working this pump does **NOT** pressurize the tank, all that is accomplished is to deliver fuel to the carburetor. Once the engine starts, air is pumped through the second hose and the fuel tank is pressurized to maintain a steady fuel supply to the carburetor. The filler cap on the fuel tank **MUST** be securely in place for this system to function properly.

A release valve installed in the fuel tank will automatically release air if the pressure should build above a predetermined psi rating.

A fuel pump may be installed on any engine equipped with this type of pressure system. Replacement of these pressure tanks has become quite expensive, therefore, the fuel pump option may be worth considering instead of replacing the pressure tank.

To install a fuel pump it is first necessary to remove the intake bypass cover plate alongside the upper or lower cylinder.

Obsolete pressure-type fuel tank. A manually-operated primer pump on the fuel tank is worked to pump fuel to the engine. After engine start, air pressure forces fuel from the tank to the engine.

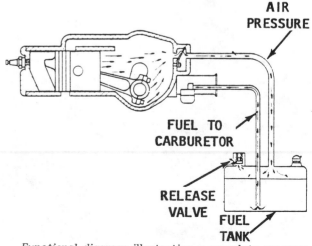

Functional diagram illustrating a complete pressure tank fuel system, from the fuel tank to the cylinder.

Next, drill the appropriate size hole for a suitable fitting to pickup the cylinder vacuum. Install the cover plate with the fitting in place. Mount the fuel pump in a convenient location, and then connect the vacuum hose. Plug the hole in the intake manifold just below the carburetor. Purchase a regular fuel tank. Because the new system will only have one hose, a new quick-disconnect fitting must also be purchased and and connected to the engine.

TROUBLESHOOTING

Observe the stamped markings on the pressure-type fuel tank adjacent to each fitting; one for air and the other for fuel. Also, observe the rib on one hose running the full length from the tank to the engine. This rib will help to identify the same hose at the tank and the engine.

With the fuel tank fill cap securely closed, insert a screwdriver into the fuel line quick-disconnect at the engine to hold the check valve open. Now, work the primer pump on the fuel tank and fuel should squirt out the quick-disconnect because the screwdriver is holding the check valve open.

If fuel is not present at the fitting, the pressure fuel tank primer pump or the check valve in the fuel pickup at the bottom of the fuel tank may be defective.

If the engine runs for short periods and then shuts down from lack of fuel, the problem may be lack of air pressure in the tank. To check for air pressure, after the engine has run for just a short period, open the fill cap very slightly. A hissing sound

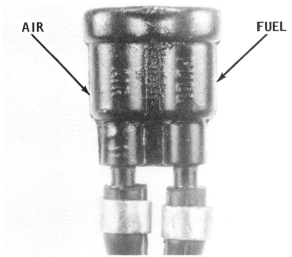

Back side of the double fuel/air connector. In this photograph, the air line is on the left and the fuel on the right.

indicating escaping air should be heard. Lack of air may be caused by defective O-rings at the connectors, or because the check valve in the intake manifold is not properly seated.

Integral Fuel Tank System

On small engines equipped with an integral fuel tank above the flywheel, or elsewhere on the engine, the filter in the tank may be plugged. To check the fuel flow, disconnect the fuel line at the carburetor and fuel should flow from the line if the shut-off valve is open. If fuel is not present, the usual cause is a plugged filter in the fuel tank. It is very difficult to determine if a porcelain-type filter is plugged. The general appearance that the filter is satisfactory may be a false indication. Therefore, if the filter is suspected, the best

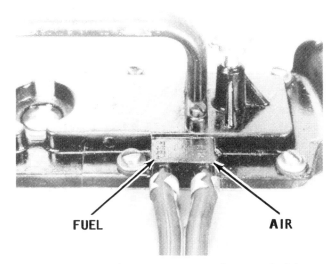

Pressure-type fuel tank with the fuel line (left), and air line (right), connected. Notice the rib on the fuel hose which extends the full length to the engine.

Pressure-type fuel tank with the air and fuel lines attached through a double connector. The fill cap must be tightly closed for the system to operate properly.

4-10 FUEL

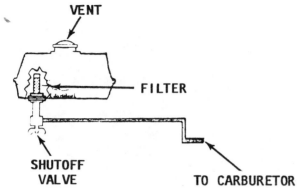

Sectional view of a engine mounted fuel tank showing the filter inside the tank and the external shut-off valve.

remedy is replacement. The cost is very modest and this one area is thus eliminated as a problem source.

Would you believe, many times lack of fuel at the carburetor is caused because the vent on the fuel tank was not opened.

To remove the filter, first remove the fuel tank filler cap, turn the engine upside down and drain the tank. Next, disconnect the fuel line to the carburetor. Remove the fuel shut-off valve. In most cases, the filter will remain with the valve and can be separated from the valve by removing the fitting. If the fitting and filter do not remain with the valve when it is removed, then remove the fitting and filter from the tank.

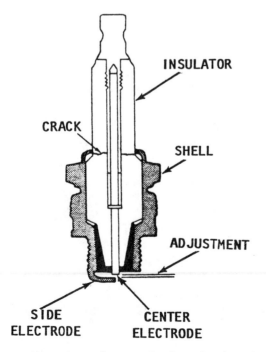

Cross-section view of a spark plug showing the principle parts with important comments for satisfactory service.

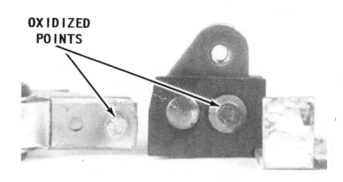

A set of points unfit for service due to oxidation.

ROUGH ENGINE IDLE

If an engine does not idle smoothly, the most reasonable approach to the problem is to perform a tune-up to eliminate such areas as: defective points; faulty spark plugs; and synchronization out of adjustment.

Other problems that can prevent an engine from running smoothly include: An air leak in the intake manifold; uneven compression between the cylinders; and sticky or broken reeds.

Of course any problem in the carburetor affecting the air/fuel mixture will also prevent the engine from operating smoothly at idle speed. These problems usually include: Too high a fuel level in the bowl; a heavy float; leaking needle valve and seat; defective automatic choke; and improper idle and high-speed needle valve adjustments.

"Sour" fuel (fuel left in a tank without a preservative additive) will cause an engine to run rough and idle with great difficulty.

Inspection of the spark plugs should become a "first" whenever the engine fails to perform properly.

TROUBLESHOOTING 4-11

Carburetor installed on most 35 hp and 40 hp engines.

EXCESSIVE FUEL CONSUMPTION

Excessive fuel consumption can result from one of three conditions, or a combination of all three.

1- Inefficient engine operation.
2- Damaged condition of the hull, including excessive marine growth.
3- Poor boating habits of the operator.

If the fuel consumption suddenly increases over what could be considered normal, then the cause can probably be attributed to the engine or boat and not the operator.

Marine growth on the hull can have a very marked effect on boat performance. This is why sail boats always try to have a haul-out as close to race time as possible.

Marine growth allowed to accumulate on the lower unit will create "drag" and seriously hamper boat performance, and corrode the metal if it is not removed.

A clean engine reflecting the owners "pride of ownership", and good care.

While you are checking the bottom take note of the propeller condition. A bent blade or other damage will definitely cause poor boat performance.

If the hull and propeller are in good shape, then check the fuel system for possible leaks. Check the line between the fuel pump and the carburetor while the engine is running and the line between the fuel tank and the pump when the engine is not running. A leak between the tank and the pump many times will not appear when the engine is operating, because the suction created by the pump drawing fuel will not allow the fuel to leak. Once the engine is turned off and the suction no longer exists, fuel may begin to leak.

If a minor tune-up has been performed and the spark plugs, points, and synchronization are properly adjusted, then the problem most likely is in the carburetor, indicating an overhaul is in order. Check for leaks at

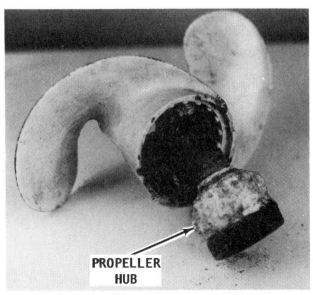

A corroded hub on a small engine propeller. Replacement of this propeller will be less expensive than the cost of a rebuild.

4-12 FUEL

the needle valve and seat. Use extra care when making any adjustments affecting the fuel consumption, such as the float level or automatic choke.

ENGINE SURGE

If the engine operates as if the load on the boat is being constantly increased and decreased, even though an attempt is being made to hold a constant engine speed, the problem can most likely be attributed to the fuel pump.

Operational description and service procedures for the fuel pump are given in Section 4-9.

4-5 JOHNSON/EVINRUDE CARBURETORS

This section provides complete detailed procedures for removal, disassembly, cleaning and inspecting, assembling including bench adjustments, installation, and operating adjustments for the three OMC carburetors installed on engines covered in this manual. The Type I carburetor is installed on the 1-1/2 to 28 hp engines; the Type II on the 30 to 40 hp models; and the Type III carburetor is used on the 9-1/2 hp model. The carburetors are further identified by the type of choke system used. One has a manual choke; the second, a heat and electric combination choke; and the third has a water choke. The following table lists the types of carburetors, the horsepower and model engines equipped with each.

CARBURETOR INSTALLATIONS

Type I
Single barrel, front draft,
with adjustable low-speed
and high-speed needle valves.

1.5 hp	1968-70	5.5 hp	1959-63
3.0 hp	1956-62	10 hp	1962-63
5.0 hp	1965-68	18 hp	1957-63
5.5 hp	1959-63	28 hp	1962-64

Type I
Single barrel, front draft,
with adjustable low-speed
and high-speed needle valve,
and filter bowl attached to carburetor.

5.5 hp	1956-58	10 hp	1956-61
7.5 hp	1956-58	15 hp	1956

Type I
Single barrel, front draft,
with adjustable low-speed
needle valve.
High-speed is fixed orifice
in float bowl.

3.0 hp	1963-68	18 hp	1964-68
5.5 hp	1964	20 hp	1966-68
6.0 hp	1965-70		

Type I
Single barrel, front draft,
with adjustable low-speed
needle valve with special packing.
High-speed is fixed orifice
in float bowl.

18 hp	1969-70	25 hp	1969-70
20 hp	1969-70		

Type II
Single barrel, front draft,
with adjustable low-speed
and high-speed needle valves.

30 hp	1956	35 hp	1957-59
33 hp	1965-70	40 hp	1960

Type II
Single barrel, front draft,
with adjustable low-speed
needle valve.
High-speed is fixed orifice
in float bowl.

40 hp   1964-1970

Type III
Single barrel, down draft,
with adjustable low-speed
needle valve.
High-speed is fixed orifice
in float bowl.

9.5 hp   1964-70

See Section 4-6 for service details for each choke systems.

For synchronizing adjustments with the ignition system, see Chapter 5.

J/E CARBURETORS

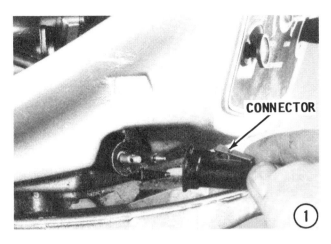

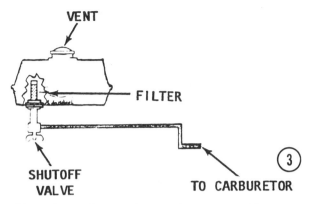

4-6 TYPE I CARBURETOR

REMOVAL AND DISASSEMBLING

1- On engines using an electric cranking motor, remove the battery leads from the battery terminals. Disconnect the fuel line from the engine or from the fuel tank at the quick-disconnect fitting.

2- Remove the hood assembly from the engine. Remove the choke and throttle linkage to the carburetor.

3- Remove the fuel line from the carburetor. This may be accomplished by either one of two methods: On engines equipped with a self-contained fuel tank, close the fuel shut-off valve, located just below the tank. Disconnect both ends, and then remove the copper fuel line between the shut-off valve and the carburetor.

4- On engines utilizing a separate fuel tank, remove the tie-strap or clamp securing the rubber hose connecting the fuel connector to the carburetor. Remove the rubber hose from the carburetor. Remove the nuts securing the carburetor to the crankcase. The carburetor may have to moved slightly forward as the nuts are loosened in order to obtain clearance for the nuts to clear the studs. After the nuts are clear, lift the carburetor from the intake manifold.

5- Remove and **DISCARD** the gasket from the intake manifold or the carburetor, if the gasket adhered to the carburetor when it was removed. A new gasket is included in a carburetor repair kit.

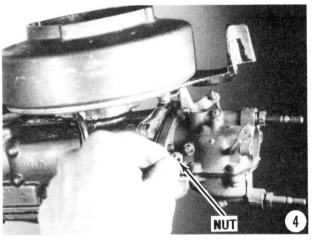

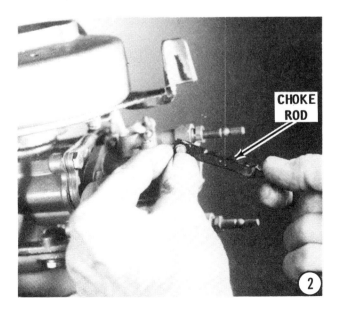

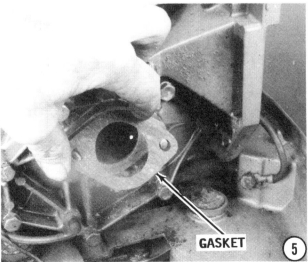

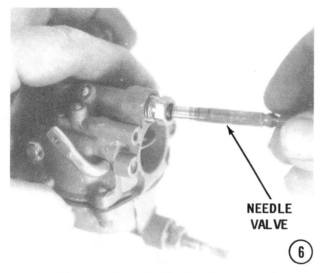

6- Observe how both the low-speed needle valve (the top one), and the high-speed needle valve (the bottom one), are secured in the carburetor body with a packing sleeve. Use a 7/16" box-end wrench and remove each sleeve. After the sleeves are removed, turn each needle valve counterclockwise until they are free of the carburetor body. If the carburetor being serviced does not have a high-speed needle valve, then it is equipped with a high-speed orifice and covered with a plug in approximately the same location in the carburetor as the high-speed needle valve. Therefore, if there is a plug, instead of the high-speed needle valve, remove the plug.

7- Use a small screwdriver, and remove the packing from the needle valve cavities in the carburetor body.

8- If installed, remove the high-speed orifice, using the proper size screwdriver.

9- Turn the carburetor upside down. If a filter bowl assembly is installed, loosen the hinge and swing it to one side. Remove the glass bowl.

10- Back off the flat knurled washer, and then remove the gasket and filter from inside the float bowl.

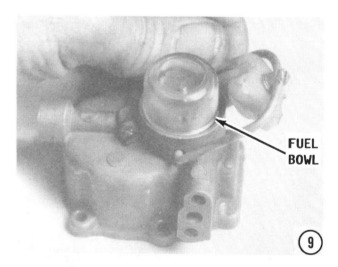

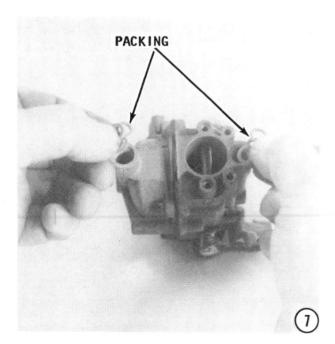

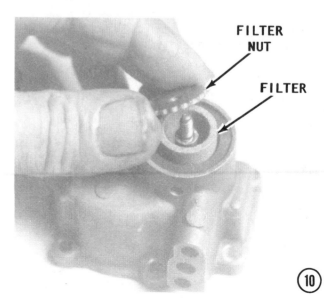

J/E CARBURETORS

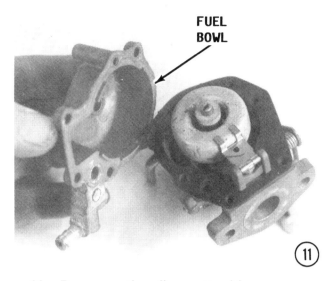

11- Remove the five attaching screws, and then lift the bowl free of the carburetor.

12- Remove the hinge pin and lift the float from the carburetor bowl cavity. As the float is lifted from the carburetor, observe the small spring attached to the needle. The needle will come out with the float assembly. Remove the needle seat from the carburetor.

GOOD WORDS

Further disassembly of the carburetor is not necessary. The nozzle in the center of the carburetor does **NOT** have to be removed in order to properly clean the carburetor. However, the nozzle can be removed with a screwdriver if it is damaged and needs to be replaced.

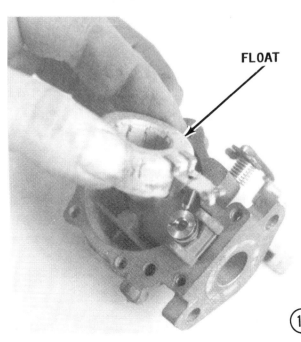

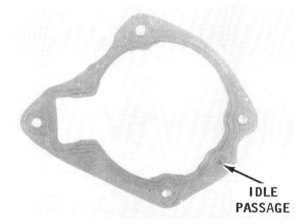

Gasket used between the bowl and the carburetor showing the opening to allow fuel to leave the bowl and enter the idle passage of the carburetor.

CLEANING AND INSPECTING

NEVER dip rubber parts, plastic parts, diaphragms, or pump plungers in carburetor cleaner. These parts should be cleaned **ONLY** in solvent, and then blown dry with compressed air.

Place all metal parts in a screen-type tray and dip them in carburetor cleaner until they appear completely clean, then blow them dry with compressed air.

Blow out all passages in the castings with compressed air. Check all parts and passages to be sure they are not clogged or contain any deposits. **NEVER** use a piece of wire or any type of pointed instrument to clean drilled passages or calibrated holes in a carburetor.

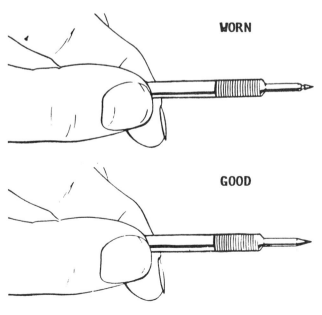

Carburetor idle adjustment screws. The top screw is worn and unfit for service. The bottom screw is new.

4-16 FUEL

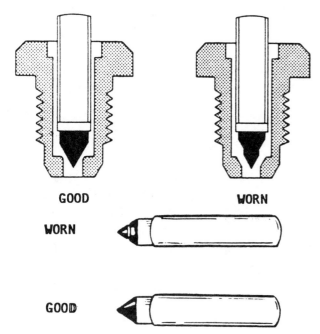

Cross-section of the needle and seat arrangement on Johnson/Evinrude carburetors, showing a worn and new needle for comparison.

Move the throttle shaft back-and-forth to check for wear. If the shaft appears to be too loose, replace the complete throttle body because individual replacement parts are **NOT** available.

Inspect the main body, airhorn, and venturi cluster gasket surfaces for cracks and burrs which might cause a leak. Check the float for deterioration. Check to be sure the float spring has not been stretched. If any part of the float is damaged, the unit must be replaced. Check the float arm needle contacting surface and replace the float if this surface has a groove worn in it.

Inspect the tapered section of the idle adjusting needles and replace any that have developed a groove.

If a high-speed orifice is installed on the carburetor being serviced, check the orifice for cleanliness. The orifice has a stamped number. This number represents a drill size. Check the orifice with the shank of the proper size drill to verify the proper orifice is used. The correct size orifice for the engine and carburetor being serviced may be obtained from the local OMC dealer.

Most of the parts that should be replaced during a carburetor overhaul are included in overhaul kits available from your local marine dealer. One of these kits will contain a matched fuel inlet needle and seat. This combination should be replaced each time the carburetor is disassembled as a precaution against leakage.

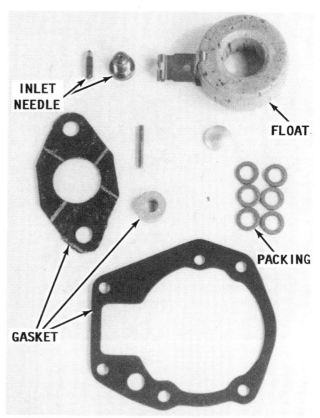

Parts necessary to properly rebuild a carburetor are included in a Johnson/Evinrude repair kit.

ASSEMBLING TYPE I CARBURETOR

13- Install a **NEW** inlet nozzle seat with a **NEW** gasket into the carburetor. Slide a **NEW** gasket down over the inlet nozzle onto the surface of the carburetor.

14- Attach a **NEW** inlet needle onto the spring included in the carburetor repair kit,

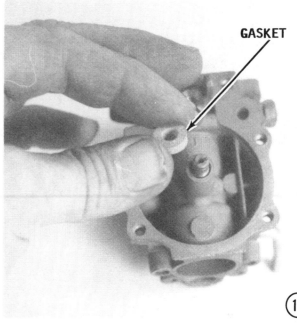

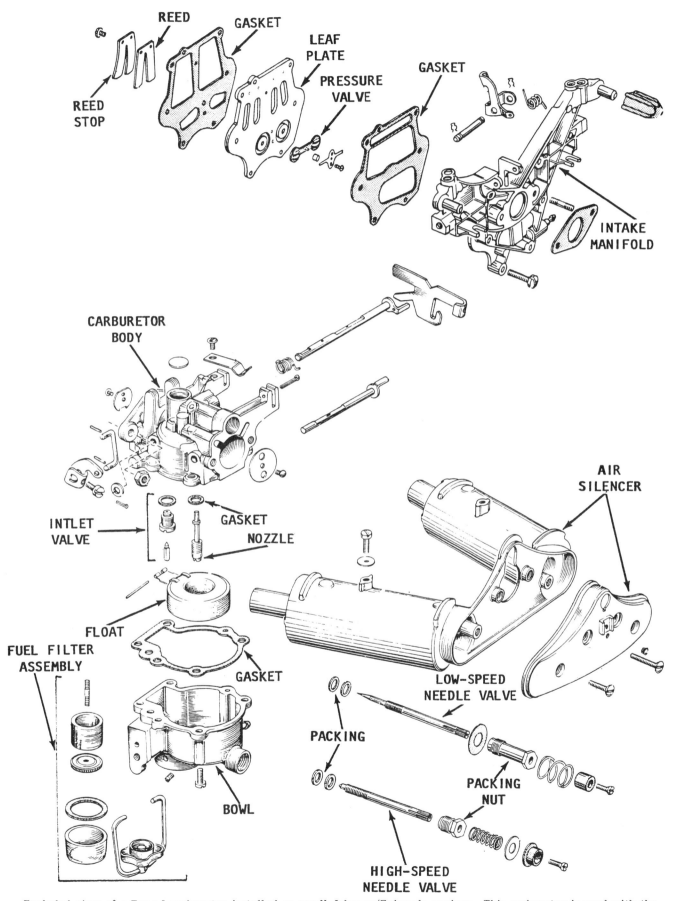

Exploded view of a Type I carburetor installed on small Johnson/Evinrude engines. This carburetor is used with the old style pressure-type fuel tank. Therefore, the filter bowl is incorporated in the carburetor.

4-18 FUEL

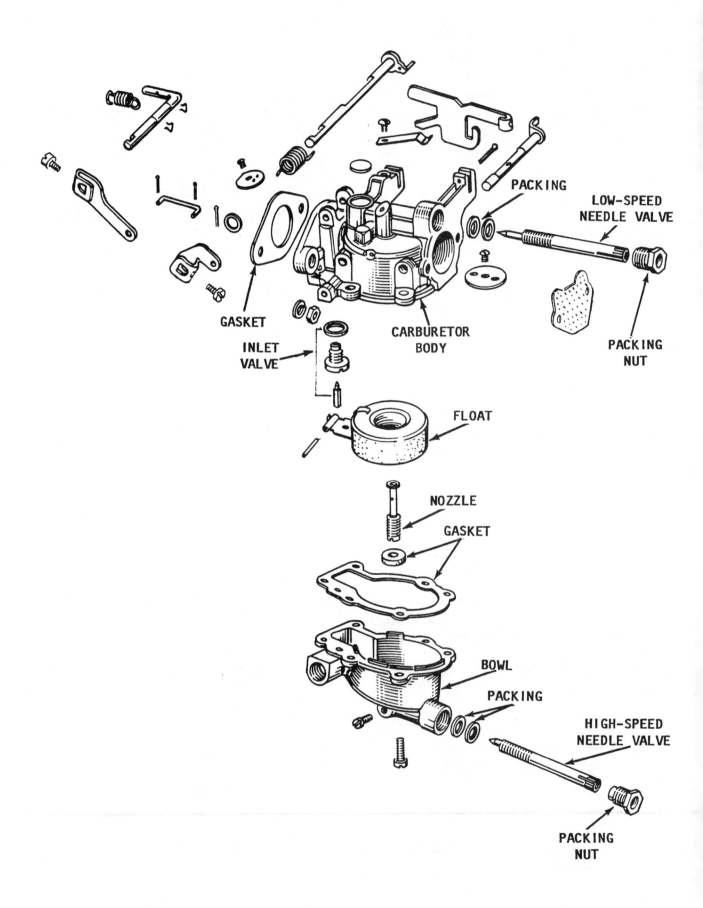

Exploded view of a Type I carburetor installed on small Johnson/Evinrude engines equipped with the fuel tank attached to the engine. The filter on these units is in the tank.

J/E CARBURETORS 4-19

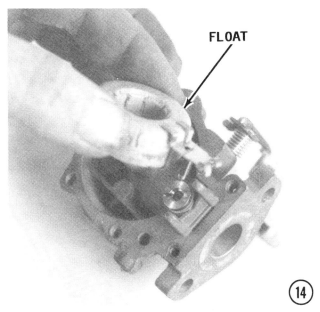

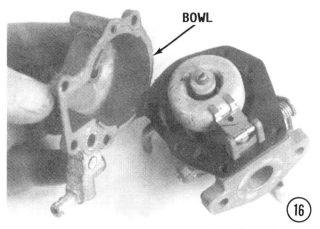

and slip the spring over the edge of the float, as shown. Apply just a drop of oil to the inlet needle and then lower the needle into the seat. Install a **NEW** hinge pin included in the kit.

15- Hold the carburetor in a horizontal position, as shown, and observe the attitude of the float. The float must be level (parallel) with the surface of the carburetor. If the float is not level, use a pair of needle-nose pliers and **CAREFULLY** bend the float tab **SQUARELY** until the float is in the correct position (level with the carburetor). Check to be sure the float is square with the carburetor cavity (one side is not further away than the other).

16- Place the bowl gasket in position on the carburetor, and then position the bowl on top of the gasket. Secure the bowl in place with the five screws.

17- If a filter is used in the float bowl, slip the filter down over the center post in the bowl. Slide the gasket into place in float bowl. Install and tighten the nut securing the filter in place.

18- Place the glass float bowl in position on top of the gasket. Swing the hinge into place and secure it with the thumb nut.

19- Insert three new packing washers into the high-speed and low-speed cavities.

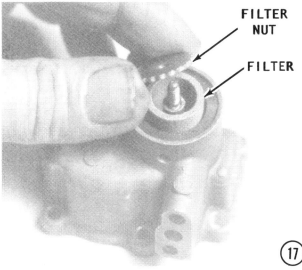

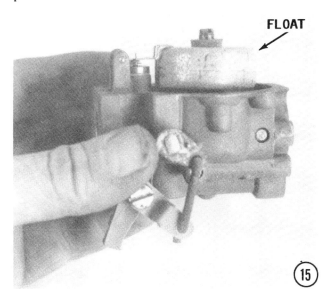

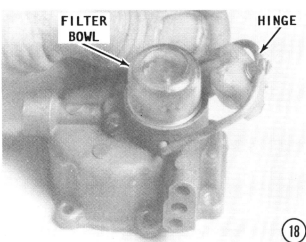

4-20 FUEL

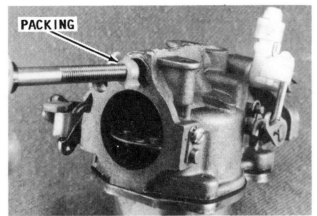

New type packing used for the low-speed needle valve. A packing nut is not used to hold the needle in adjustment.

20- Place a white plastic washer into the high- and low-speed cavities. If a high-speed orifice is used, install the orifice with the proper size screwdirver until the orifice just **BARELY** seats. Install a new gasket on the orifice plug, and then the plug.

21- Start the two sleeve nuts into the carburetor, but **ONLY** far enough to allow the needle valves to be installed.

22- Install the high-speed needle valve into the bottom cavity by rotating it **CLOCKWISE**. Allow the needle valve to seat **LIGHTLY**, then back it out **(COUNTERCLOCKWISE)** 3/4 turn. Now, tighten the sleeve nut only until it is difficult to turn the needle valve by hand.

23- Install the low-speed needle valve into the upper cavity in the same manner as the high-speed needle valve in Step 22. After the valve is seated **LIGHTLY**, back it out **(COUNTERCLOCKWISE)** 1-1/2 turns. Now, tighten the packing nut until there is drag on the needle valve, but the valve may still be rotated by hand, but with just a little difficulty.

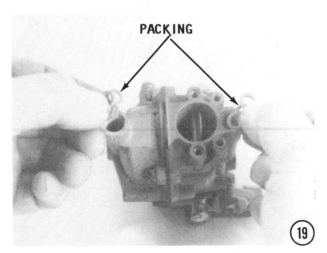

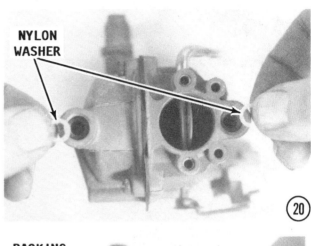

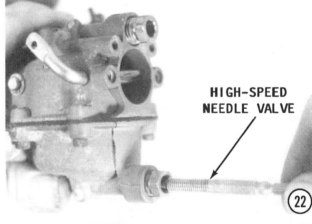

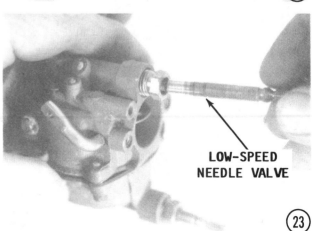

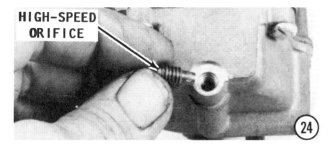

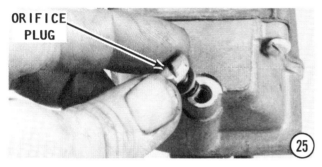

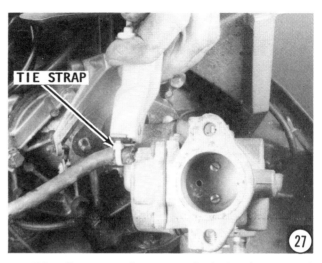

29- Connect the manual choke lever and the front cover onto the carburetor. Thread the knobs onto the low- and high-speed needle valves.

24- Install the high-speed orifice, if one is used, into the float bowl. **ALWAYS** take time to use the proper size screwdriver to prevent damaging the orifice. Tighten the orifice only until it just seats.

25- Install the high-speed orifice plug using a **NEW** gasket. Tighten the plug securely.

26- Check to be sure the surface of the intake manifold is clean and free of any old gasket material. Place a **NEW** gasket in position on the intake manifold.

27- Connect the fuel line to the carburetor, or on engines equipped with a self-contained fuel tank, replace the line between the shut-off valve and the carburetor.

28- Slide the carburetor over the intake manifold mounting studs, and then secure it in place with the two nuts. Tighten the nuts **ALTERNATELY** to avoid warping the carburetor body.

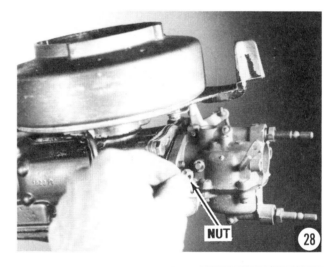

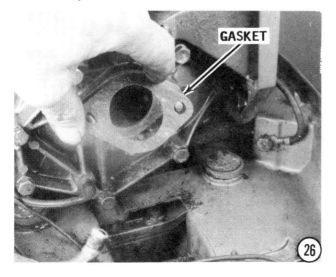

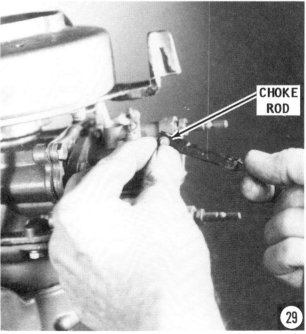

4-22 FUEL

30- Check the synchronization of the fuel and ignition systems according to the procedures outlined in Chapter 5. Mount the engine in a body of water. Connect the fuel line to a fuel source. Prime the engine. If the engine is equipped with a self-contained fuel tank, open the fuel shut-off valve and allow the carburetor to fill with fuel.

CAUTION: Water must circulate through the lower unit to the engine any time the engine is run to prevent damage to the water pump in the lower unit. Just five seconds without water will damage the water pump.

31- Start the engine and allow it to warm to operating temperature. Shift the engine into **FORWARD** gear. Adjust the low-speed idle by turning the low-speed needle valve **CLOCKWISE** until the engine begins to misfire or the rpm drops noticeably. From this point, rotate the needle valve **COUNTERCLOCKWISE** until the engine is operating at the highest rpm. Advance the throttle to the wide open position (WOT). Adjust the high-speed by rotating the high-speed needle valve **CLOCKWISE** until the number of rpm begins to drop, then rotate the high-speed valve **COUNTERCLOCKWISE** until the highest rpm is reached. Return the throttle to idle speed. Adjust the idle speed a second time as described earlier in this step. Again, advance the throttle to the WOT position and check the high-speed adjustment. Return the engine to idle speed. If the engine coughs and operates as

if the fuel is too lean, but the idle and high-speed adjustments have been correctly made, then recheck the synchronization between the fuel and ignition systems. Now, shut off the fuel supply and allow the engine to run until it first begins to misfire from lack of fuel. Retard the spark and shut the engine down. Tighten the sleeve nut securely to prevent the needle valves from changing position through engine vibration while it is operating, but still allow the needle valves to be adjusted by hand using the knob on the end of the valve.

32- The idle stop is located on the port side of the engine, on the outside of the cowling. Adjust the nylon screw inward or outward to obtain the desired idle speed.

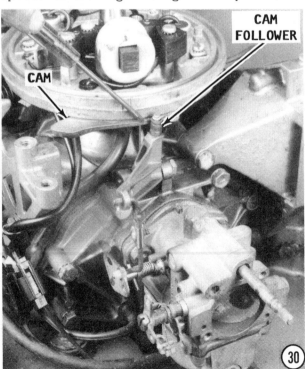

4-7 CHOKE SYSTEM SERVICE

FIRST THESE WORDS

Johnson/Evinrude carburetors may be identified by the type of choke system used. One has a manual choke; the second, a heat and electric combination choke; and the third has a water choke. Detailed instructions will be given in this section for removal of the three different types of choke systems. The following section provides procedures for rebuilding the carburetors. The next section covers installation of the choke systems to the carburetor.

The larger engines covered in this manual (25 hp thru 40 hp) are equipped with electric starter motors and a generator system. In most cases, these items must be moved out of the way to provide clearance for the carburetor. **HOWEVER,** they need only be moved and not removed. Therefore, only attaching hardware need be removed in order to move the item for carburetor clearance. Electrical connections may be left intact. When working on an engine equipped with a combination heat/electrical choke system, the heat tube between the exhaust manifold and the carburetor must be removed **BEFORE** the starter motor can be moved.

EACH carburetor has only one choke system installed. Perform one of the following three procedures depending on the choke system installed on the carburetor of the engine being serviced.

HEAT/ELECTRIC CHOKE — REMOVAL

1- Observe the compression nut under the plate securing the heat tube to the exhaust manifold. Loosen the nut and slide it back on the heat tube. At the other end of the heat tube is another compression nut securing the tube in the choke chamber of the carburetor. Loosen the compression nut and slide it back on the tube. Remove the heat tube from the engine.

2- Disconnect the fuel line at the carburetor. Remove the two nuts from the inlet manifold mounting studs and remove the carburetor. It may be necessary to shift the carburetor slightly as the nuts are loosened in order to obtain clearance for the nuts to be removed from the studs.

3- Observe the marks on the cover plate and on the base. Notice the cover plate has only one mark, but the base has several marks. Identify the relationship of the cover plate and the marks on the base to ensure the cover plate will be installed back into its original position. Remove the three screws securing the cover plate to the choke base.

4- Lift the cover and gasket from the choke base. Observe the cover has a spring on the inside. This spring performs a very important role in the choke system. As the heat from the exhaust manifold moves

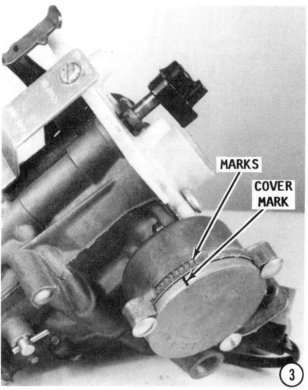

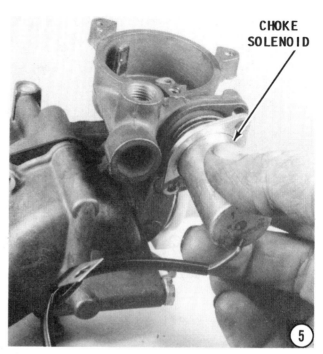

through the tube and around the spring, the spring expands and releases its pressure on the choke. Also notice how the spring is attached to the choke lever. The spring is attached to a plunger which actuates the electrical part of the choke system.

5- Remove the two screws securing the electric choke solenoid to the carburetor. Remove the solenoid and take care to retain the spring inside the solenoid. Remove the choke solenoid gasket.

6- Using a pair of needle-nose pliers, reach inside the choke base and remove the spring from the choke lever. Lower the spring and plunger from the carburetor. Further disassembly of the combination heat/electrical choke is not necessary. If the choke is to be assembled without rebuilding the carburetor, proceed directly to Section 4-8. If the carburetor is to be rebuilt at this time, proceed to Section 4-7.

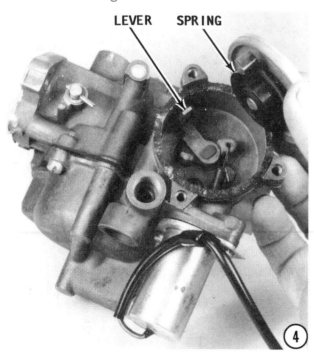

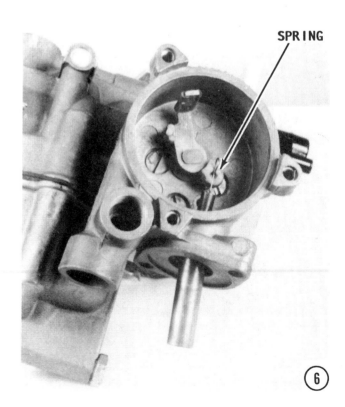

ALL ELECTRIC CHOKE — REMOVAL

The electric choke is mounted on the bottom side of the carburetor.

1- Disconnect the electrical wire from the choke solenoid. Disconnect the fuel line from the carburetor. Remove the two nuts securing the carburetor to the intake manifold. Lift the choke from the engine. Scribe a mark on the side of the bracket to ensure the choke solenoid will be installed in its orginal position.

2- Remove the cotter key, washers, and pin from the choke plunger and pin.

3- Remove the two screws securing the clamp to the carburetor. Remove the choke solenoid from the carburetor. The choke solenoid **CANNOT** be serviced. The boot should be in good condition. It may be removed by sliding it off the solenoid. Observe the two washers and spring under the boot. Take care not to lose these three items. Clean the solenoid, and then slide the boot back in place. The solenoid is then ready to be installed.

4- Remove the cotter pins from each end of the rod extending from the upper section to the lower section of the carburetor, and remove the rod. This rod works the choke in the upper body of the carburetor when the electric choke is activated. Further disassembly of the all electric choke is

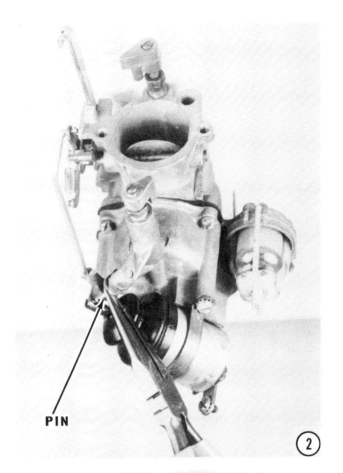

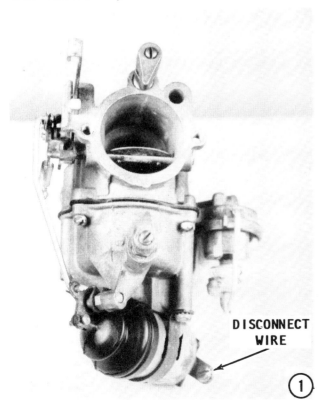
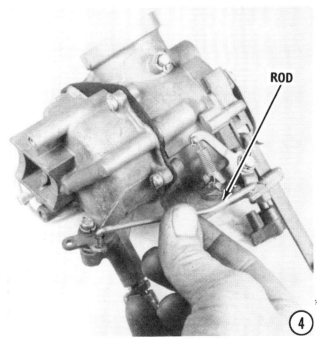

4-26 FUEL

not necessary. If the choke is to be assembled without rebuilding the carburetor, proceed directly to Section 4-8. If the carburetor is to be rebuilt at this time, proceed to Section 4-7.

WATER CHOKE — REMOVAL

An understanding of exactly how the water choke functions is essential to ensure the service procedures are properly performed. Hot water from the engine is allowed to circulate inside a chamber of the choke. As the hot water heats the spring, the metal expands and the spring releaves its tension on the choke. Once the tension is released, the choke opens.

The hot circulating water is routed to the port side of the engine and is discharged into the exhaust chamber. An additional feature is included in the automatic choke assembly. A spring-loaded diaphragm and plunger assembly, activated by intake manifold pressure, closes the choke the instant the engine is shut down, whether the engine is hot or cold. The plunger will release its hold on the choke only when the engine is again started. At low manifold pressure, with the engine operating, the choke is released to open the choke and the thermal (water) system is in control. This system operates independently of the water choke system.

A check valve and orifice is installed in the water system to prevent flooding of the diaphragm when the manifold pressure fluctuates during slow speed operation. A puncture, or other damage, to the diaphragm will cause the choke to be held in the closed position at all times, when the vacuum system is in the automatic position. However, the choke can still be opened by using the manual lever.

During installation of the carburetor, considerable attention must be exercised to install the proper hose to the correct fitting. If the hoses are crossed, water will be discharged into the intake manifold and the engine, resulting in a series of horrendous problems. This water will then be released from the cylinder through the spark plug openings when the spark plugs are removed and the engine is cranked in an effort to start it.

THEREFORE, take time to identify the hoses to ensure each is correctly connected to the proper carburetor fitting.

1- Remove the hoses connected to the choke, one from the top, and the other two from the bottom. The hose connected to the bottom of the choke and closest to the carburetor is the water outlet line. The other hose on the bottom is the vacuum line. The hose connected to the top of the choke is the water inlet line. Disconnect the fuel line at the carburetor. Remove the two nuts securing the carburetor to the intake manifold, and then remove the carburetor.

WATER CHOKE   4-27

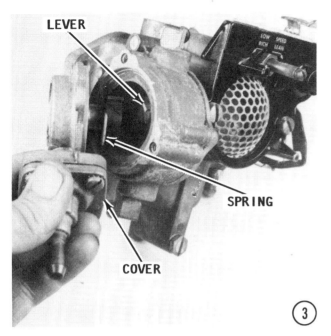

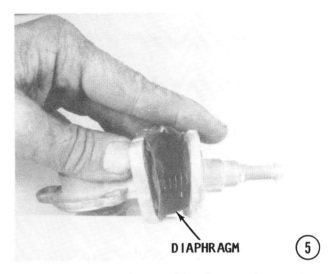

2- Observe the mark on the choke cover and a matching mark on the choke base. These marks must be aligned in the same position when the choke is assembled to the carburetor.

3- Remove the three screws securing the choke cover to the base. Lift the cover from the carburetor and take notice of the spring on the inside of the cover. This spring releaves pressure on the choke as heated hot water from the engine circulates through the chamber.

4- After the cover has been removed, it is not necessary to remove the choke base from the carburetor. Remove the four screws securing the choke head, and then remove the head.

5- Notice the diaphragm. Under the diaphragm is a spring. Remove the diaphragm and spring from the choke head.

6- Remove the small valve and seat located under the spring.

Further disassembly of the water choke valve is not necessary. If the choke is to be assembled without rebuilding the carburetor, proceed directly to Section 4-8. If the carburetor is to be rebuilt at this time, proceed to Section 4-7.

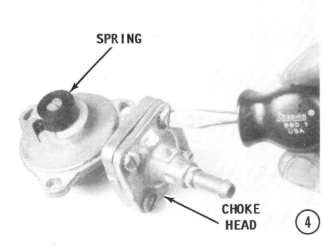

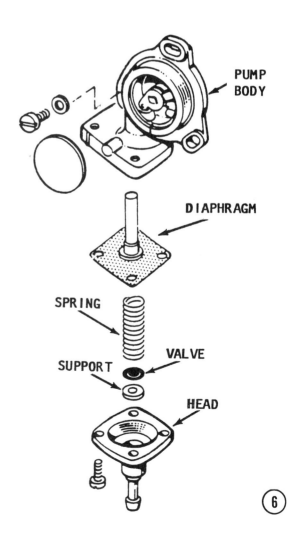

4-28 FUEL

4-8 TYPE II CARBURETOR

DISASSEMBLING

The following steps outline service procedures for the Type II carburetor. Except for the choke system, this carburetor is very similar to the Type I carburetors covered in the previous section.

1- Remove the knobs from the low-speed needle valve. This is accomplished by holding the knob firmly and at the same time backing out the retaining screw from the center of the knob. Once the screw is removed, the knob may be pulled free of the

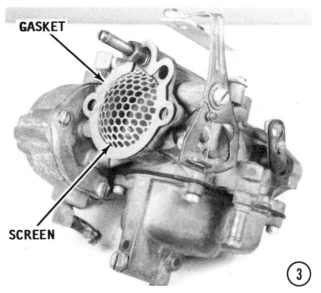

needle valve. If the carburetor being serviced has a front shield installed, remove the top screw.

2- Remove the two screws on the front side of the carburetor cover. Remove the cover from the front of the carburetor.

3- Remove the small screen and gasket.

4- Loosen the packing nut and rotate the low-speed needle valve counterclockwise to remove it from the carburetor. After the needle valve has been removed, back-out the packing nut.

5- The low-speed needle valve has a removable sleeve. To remove this sleeve, use an **OLD** needle valve as follows: First screw the old valve into the sleeve. Next clamp the needle valve in a vise. Now, tap on the carburetor and remove the packing and sleeve. The packing is installed in front of the sleeve.

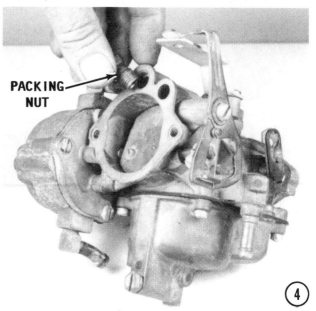

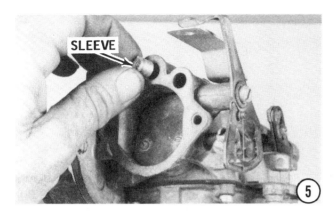

6- Remove the 7/16" nut from the bowl at the bottom of the carburetor. Using the proper size screwdriver, remove the high-speed needle valve from the bowl.

GOOD WORDS:
The following steps are to be performed if the carburetor being serviced has an adjustable high-speed needle valve.

7- Loosen the packing nut securing the high-speed needle valve to the carburetor bowl. Turn the high-speed needle valve **COUNTERCLOCKWISE** until it is free, then remove it from the carburetor bowl. Remove the carburetor flange nut from the bowl.

8- With a small screwdriver, reach into the bowl cavity and remove the packing glands from the carburetor.

9- Remove the four screws securing the bowl to the carburetor, and then remove the bowl.

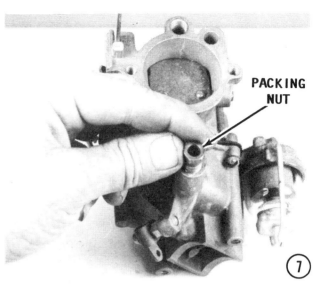

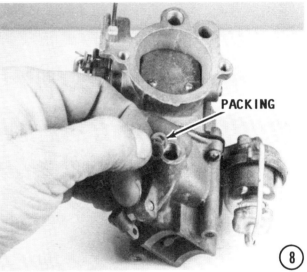

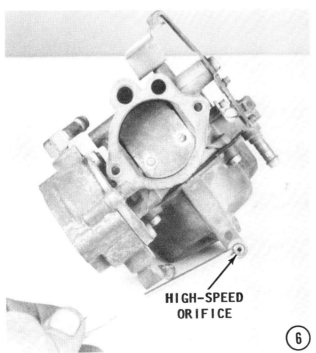

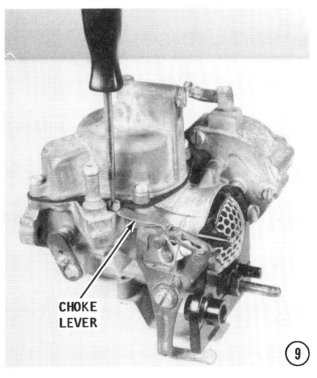

4-30 FUEL

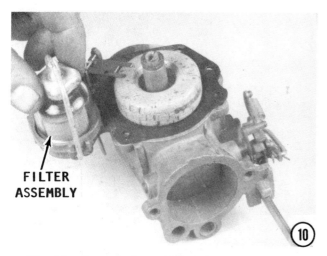

10- If the carburetor has an attaching fuel filter bowl, turn the locking disc holding the glass bowl to the carburetor **COUNTERCLOCKWISE**, and then move the hanger clear of the bowl. Remove the bowl.

11- Remove the center screw, washer, filter, and gasket from the carburetor.

12- Remove the bowl gasket from the carburetor. Remove the gasket from the high-speed nozzle. Remove the high-speed nozzle from the carburetor using the proper size screwdriver to prevent possible damage to the nozzle. If difficulty is experienced in removing the high-speed nozzle, leave it in place. When the carburetor is immersed in the carburetor cleaner the high-speed nozzle will be cleaned suitable for further service.

13- Work the hinge pin free of the float and then remove the float.

14- Remove the inlet needle valve from the seat. Remove the seat and gasket from the carburetor.

Further disassembly of the carburetor is not necessary.

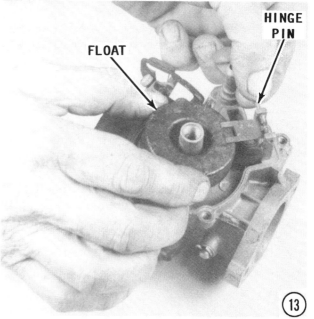

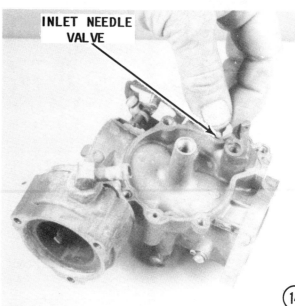

TYPE II CARBURETOR   4-31

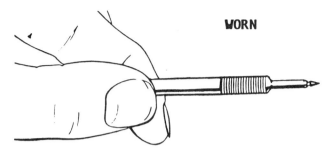

A worn needle valve unfit for further service.

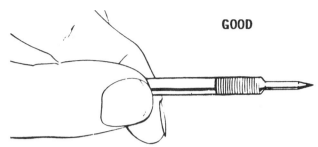

Compare this new needle valve with the worn needle to the left.

CLEANING AND INSPECTING

NEVER dip rubber parts, plastic parts, diaphragms, or pump plungers in carburetor cleaner. These parts should be cleaned **ONLY** in solvent, and then blown dry with compressed air.

Place all metal parts in a screen-type tray and dip them in carburetor cleaner until they appear completely clean, then blow them dry with compressed air.

Blow out all passages in the castings with compressed air. Check all of the parts and passages to be sure they are not clogged or contain any deposits. **NEVER** use a piece of wire or any type of pointed instrument to clean drilled passages or calibrated holes in a carburetor.

Move the throttle shaft back-and-forth to check for wear. If the shaft appears to be too loose, replace the complete throttle body because individual replacement parts are **NOT** available.

Inspect the main body, airhorn, and venturi cluster gasket surfaces for cracks and

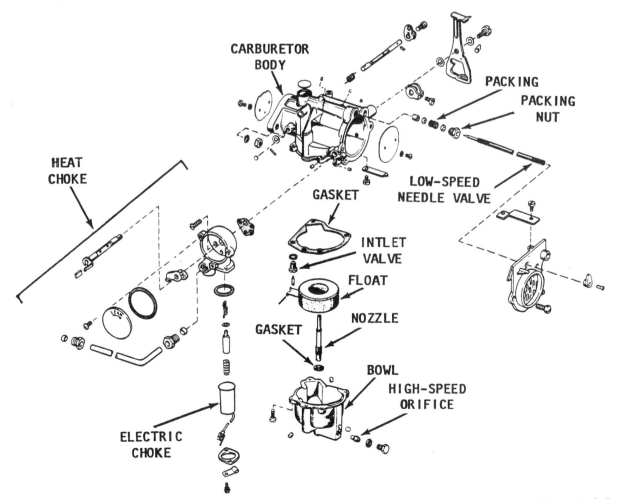

Exploded view of a Type II carburetor installed on larger horsepower engines equipped with a heat/electric choke.

4-32   FUEL

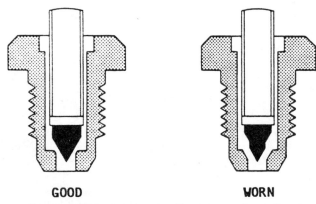

Cross-section drawing to allow comparison of a new needle and seat (left), with one badly worn (right).

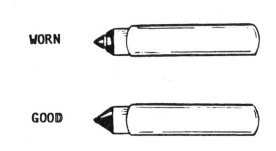

Comparison of a worn and new needle valve. The top needle is unfit for further service.

burrs which might cause a leak. Check the float for deterioration. Check to be sure the float spring has not been stretched. If any part of the float is damaged, the unit must be replaced. Check the float arm needle contacting surface and replace the float if this surface has a groove worn in it.

Inspect the tapered section of the idle and high-speed adjusting needles and replace any that have developed a groove.

If a high-speed orifice is installed on the carburetor being serviced, check the orifice for cleanliness. The orifice has a stamped

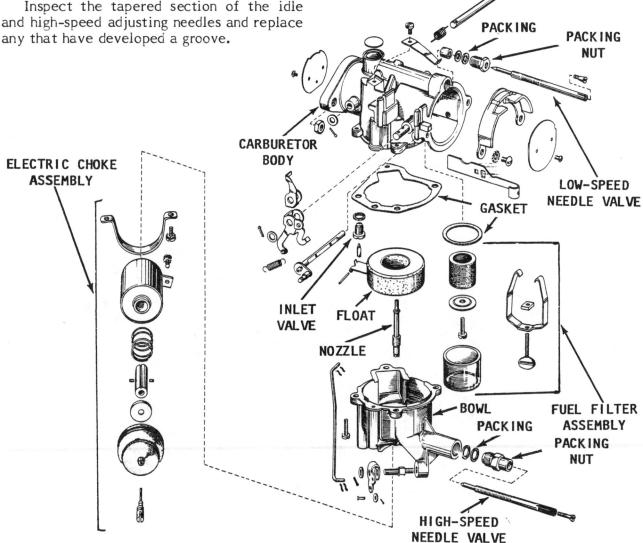

Exploded view of a Type II carburetor installed on larger horsepower engines equipped with the all-electric choke.

TYPE II CARBURETOR   4-33

number. This number represents a drill size. Check the orifice with the shank of the proper size drill to verify the proper orifice is used. The local OMC dealer will be able to provide the correct size orifice for the engine and carburetor being serviced.

Most parts that should be replaced during a carburetor overhaul are included in overhaul kits. These kits are available from your local marine dealer. One of these kits will contain a matched fuel inlet needle and seat. This combination should be replaced each time the carburetor is disassembled as a precaution against leakage.

ASSEMBLING TYPE II CARBURETOR

Purchase of a carburetor repair kit is almost a must when servicing this unit. All parts required for a complete rebuild, including the necessary gaskets, are contained in the kit. Most repair kits contain more parts and gaskets than are needed because the kit may be used to service a wide range of carburetor models.

1- Install the **NEW** inlet seat and gasket into the carburetor base. Take care to use the proper size screwdriver as a precaution against damaging the inside surface of the the seat. Apply just a drop of oil into the seat, to prevent the needle from sticking when it is installed. Insert the inlet needle valve into the seat.

2- Install the **NEW** float and hinge pin from the repair kit with the tang of the float facing **DOWNWARD** toward the inside of the carburetor.

3- If the inlet nozzle was removed, install the nozzle into the center of the carburetor and secure it tightly with the proper size screwdriver. Exercise care not to burr the edges of the nozzle. The end of

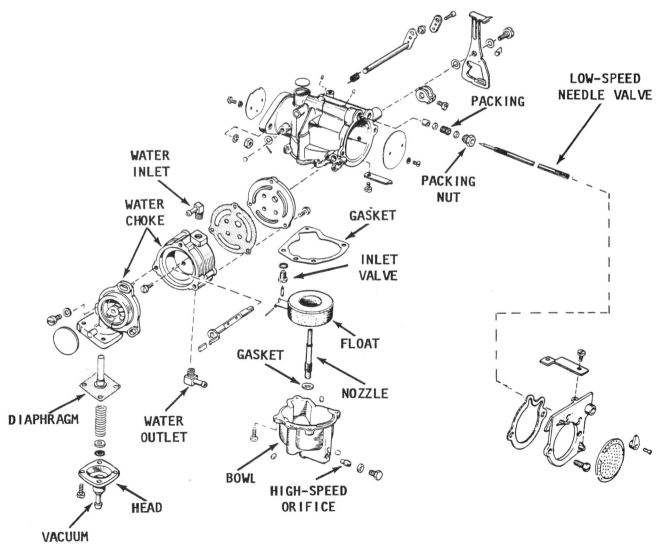

Exploded view of a Type II carburetor installed on larger horsepower engines equipped with a hot water choke.

4-34 FUEL

the nozzle seats in the bowl of the carburetor and burrs on the nozzle will result in fuel leakage, or the bowl will **NOT** fit properly onto the carburetor.

4- Install a **NEW** gasket over the nozzle. Force the gasket down onto the carburetor stem.

5- Place a **NEW** gasket in position on the the carburetor base. **NEVER** attempt to install a used gasket at this location. As the gasket is used, the low-speed hole has a tendency to shrink slightly and prevent sufficient fuel from passing through. Therefore, the need for installation of a new gasket each time the carburetor is serviced.

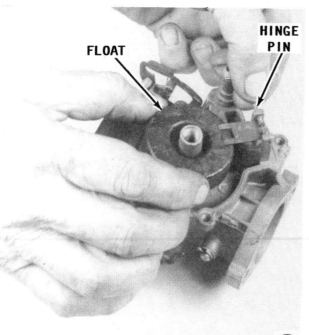

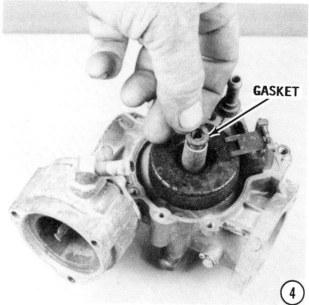

TYPE II CARBURETOR   4-35

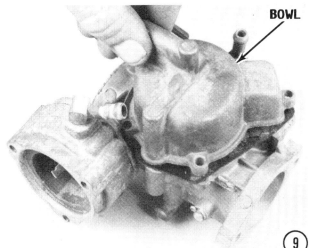

6- If the carbureotor being serviced has a filter bowl attached to the carburetor, install a **NEW** filter bowl gasket, **NEW** filter, washer and screw at this time.

7- Place the fitler bowl over the filter and secure it in place with the clamp.

8- Hold the carburetor in a horizontal position, as shown, and observe the attitude of the float. The float must be level (parrallel) with the surface of the carburetor. If the float is not level, use a pair of needle-nose pliers and **CAREFULLY** bend the foat tab **SQUARELY** until the float is in the correct position (level with the carburetor). Check to be sure the float is square with the carburetor cavity (one side is not further away than the other).

9- Lower the bowl onto the carburetor and secure it with the four screws.

10- Install the bushing into the low-speed needle valve opening.

11- Insert the two fiber washers and the one nylon washer into the low-speed cavity.

12- Start the low-speed needle valve packing nut into the opening. **DO NOT** tighten the nut at this time. Thread the low-speed needle valve into the low-speed

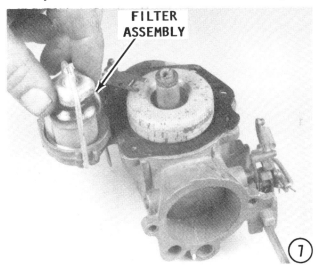

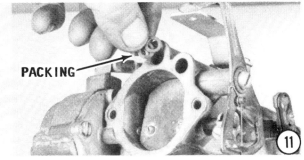

4-36 FUEL

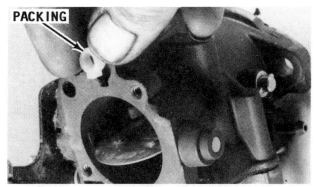

Installing a new type packing. A packing nut is not used to hold the needle adjustment.

opening. Continue threading it into the opening until it barely seats. After the needle valve seats, back it out 1-1/2 turns **COUNTERCLOCKWISE**. Now, tighten the packing nut until there is drag on the needle valve, but the valve may still be rotated by hand, but with just a little difficulty.

13- Install the high-speed orifice into the carburetor bowl, using the proper size screwdriver.

14- Install a **NEW** drain plug gasket, and then thread the drain plug into the carburetor bowl cavity.

SPECIAL NOTE:

The following step is to be performed if the carburetor being serviced has an adjustable high-speed needle valve.

15- Install the three packing washers into the carburetor bowl.

16- Start the high-speed needle valve packing nut into the high-speed opening. **DO NOT** tighten the nut at this time.

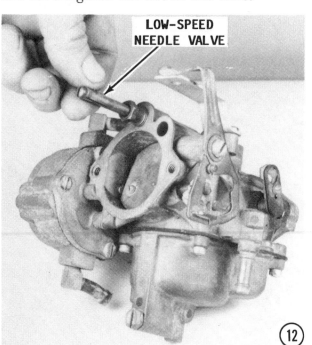

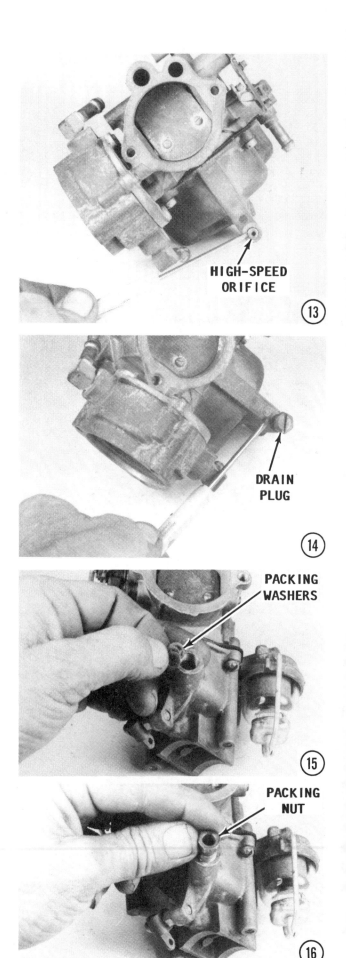

TYPE II CARBURETOR 4-37

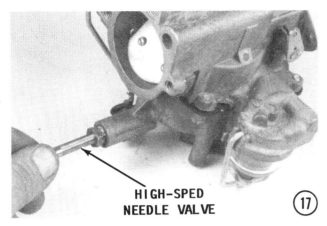

HIGH-SPED NEEDLE VALVE

17- Thread the high-speed needle valve into the high-speed opening. Continue to thread it into place until it **BARELY** seats. After the needle valve seats, back it out 3/4 turn **COUNTERCLOCKWISE**. Now, tighten the packing nut until there is drag on the needle valve, but the valve may still be rotated by hand, but with just a little difficulty.

18- Install a **NEW** gasket and screen onto the front of the carburetor.

19- If a front cover is used on the unit being serviced, install the front cover and secure it in place with the two attaching screws.

20- Install the spring steel lever into the indent of the manual choke on the bottom side of the carburetor. This lever holds the choke in the neutral position and prevents loose movement of the handle.

21- Install the low-speed needle valve knob by sliding it onto the valve stem, and then secure it in place with the screw.

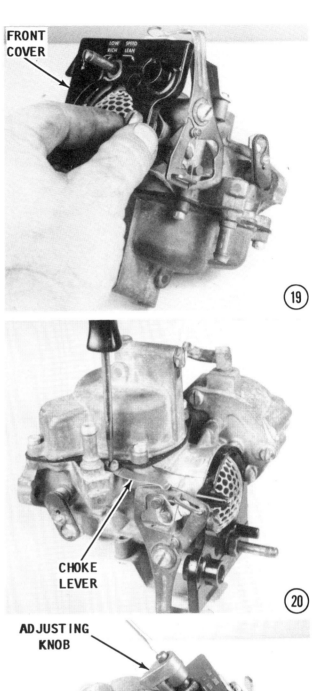

FRONT COVER

CHOKE LEVER

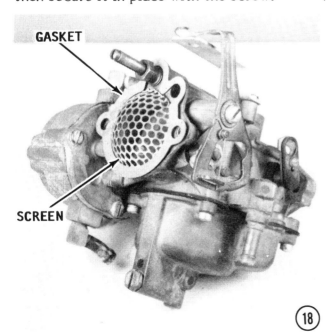

GASKET

SCREEN

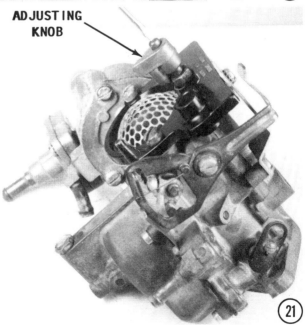

ADJUSTING KNOB

4-38 FUEL

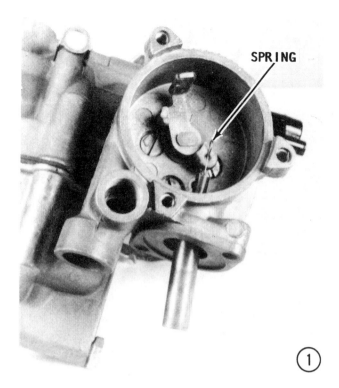

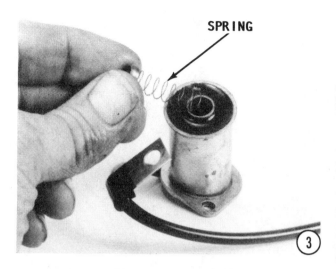

Heat/Electrical Choke Installation

1- Install the choke plunger and spring by hooking the spring into the choke lever.

2- Insert a **NEW** gasket into the cavity of the choke base.

3- Insert the spring into the bore of the choke solenoid.

4- Work the choke solenoid up the plunger and secure the plunger with the two retaining screws.

5- Place a **NEW** gasket onto the surface of the carburetor. Install the choke cover onto the carburetor. Observe how the spring has a clip and the choke lever has a protrusion. The clip **MUST** fit around the lever and protrusion.

6- Align the mark on the cover plate with the matching mark on the carburetor. Secure the cover plate in this position with the three attaching screws.

4-9 ASSEMBLING CHOKES TO TYPE II CARBURETORS

Perform only one of the three following choke installation procedures, depending on the type of unit being serviced. After the choke and carburetor have been properly installed onto the engine, proceed directly to the adjustment portion of this section to operate the engine for testing and carburetor adjustments under load.

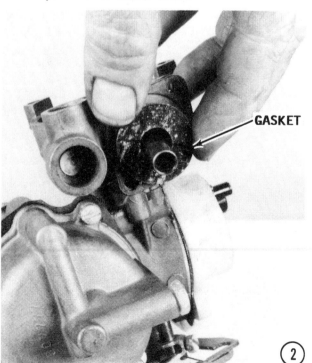

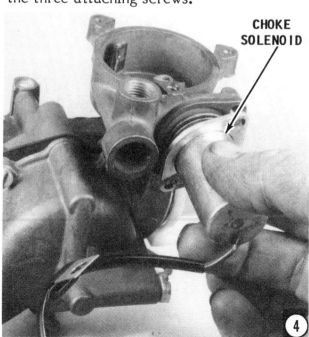

TYPE II CARBURETOR 4-39

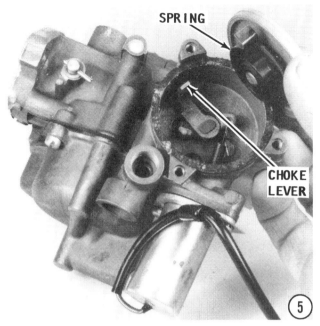

7- Clean the surface of the intake manifold thoroughly. Check to be sure all old gasket material has been removed. Place a **NEW** gasket over the studs and into place on the manifold. Slide the carburetor down over the studs and secure it in place with the two nuts. Tighten the nuts **EVENLY** and **ALTERNATELY**. Connect the fuel line to the carburetor. Connect the choke quick-disconnect on the choke wire leading to the dash. Install the starter and generator, if the engine being serviced is equipped with these two units.

8- Connect the heat tube to the engine and secure it in place with the compression nut at each end.

9- Position the heat shield over the heat tube and secure it in place with the three attaching screws. Proceed directly to the adjustment portion of this section to operate the engine for testing and carburetor adjustments under load.

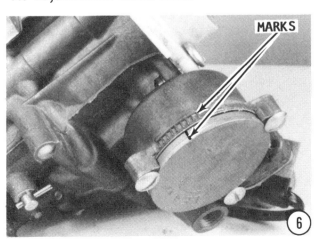

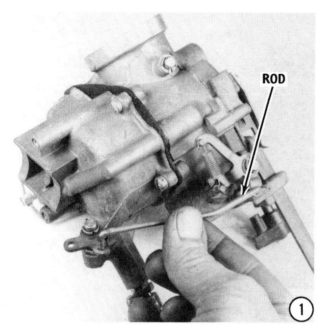

ALL Electric Choke Installation

1- Install the rod extending from the upper choke assembly to the lower hinge of the choke. Secure the rod in place with a cotter pin at each end.

2- Insert the choke assembly into the cavity of the float bowl with the electrical connector facing **DOWNWARD** to permit installation of the electrical wires after the carburetor is installed. Now, bring the clamp over the top of the choke to secure the choke assembly in place.

3- Install the pin through the end of the shaft of the choke solenoid and into the lever attached to the carburetor. Install the washer and cotter pin.

4- Clean the surface of the intake manifold thoroughly. Check to be sure all old gasket material has been removed. Place a **NEW** gasket over the studs and into place on

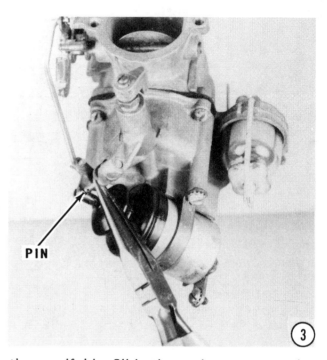

the manifold. Slide the carburetor onto the intake manifold studs and secure it in place with the attaching nuts. Tighten the nuts **EVENLY** and **ALTERNATELY**. Connect the fuel line to the carburetor. Attach the electrical wires to the choke solenoid. Install the starter motor and generator if the engine being serviced is equipped with these two units. proceed directly to the adjustment portion of this section to operate the engine for testing and carburetor adjustments under load.

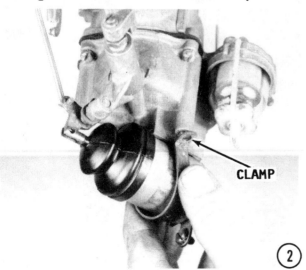

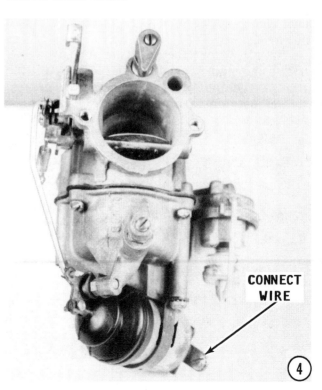

TYPE II CARBURETOR 4-41

Water Choke Installation

1- Install the check valve support into the head.

2- Install the small diaphragm into the vacuum opening of the choke body.

3- Install the spring onto the top of the support.

4- Lower the stem down through the spring and diaphragm and onto the base.

5- Install the cover and secure it in place with the attaching screws. Tighten the screws **ALTERNATELY** as a precaution against warping the cover.

6- Take the cover assembly and install a new gasket onto the base of the carburetor. Work the spring clip over the protrusion of the choke lever and lower it into the cavity.

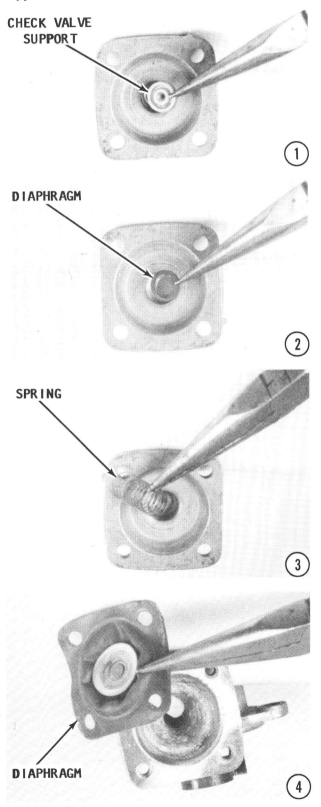
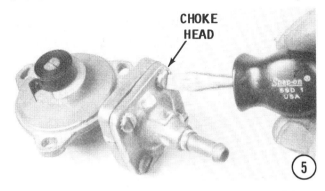
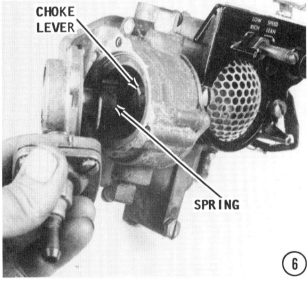

7- Align the mark on the cover with the mark on the body, and then secure the cover in this position with the three attaching screws.

8- Clean the surface of the intake manifold thoroughly. Check to be sure all old gasket material has been removed. Place a **NEW** gasket over the studs and into place on the manifold. Slide the carburetor onto the intake manifold studs and secure it in place with the attaching nuts. Tighten the nuts **EVENLY** and **ALTERNATELY**. Connect the fuel hose to the carburetor. Connect the vacuum and water hoses to the carburetor paying careful **ATTENTION** to connect the proper hose to the correct fitting according to the identification given to the hoses during disassembly. If identification was not made during disassemby, follow the hose to the other end to ensure proper connection to the carburetor.

a- Connect the water hose to the top fitting of the water choke.

b- Connect the outlet water hose to the fitting closest to the carburetor. The other end of this line is connected to a fitting on the exhaust chamber on the port side of the engine.

c- Connect the vacuum line to the vacuum diaphragm fitting. The other end of this line is connected to the intake manifold.

If the hoses are not connected correctly, water will be injected into the cylinder when the engine is cranked, resulting in a series of horrendous problems.

Install the starter motor and generator if the engine being serviced is equipped with these two units. Now, perform the procedures outlined in the next paragraphs to test and make the carburetor adjustments under a load condition.

TYPE II CARBURETOR ADJUSTMENTS

GOOD WORD

Under all conditions, the ignition and fuel system **MUST** be synchronized before the fine adjustments to the carburetor are made. See Chapter 5. After the synchronization has been completed, proceed with the following work.

9- Mount the engine in a test tank or body of water. If this is not possible, connect a flush attachment and garden hose to the lower unit. **ONLY** the low-speed adjustment may be made using the flush attachment. If the engine is operated above idle speed with no load on the propeller, the engine could **RUN-AWAY** resulting in serious damage or destruction of the unit.

CAUTION: Water must circulate through the lower unit to the engine any time the engine is run to prevent damage to the water pump in the lower unit. Just five seconds without water will damage the water pump.

Start the engine and allow it to warm to operating temperature. Adjust the low-speed idle by turning the low-speed needle valve **CLOCKWISE** until the engine begins to misfire or the rpm drops noticeably. From this point, rotate the needle valve **COUNTERCLOCKWISE** until the engine is operating at the highest rpm.

If the engine is equipped with an adjustable high-speed needle valve, shift the engine into **FORWARD** gear, and then advance the throttle to the wide open position (WOT). **NEVER** attempt to make this adjustment with a flush attachment and garden hose attached to the lower unit. Adjust the high-speed by rotating the high-speed needle valve **CLOCKWISE** until the number of rpm begins to drop, then rotate the high-speed needle valve **COUNTER-CLOCKWISE** until the highest rpm is reached. Return the throttle to idle speed. Adjust the idle speed a second time as described earlier in this step. Again, advance the throttle to the WOT position and check the high-speed ad-

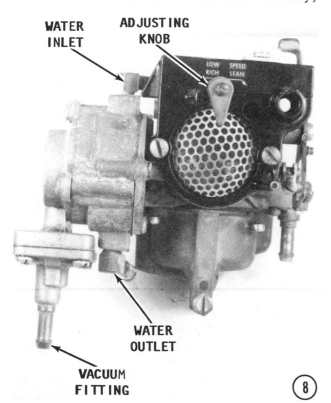

justment. Return the engine to idle speed. If the engine coughs and operates as if the fuel is too lean, but the idle and high-speed adjustments have been correctly made, then recheck the synchronization between the fuel and ignition systems. Now, shut off the fuel supply and allow the engine to run until it first begins to misfire from lack of fuel. Retard the spark and shut the engine down. Tighten the sleeve nut securely to prevent the needle valves from changing position through engine vibration while it is operating, but still allow the needle valves to be adjusted by hand using the knob on the end of the valve.

The idle stop is located on the port side of the engine, on the outside of the cowling. Adjust the nylon screw inward or outward to obtain the desired idle speed.

4-10 TYPE III CARBURETOR

This carburetor is installed on the 9-1/2 hp engine. The unit has an adjustable low-speed needle valve and fixed high-speed needle valve. The only changes that have been made to this particular model carburetor over the years was a redesign of the idle needle valves. On early model 9.5 hp engines, the reeds set directly below the carburetor. On later models, the reeds were removed from this location and installed between the engine block and the intake manifold.

The early model idle needle valves extended out of the carburetor with an O-ring, spring, and E-clip. The needle valve was controlled from the front of the engine by means of a flexible cable. On this model carburetor, the cable and control knob must be removed before the carburetor is removed from the engine.

Later model carburetors are equipped with a long packing nut and linkage. An adjustable knob located on the front of the engine controls movement of the valve. When servicing the late model carburetors, the low-speed needle valve adjustment knob and linkage must be removed before the carburetor is removed from the engine.

REMOVAL

1- Remove the choke rod that extends over the top of the carburetor by snapping the choke rod out of the nylon snap. Disconnect the fuel line from the carburetor.

2- If the carburetor has a packing nut with the needle through the nut and a nylon adjustment knob, remove the knob or remove the linkage from the knob.

3- If the carburetor has a flexible line to the front of the engine, remove the knob on the front of the engine and then turn the flexible cable **COUNTERCLOCKWISE** until the needle valve is removed from the carburetor.

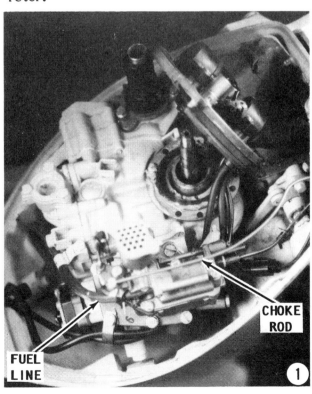

4-44 FUEL

SPECIAL NOTE:
As the needle valve is being removed, take care to retain the washer, O-ring, and spring installed between the E-ring on the valve and the carburetor. The washer and spring will be used again. The O-ring may be discarded.

4- Remove the five screws securing the carburetor to the intake manifold. Notice that four of the screws have slots and one is a countersunk screw. Lift the carburetor from the engine.

5- Remove the four screws securing the float bowl, and then remove the float bowl.

6- Remove and **DISCARD** the bowl gasket.

7- Remove the hinge pin, and then lift the float assembly from the carburetor body.

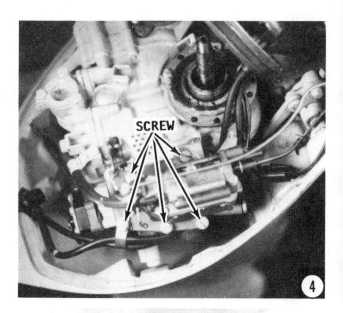

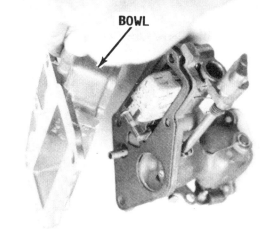

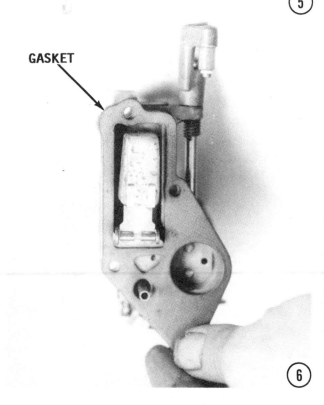

TYPE III CARBURETOR 4-45

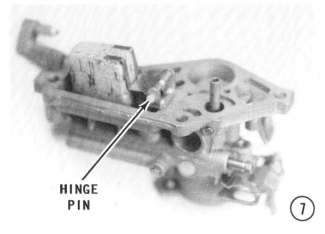

HINGE PIN

7

8- Reach inside the inlet seat and remove the inlet needle. Remove the inlet needle seat.

9- Remove the drain plug from the bottom of the float bowl. Use the proper size screwdriver and remove the high-speed orifice from the float bowl. Loosen the low-speed needle valve packing nut by turning it **COUNTERCLOCKWISE**. Now, remove the low-speed needle valve. Remove the packing nut and washers.

CLEANING AND INSPECTING

NEVER dip rubber parts, plastic parts, nylon parts, diaphragms, or pump plungers in carburetor cleaner. These parts should be cleaned **ONLY** in solvent, and then blown dry with compressed air.

Place all metal parts in a screen-type tray and dip them in carburetor cleaner until they appear completely clean, then blow them dry with compressed air.

Blow out all passages in the castings with compressed air. Check all parts and

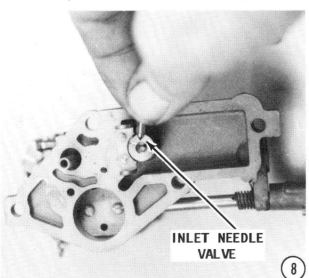

INLET NEEDLE VALVE

8

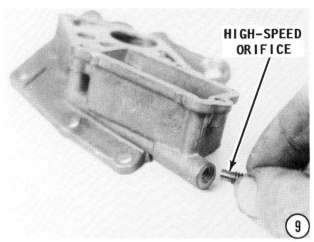

HIGH-SPEED ORIFICE

9

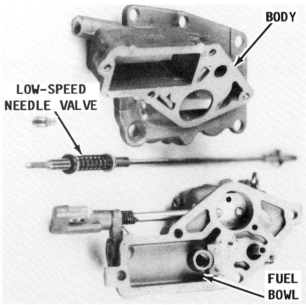

BODY

LOW-SPEED NEEDLE VALVE

FUEL BOWL

Major parts of carburetor installed on a Johnson/Evinrude 9.5 horsepower engine.

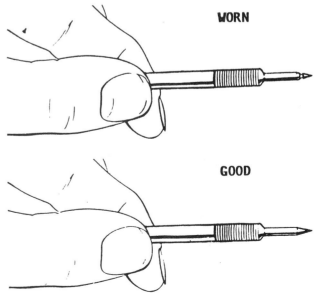

WORN

GOOD

Comparison of worn and new carburetor adjustment screws. The upper screw is unfit for further service.

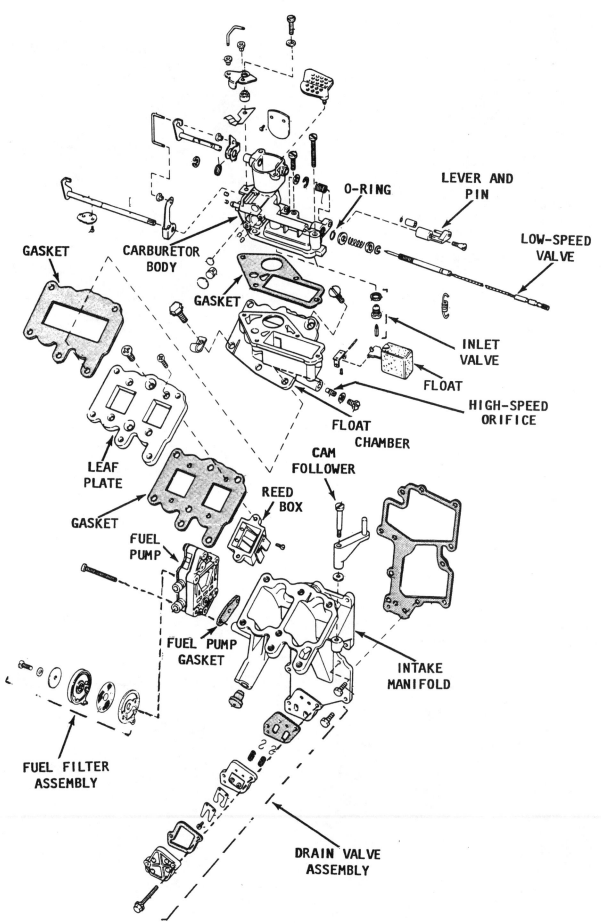

Exploded view of a Type III carburetor installed on 9.5 horsepower engines.

TYPE III CARBURETOR   4-47

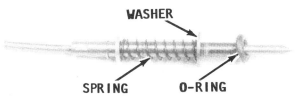

Low-speed needle valve and associated parts used in the carburetor of a 9.5 horsepower engine.

Parts included in a carburetor repair kit for the Johnson/Evinrude 9.5 horsepower engine.

passages to be sure they are not clogged or contain any deposits. **NEVER** use a piece of wire or any type of pointed instrument to clean drilled passages or calibrated holes in a carburetor.

Move the throttle shaft back-and-forth to check for wear. If the shaft appears to be too loose, replace the complete throttle body because individual replacement parts are **NOT** available.

Inspect the main body, airhorn, and venturi cluster gasket surfaces for cracks and burrs which might cause a leak. Check the float for deterioration. Check to be sure the float spring has not been stretched. If any part of the float is damaged, the unit must be replaced. Check the float arm needle contacting surface and replace the float if this surface has a groove worn in it.

Inspect the tapered section of the idle adjusting needles and replace any that have developed a groove.

If a high-speed orifice is installed on the carburetor being serviced, check the orifice for cleanliness. The orifice has a stamped number. This number represents a drill size. Check the orifice with the shank of the proper size drill to verify the proper orifice is used. The local OMC dealer will be able to provide the correct size orifice for the engine and carburetor being serviced.

Most of the parts that should be replaced during a carburetor overhaul are included in overhaul kits available from your local marine dealer. One of these kits will contain a matched fuel inlet needle and seat. This combination should be replaced each time the carburetor is disassembled as a precaution against leakage.

ASSEMBLING TYPE III CARBURETOR

1- Install the high-speed orifice into the float body using the proper size screwdriver to prevent burring the edges of the orifice.

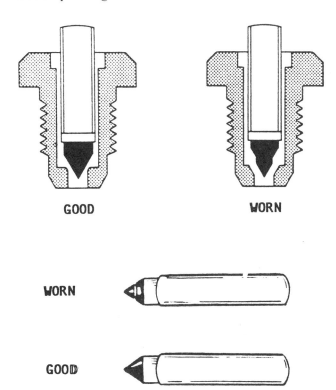

Needle and seat arrangement, showing a worn and new needle for comparison.

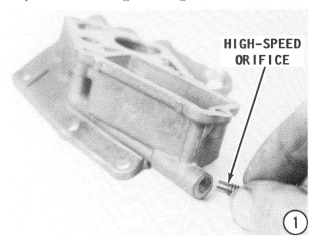

4-48 FUEL

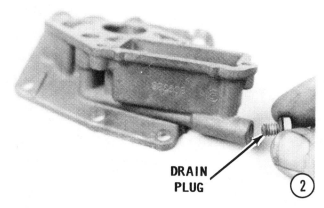

Any damage to the orifice will result fuel leakage and poor engine performance.

2- Install the drain plug with a **NEW** gasket and tighten it securely with a 7/16" wrench.

3- Install the inlet needle seat and gasket. **TAKE CARE** to use the proper size screwdriver to install the seat. If the inside diameter of the seat is damaged the needle valve will leak fuel causing a flooding condition in the carburetor.

4- Apply just a drop of oil into the seat, and then insert the inlet needle into seat.

5- Position a **NEW** float over the needle, and then slide a **NEW** hinge pin into place.

6- Hold the carburetor in a vertical position (up-and-down), and observe the float. The float **MUST** be parallel (align evenly), with the surface of the carburetor body. If the float is not parallel **CAREFULLY** bend the float ever so slightly, as shown, until the correct positioning is obtained.

7- Slide the float bowl gasket over the float and nozzle into position on the carburetor base.

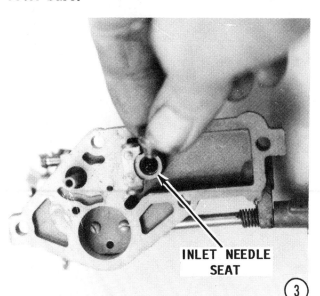

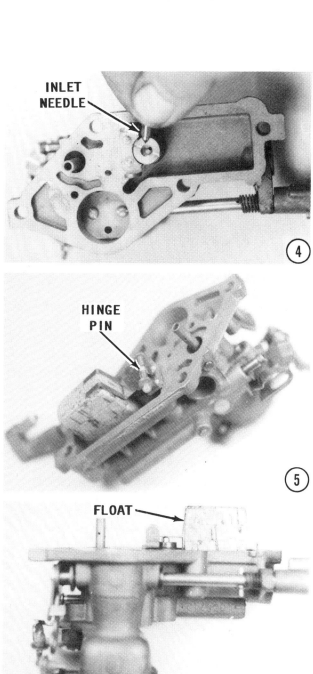

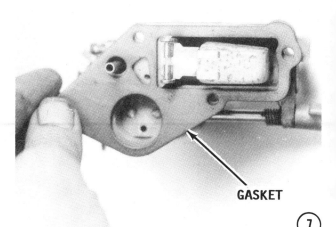

TYPE III CARBURETOR 4-49

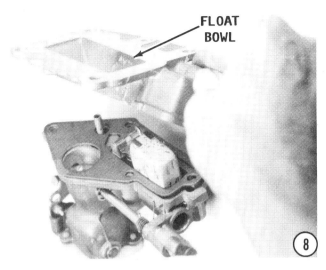

8- Lower the float bowl down over the top of the carburetor body and secure it in place with the four attaching screws.

SPECIAL WORDS

If the carburetor being serviced has packing nuts and washers for the needle valves, perform Step 9. If the carburetor has the flexible line extending from the valve to the front of the engine, the needle valve will be installed **AFTER** the carburetor is in place on the engine.

9- Install the packing nut washers into the needle valve openings. Thread the packing nut into the opening but **DO NOT** tighten them at this time. Thread the low-speed needle valve into place until it just **BARELY** seats. From this position, back it out (COUNTERCLOCKWISE) 1-1/2 turns as a preliminary rough adjustment.

10- Check the surface of the intake manifold to be sure it has been thoroughly cleaned and is free of any old gasket material. Place a **NEW** gasket in position on the manifold. Set the carburetor into place on the manifold and secure it with the five

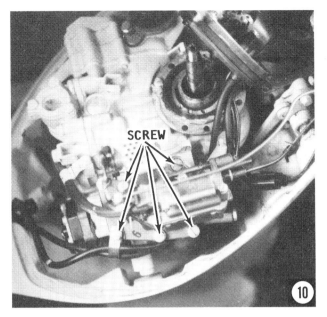

attaching screws. **OBSERVE** that one of the screws is a countersunk type. This screw **MUST** be installed into the countersunk hole.

11- If the carburetor being serviced has the flexible low-speed needle valve arrangement, check to be sure the snap ring is in place and then install the spring, washer and **NEW** O-ring onto the needle. Apply just a drop of oil onto the O-ring to ease installation of the needle valve. Thread the low-speed needle valve into the carburetor until it just **BARELY** seats. From this poistion, back it out (COUNTER CLOCKWISE) 1-1/2 turns as a preliminary rough adjustment. Install the choke rod by snapping it into place in the nylon retainer. Connect the fuel line to the carburetor.

GOOD WORDS

It is best to synchronize the fuel and ignition systems at this time. See Chapter 5. After the synchronization has been completed, proceed with the following work.

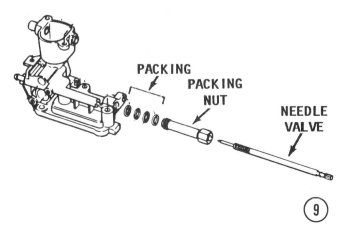

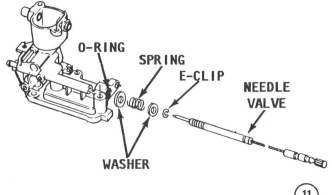

TYPE III CARBURETOR ADJUSTMENTS

12- Mount the engine in a test tank or body of water. If this is not possible, connect a flush attachment and garden hose to the lower unit. **NEVER** operate the engine above idle speed using the flush attachment. If the engine is operated above idle speed with no load on the propeller, the engine could **RUN-A-WAY** resulting in serious damage or destruction of the unit.

CAUTION: Water must circulate through the lower unit to the engine any time the engine is run to prevent damage to the water pump in the lower unit. Just five seconds without water will damage the water pump.

Start the engine and allow it to warm to operating temperature. Adjust the low-speed idle by turning the low-speed needle valve **CLOCKWISE** until the engine begins to misfire or the rpm drops noticeably. From this point, rotate the needle vavle **COUNTERCLOCKWISE** until the engine is operating at the highest rpm. If the engine coughs and operates as if the fuel is too lean, but the idle and high-speed adjustments have been correctly made, then recheck the synchronization between the fuel and ignition systems.

On engines with the flexible low-speed extension to the front of the engine, the spring maintains tension on the needle and adjustment will not be lost because of vibration during operation. On engines with the packing nut arrangement, the nut must be

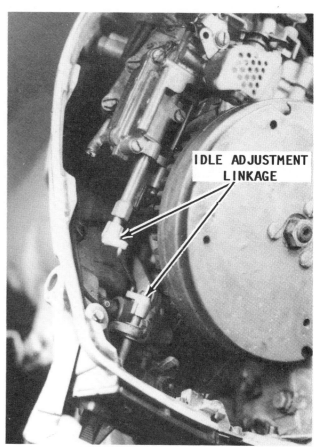

Idle adjustment linkage on Johnson/Evinrude 9.5 horsepower engines.

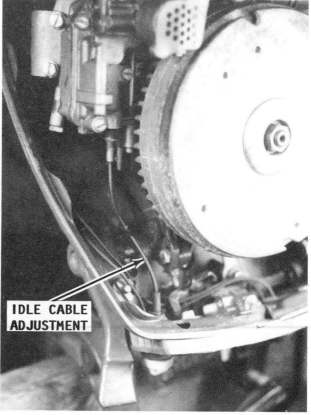

Flexible cable arrangement on the 9.5 horsepower engine.

FUEL PUMP 4-51

tightened securely to hold the adjustment. **HOWEVER,** do not tighten it to the point where an adjustment cannot be made by hand.

4-11 FUEL PUMP SERVICE

A considerable number of fuel pump designs and sizes have been installed on the larger Johnson/Evinrude engines covered in this manual. Only one can be rebuilt and detailed procedures with illustrations are given in this section.

This fuel pump has three nipples providing the means of connecting fuel and vacuum lines. The vacuum line is connected to one nipple. The other end of this hose is connected to the vacuum side of the engine. The inlet hose (from the fuel tank), is connected to the second nipple. The outlet hose (to the carburetor), is connected to the third nipple.

Minor changes have been incorporated into the fuel pump over the years. These changes will be identified in the text.

All other fuel pumps must be replaced as a unit. However, the pump cover can be

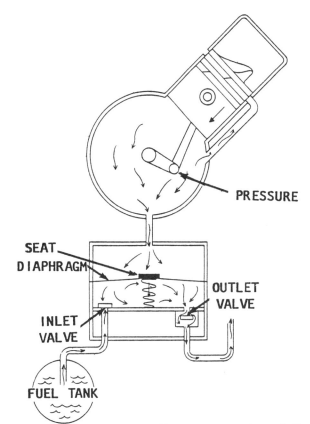

*Functional diagram to illustrate operation of the fuel pump during **DOWNWARD** movement of the piston, causing pressure in the crankcase. Notice how the pump inlet valve is closed and the outlet valve is open to allow fuel to be transferred to the carburetor.*

removed, the filter screen cleaned or replaced, and a **NEW** cover gasket installed.

The accompanying illustrations show only a couple of these fuel pumps, including the unit that can be rebuilt.

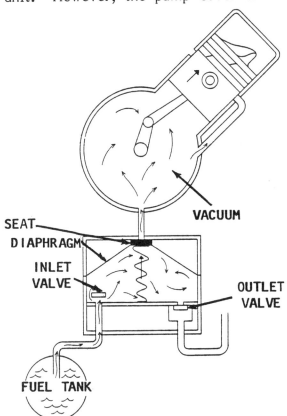

*Functional diagram to illustrate operation of the fuel pump during **UPWARD** movement of the piston causing a vacuum condition in the crankcase. Notice how the intake valve in the fuel pump is open and the outlet valve is closed.*

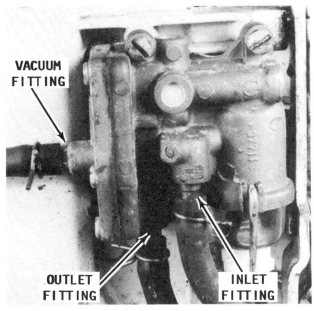

Fuel pump with the vacuum, fuel outlet, and fuel inlet fittings visible. This is the only Johnson/Evinrude pump that can be rebuilt.

4-52 FUEL

Side of the engine with the vacuum opening shown. The fuel pump is mounted over this opening to receive the vacuum/pressure from the crankcase for operation.

TROUBLESHOOTING

If the spark plug of the cylinder to which the vacuum line is connected becomes wet fouled, the cause may very well be a ruptured fuel pump diaphragm. This reasoning is sound for both types of fuel pumps. Sometimes the pump that is not serviceable is bolted directly to the side of the engine block. On other installations this non-serviceable pump has a vacuum hose connected to the cylinder block. When the pump is bolted directly to the block, the pump receives vacuum directly from the cylinder through a hole in the back side of the pump and a matching hole in the block.

A good test for the pump that is connected to the intake manifold with a vacuum hose, is to disconnect the vacuum line

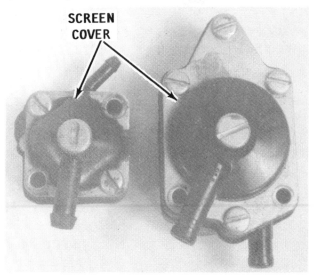

Two different fuel pumps installed on OMC engines. Service on these two models is limited to removing the cover and cleaning the screen.

from the engine, operate the squeeze bulb in the fuel line until it is firm, and then to carefully observe the end of the vacuum hose to detect any fuel leakage. The smallest amount of fuel from the hose indicates a damaged diaphragm. The pump must be rebuilt to restore satisfactory service of the pump.

To test the non-serviceable pump, remove the pump from the engine, operate the squeeze bulb until it is firm, and then carefully observe the vacuum hole in the back side of the pump for any indication of fuel. The smallest amount of fuel indicates a damaged diaphragm. In this case, the pump **MUST** be replaced.

PUMP REMOVAL AND REPAIR

1- Identify each hose and its location, then disconnect the vacuum hose and two fuel hoses from the fuel pump. Remove the three attaching screws securing the pump to the engine. Two screws are visible on top of the pump and the third is hidden behind the fuel inlet nipple.

2- Mount the fuel pump in a vise, as shown. Loosen the plastic screw securing the filter bowl to the pump.

3- Swing the hinge down and lift the fuel bowl free of the pump. Remove and **DISCARD** the bowl gasket and filter.

FUEL PUMP 4-53

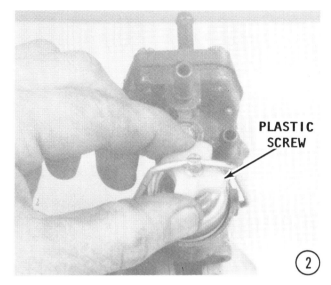

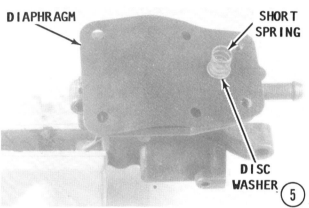

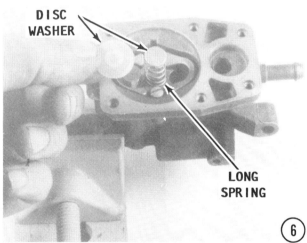

4- Shift the pump position in the vise, as shown. Remove the six screws securing the pump cover.

5- Remove the cover. Take care not to lose the disc washer and spring from the top of the diaphragm. Remove the disc washer, spring and diaphragm. **SAVE** the small disc washer because it will be used again.

6- Some model pumps may have a small disc washer and long spring installed under the diaphragm. Other models may have a large nylon washer and spring. **DISCARD** the small disc washer because it has been replaced with the larger nylon washer and is included in the pump repair kit.

7- Remove the two screws from the check valve retainer. Reach into the pump and remove the check valve retainer.

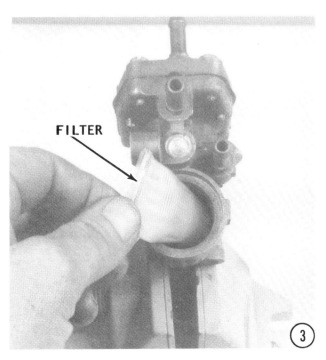

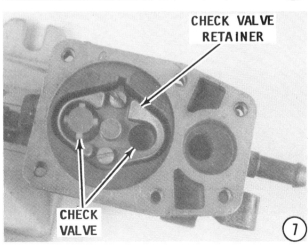

8- CAREFULLY observe how one check valve is facing downward and the other valve is facing upward. Also notice the groove in the fuel pump body. A small boss on the retainer fits into the groove as a prevention against installing the retainer incorrectly. Remove the two check valves and the check valve gasket. Further disassembly of the pump is not necessary.

CLEANING AND INSPECTING

Wash the fuel pump body thoroughly and then blow it dry with compressed air. All internal parts necessary to rebuild the pump, including diaphragm, check valves, gaskets, etc., will be included in the pump repair kit. At one time, these kits were available from the local OMC dealer at modest cost. However, OMC has discontinued packaging the parts in kit form. Therefore, unless the dealer still has one of the old kits in stock, the fuel pump parts must be purchased individually.

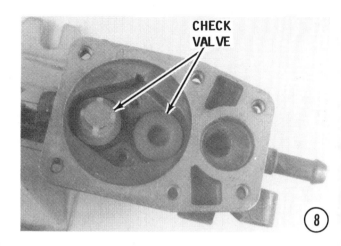

PUMP ASSEMBLING AND INSTALLATION

1- Insert one of the **NEW** check valves through the check valve gasket.

2- Install the gasket and check valve into the pump body with the valve facing **DOWNWARD**, as shown.

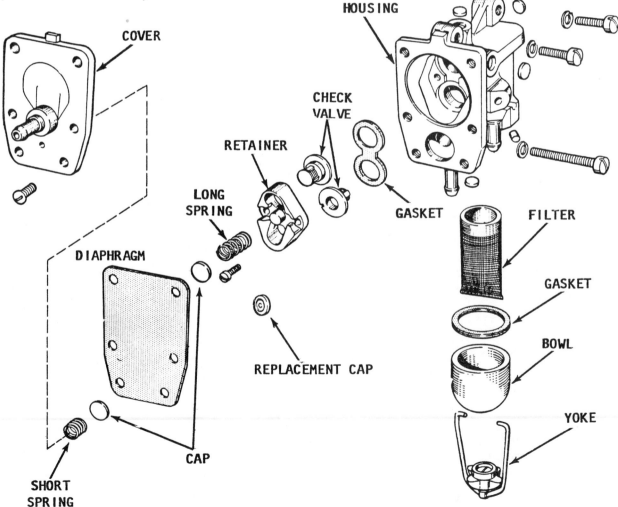

Exploded drawing of the only fuel pump that can be rebuilt.

FUEL PUMP 4-55

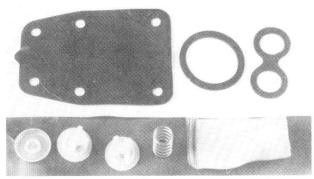

Parts included in a fuel pump repair kit for those model pumps that can be rebuilt.

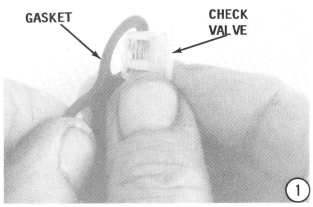

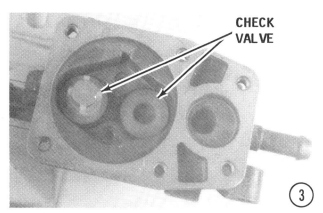

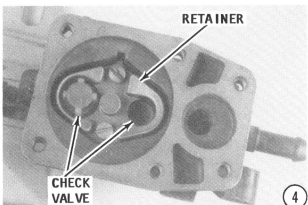

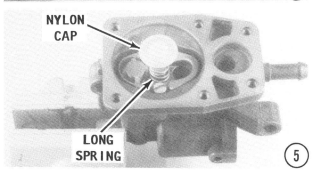

3- Position the other check valve on top of the gasket facing **UPWARD**. Slide the retainer down over the check valves with the boss on the retainer in the groove of the pump body. Observe the large and small hole in the retainer. The large hole **MUST** be positioned over the check valve facing **UPWARD**.

4- Secure the retainer in place with the two attaching screws.

5- Install the **NEW LONG** spring over the boss of the retainer, and then place the nylon disc washer on top of the spring that was provided in the kit. **NEVER** use the small disc on top of the long spring.

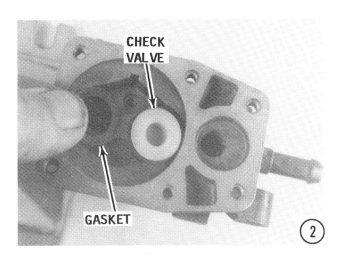

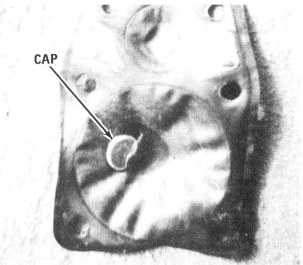

Damaged old-style diaphragm with the small cap installed on early model engines. This small cap has been replaced with a much larger one. Notice how this cap worked its way free of the spring and came through the diaphragm.

(6)

GOOD WORD:
The following steps may only be properly accomplished by exercising patience and a little time.

6- Mount the pump in a vise in a vertical position, as shown. Lay the diaphragm over the top of the nylon disc washer and onto the pump body. Notice how the spring holds the disc up and partially lifts the diaphragm from the pump. This is a normal condition.

7- Insert the small disc and the short spring into the cavity, as shown. This spring and disc helps to cushion the vacuum impulses from the engine.

8- Ease the fuel pump cover down over the diaphragm and then thread the six cover attaching screws into the pump body. As each screw is started, pull on the edge of the diaphragm to align the screw holes in the diaphragm with the matching holes in the pump body. Tighten the attaching screws securely.

9- Slide the filter element into the fuel. The end of the filter element must slide over the indexing peg in the bottom of the pump. Force the element onto the peg until it is fully seated.

10- Place a **NEW** gasket into position in the fuel pump, as shown.

11- Place the fuel pump bowl over the filter element and into position on the gasket.

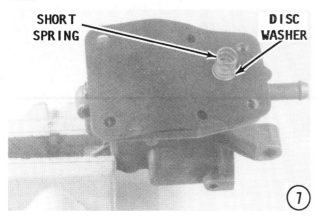

(7)

(8)

(9)

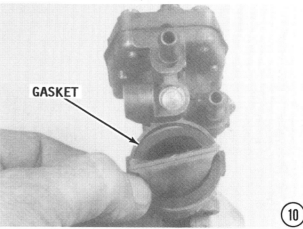

(10)

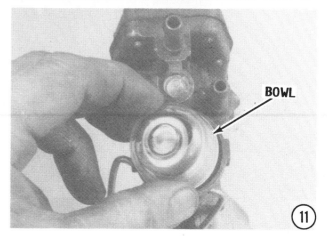

(11)

FUEL TANK AND LINE 4-57

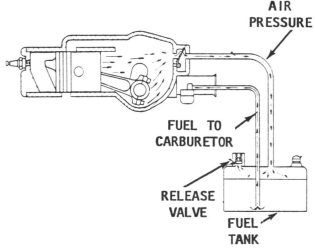

Functional diagram illustrating a complete pressure tank fuel system from the fuel tank to the cylinder.

12- Swing the hinge up over the pump bowl, and then secure it in place by tightening the plastic thumb screw.

13- Mount the fuel pump onto the engine and secure it with the two screws on top and one behind the fuel inlet nipple. **CAREFULLY** connect the vacuum hose and the two fuel hoses to their proper nipples, as identified during disassembling. If identification was not made follow the hoses as described. Connect the vacuum hose (from the engine) to the nipple on the pump cover. Connect the inlet hose (from the fuel tank) to the inside nipple (the one closest to the engine). Connect the outlet hose (to the carburetor) to the remaining nipple.

4-12 FUEL TANK AND LINE SERVICE

The procedures outlined in this section cover service of original equipment produced by Johnson/Evinrude.

Pressurized Fuel System

A pressure-type fuel tank was used with the early model engines not equipped with a fuel pump. Two hoses were connected between the tank and the engine. One hose served as the fuel transfer line and the other supplied pressurized air to the tank. For the system to operate, the fuel fill cap on the tank must be completely closed, making the tank air-tight.

The system is primed by operating the primer pump located on top of the fuel tank.

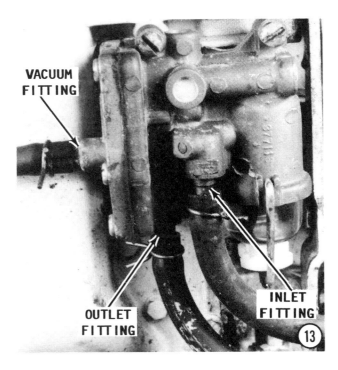

Obsolete pressure-type fuel tank. A manually-operated primer pump on the fuel tank is worked to pump fuel to the engine. After engine start, air pressure forces fuel from the tank to the engine.

The primer pump is operated until the carburetor is full of fuel. The engine is then started. Once the engine is operating, even at idle speed, pressurized air is fed from the intake manifold through a check valve and the hose to the fuel tank. As pressure is increased inside the fuel tank, the fuel is forced through the fuel hose to the carburetor.

Each tank should build and retain the following pressures:

Model CD	2 to 4-1/2 lbs.
Model AD	2 to 5 lbs.
Model QD	2 to 5 lbs.
Model FD	2 to 5-1/2 lbs.
Model RD	2 to 4-1/2 lbs.

When the engine is shut down and the fuel hose is disconnected from the engine, the fuel tank cap should be opened slightly to allow the air pressure in the tank to escape. It is not a good practice to allow the tank to remain pressurized.

Any engine equipped with this pressurized fuel system can be updated by installing a fuel pump. The cost of replacing the pressurized-type fuel tank actually exceeds the cost involved in making the conversion. The necessary work involves drilling a hole in the bypass cover to mount the pump and provide the pump with vacuum from one of the cylinders. Installation of new hoses to the carburetor is not a difficult or expensive task.

The accompanying illustration clearly shows the two hoses connected to the fuel tank. The hose on the left transfers fuel to the carburetor and the hose on the right is the air pressure line. To assist in identifying the fuel line, the fuel hose has a rib extending its full length from the tank to the carburetor. The primer pump is also clearly shown in the illustration.

DISASSEMBLING

WARNING

TAKE EXTREME CARE during work with the fuel tank, because highly flamable fumes are present and the danger of fire or explosion is present. Demand and observe **NO SMOKING** or open flame in the work area. Clean the tank in the out-of-doors.

1- Check to be sure the knob on the pump shaft is able to rotate on the shaft. The knob is secured to the shaft with a cotter pin. If the knob does not turn during operation, the diaphragm in the pump will be damaged or ruptured.

Explanation

The following description of pump operation may be helpful in explaining how damage to the diaphragm may be caused by failure of the button to turn.

When a person operates the pump, the action is not in a straight down direction. The natural tendancy is to turn the thumb or hand while operating the plunger. The shaft is connected directly to the diaphragm. Therefore, if the knob fails to turn, the shaft will be torn away from the diaphragm.

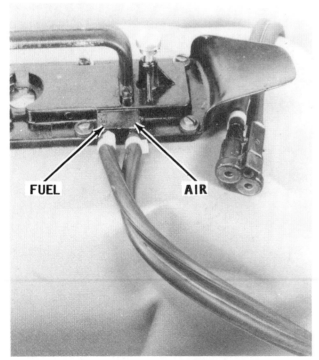

Fuel and air pressure lines attached to a pressure-type fuel tank through the double connector. The fuel line has a rib extending its full length.

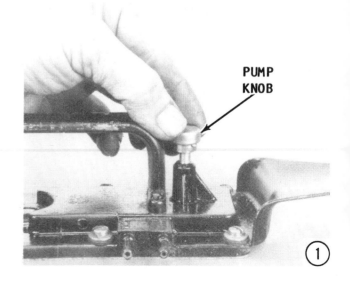

FUEL TANK AND LINE 4-59

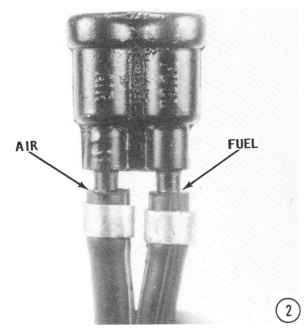

2- Check to be sure the proper hose is connected to the correct fitting at the carburetor. The hose on the left in the accompanying illustration is the air pressure line and the hose on the right is the fuel supply line. Notice the rib extending the full length of the fuel hose.

3- Remove the fuel tank cap. Disconnect the retaining chain from inside of the fuel cap. The chain may hang free inside the fuel tank.

4- If the chain cannot be disconnected from the cap by removing the retaining screw, then remove the cotter pin from inside the tank. The chain and cap can then be removed as an assembly.

5- Remove the cotter pin from the push-knob on the fuel pump plunger.

6- Remove the push knob from the pump plunger. If the knob is "frozen" and cannot be removed easily, hold the plunger with a pair of pliers and work the knob free with a second pair of pliers.

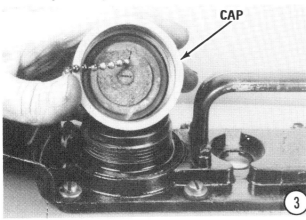

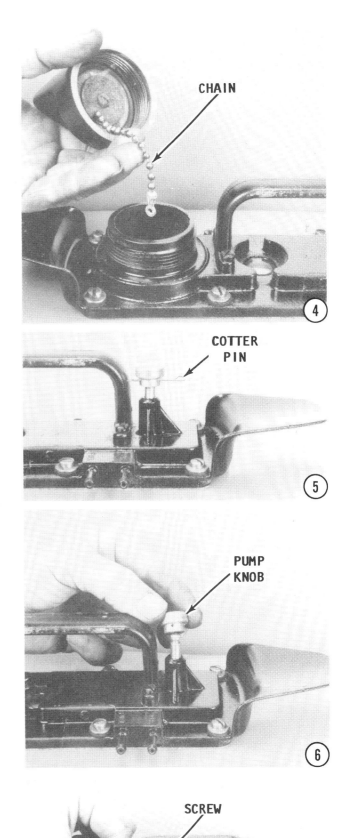

4-60 FUEL

7- Remove the retaining scews from around the pump housing. Take care to save the washer on each screw for use during installation.

8- Lift the pump and cover from the fuel tank, as shown.

9- Remove and discard the gasket from the fuel tank. Clean any old gasket material from the fuel tank surface.

10- Secure the pump housing in a vise, as shown. This position will provide freedom to work on the pump and pickup assembly. Remove the nut and separate the pickup from the pump body. The pickup and screen are sold only as an assembly.

11- Test operation of the check valve by attempting to suck air up through the tube. The attempt should be successful. Attempt to blow air down through the tube. The attempt should fail. The check valve should allow air to pass up through the tube but prevent the movement of air in the opposite direction. If the check valve fails the tests, clean the assembly in carburetor cleaner and then wash it thoroughly with soap and water to prevent acid burn to the lips and mouth when the test is repeated. Perform the test a second time. If the unit fails to pass the second test, the entire pickup assembly **MUST** be replaced.

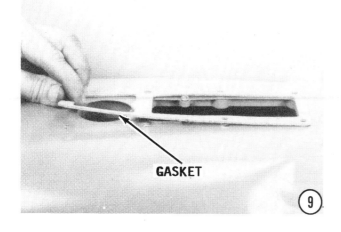

12- Remove the eight screws from the pump body and lift the body free of the pump. Take care not to damage or loose the two springs, two washers, disc, and the diaphragm. Observe the relationship of parts: The spring on top of the diaphragm; the rubber washer and spring to the right; the small disc in the center, as shown in the accompanying illustration.

13- Remove the spring from the top of the diaphragm. Lift the rubber washer and spring from the right side of the diaphragm.

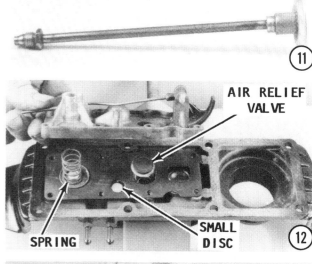

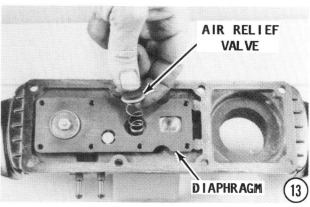

FUEL TANK AND LINE

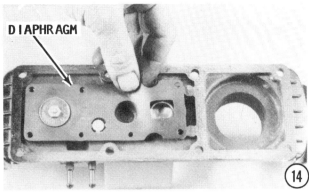

14- Lift the diaphragm free of the pump body with the pump plunger passing out through the housing.

15- Remove the spring from the housing located under the diaphragm.

16- Remove the small disc from the center of the housing.

17- Remove the small nut and washer from the old diaphragm.

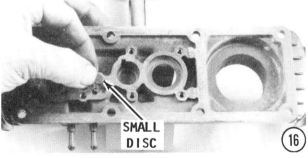

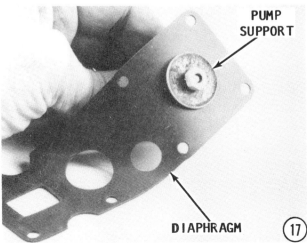

Pickup and filter screen used in a pressure-type fuel tank. The screen should be cleaned regularly with solvent and then blown dry with compressed air.

CLEANING AND INSPECTING

WARNING
TAKE EXTREME CARE during work with the fuel tank, because highly flamable fumes are present and the danger of fire or explosion is present. Demand and observe **NO SMOKING** or open flame in the work area. Clean the tank in the out-of-doors.

Wash the interior of the fuel tank with solvent. Agitate the solvent violently while rapidly changing position of the tank to remove any foreign material in the tank. After the tank has been cleaned with the solvent, rinse it thoroughly with clear water and then dry and remove any moisture from the inside.

ASSEMBLING

1- Slide the large concave washer onto the pump plunger. Insert the pump plunger through the diaphragm.

2- Slide a flat washer onto the plunger on the opposite side of the diaphragm from the concave washer. Thread the nut onto

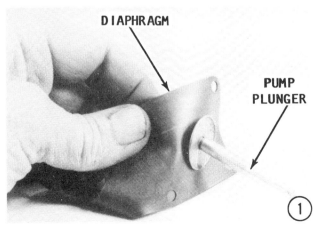

4-62 FUEL

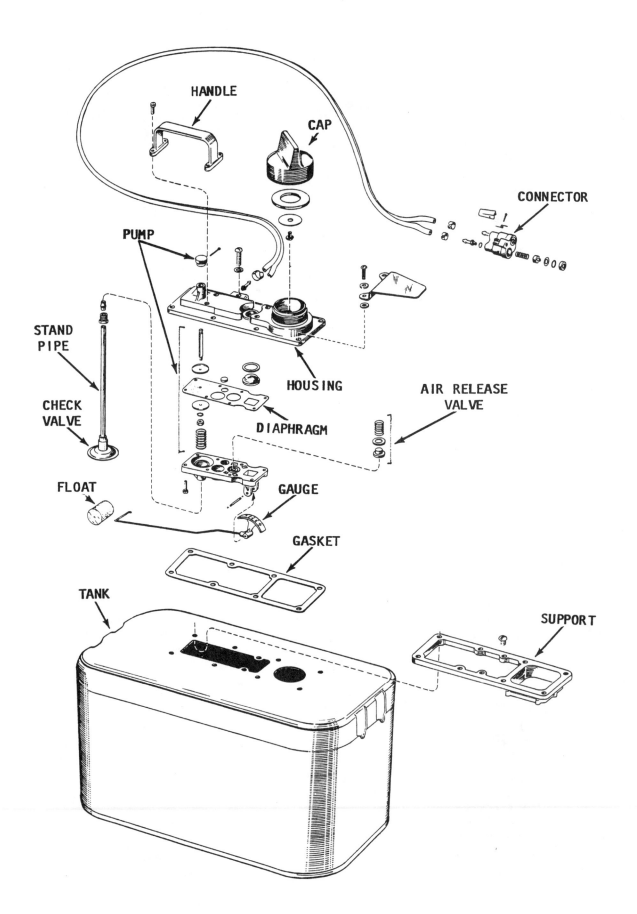

Exploded view of a pressure-type fuel tank with major parts identified.

FUEL TANK AND LINE 4-63

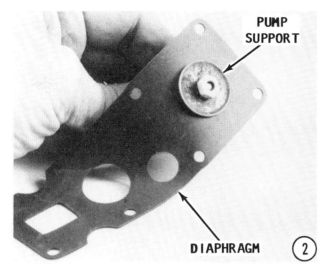

the plunger and tighten it just snugly. **DO NOT** overtighten the nut, to prevent possible damage to the diaphragm.

3- Insert the small disc washer into the center hole of the housing.

4- Insert the glass sight gauge in the housing and check to be sure it is properly seated.

5- Install the spring over the boss on the plunger.

6- Work the pump plunger down through the spring and the housing with the diaphragm seated on the housing. Install the relief spring and washer into the housing and hole of the diaphragm.

7- Install the spring over the top of the pump washer. Check to be sure the housing is level to allow the spring to remain in position when the pump housing is installed.

8- Position the pump housing down over the spring and relief valve.

9- Start to thread the eight screws securing the pump housing to the pump body, and then tighten them evenly and alternately.

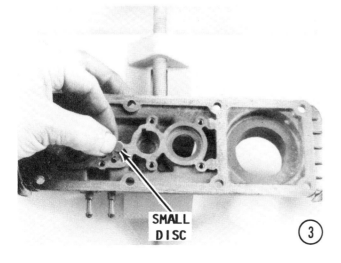

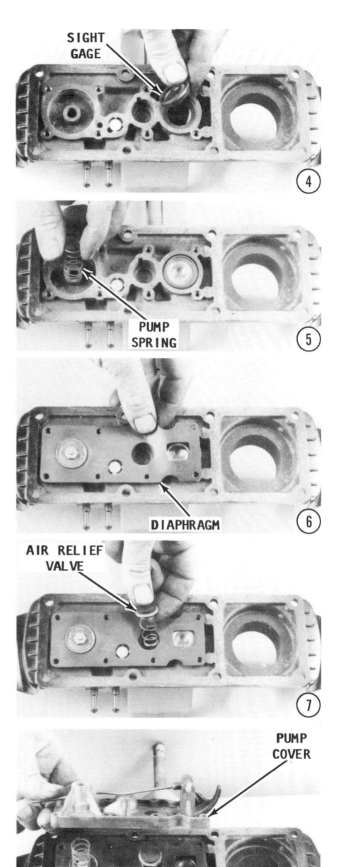

4-64 FUEL

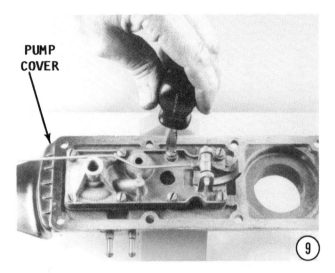

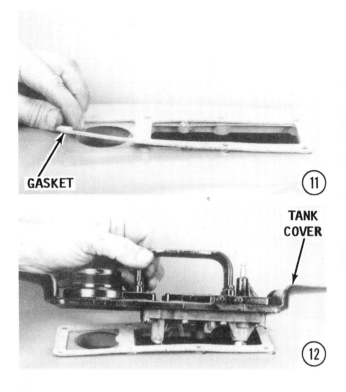

10- Place the pickup check valve assembly in position, and then start the compression nut **BY HAND** to prevent cross-threading. Tighten the nut securely with the proper size wrench.

11- Position a **NEW** gasket in place around the fuel pump opening on the tank.

12- Lower the complete assembly into the fuel tank. Align the retaining screw holes with the matching holes in the gasket and tank.

13- Securing the pump in place with the attaching screws. Use **NEW** tiny gaskets on each screw to prevent fuel linkage through the screw holes.

14- Slide the push button onto the pump plunger.

15- Secure the push button in place with a **NEW** cotter pin. After the pin is in place and the ends have been bent back in the usual manner, clip off the ends of the pin to prevent scratching a finger or hand when the pump is operated. Check to be sure the buttom will rotate in a full circle, without binding.

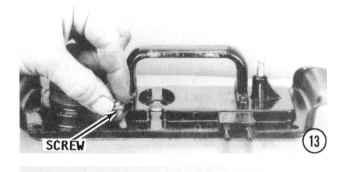

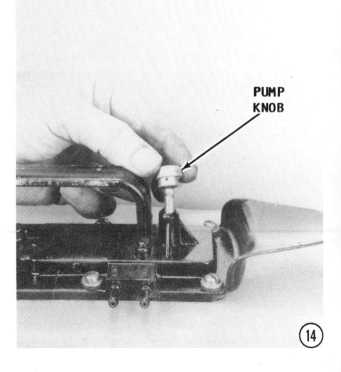

FUEL TANK AND LINE   4-65

16- Install the fuel cap and chain. If the chain and cap were removed, use a **NEW** cotter pin through the flange in the tank to secure the chain in place.

17- Install the fuel cap. If only the cap was removed, attach the chain to the cap with the retaining screw.

18- After the work is completed, add a quantity of fuel/oil mixture to the tank and test the system for proper operation and no leaks. Insert a small screwdriver into the end of the fuel connector at the end of the line to open the check valve. Operate the fuel tank pump and and be prepared to catch fuel being discharged from the end of the hose.

Fuel Line Service

The only service work to be performed on the fuel lines is replacement of the O-rings in the fuel line connectors, and replacement of the hoses. New O-rings and fuel lines are available at the local OMC dealer.

19- Use two ice picks or similar tool, and push down the center plunger of the connector and work the O-ring out of the hole. Repeat the procedure to remove the O-ring from the other check valve.

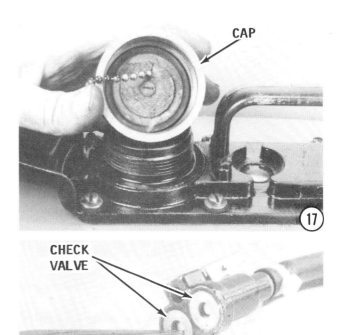

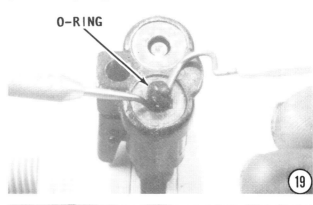

20- Apply just a drop of oil into the hole of the connector. Apply a thin coating of oil to the surface of the O-ring. Pinch the O-ring together and work it into the hole while simultaneously using a punch to depress the plunger inside the connector.

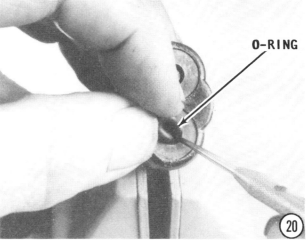

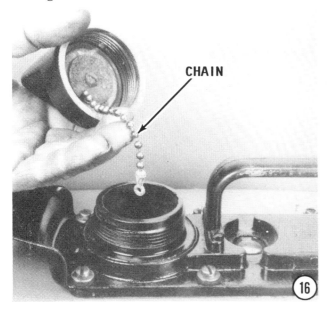

4-66 FUEL

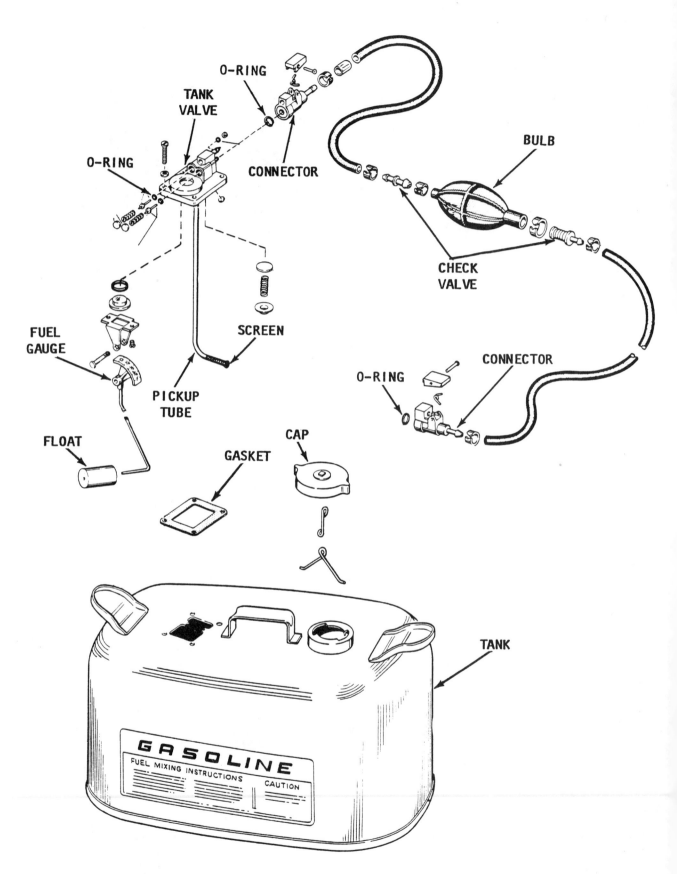

Exploded drawing of a modern non-pressurized fuel tank and fuel line.

FUEL TANK AND LINE 4-67

4-13 LATE MODEL FUEL TANK SERVICE

1- Late model fuel tanks (since about 1959), are not pressurized. A squeeze bulb is used to move fuel from the tank to the carburetor until the engine is operating. Once the engine starts, the fuel pump, mounted on the engine, transfers fuel from the tank to the carburetor.

2- The pickup unit in the tank is sold as a complete unit, but without the gauge and float.

3- To replace the pickup unit, first remove the four screws securing the unit in the tank. Next, lift the pickup unit up out of the tank.

4- Remove the two Phillips screws securing the fuel gauge to the bottom of the pickup unit and set the gauge aside for installation onto the new pickup unit.

If the pickup unit is not to be replaced, clean and check the screen for damage. It is possible to bend a new piece of screen material around the pickup and solder it in place without purchasing a complete new unit.

Attach the fuel gauge to the new pickup unit and secure it in place with the two Phillips screws.

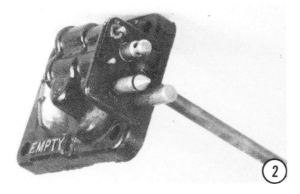

5- Clean the old gasket material from fuel tank and old pickup unit (if the old pickup unit is to be installed for further service). Work the float arm down through the fuel tank opening, and at the same time the fuel pickup tube into the tank. It will probably be necessary to exert a little force on the float arm in order to feed it all into the hole. The fuel pickup arm should spring into place once it is through the hole. Secure the pickup and float unit in place with the four attaching screws.

6- The primer squeeze bulb can be replaced in a short time. A squeeze bulb assembly, complete with the check valves installed, may be obtained from the local OMC dealer.

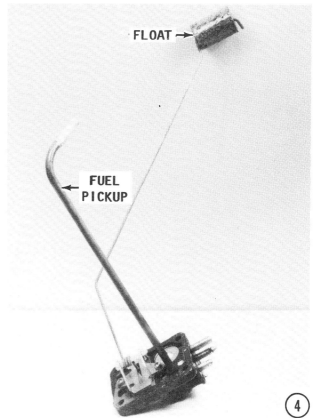

4-68 FUEL

An arrow is clearly visible on the squeeze bulb to indicate the direction of fuel flow. The sqeeze bulb **MUST** be installed correctly in the line because the check valves in each end of the bulb will allow fuel to flow in **ONLY** one direction. Therefore, if the squeeze bulb should be installed backwards (in a moment of haste to get the job done), fuel will not reach the carburetor.

7- To replace the bulb, first unsnap the clamps on the hose at each end of the bulb. Next, pull the hose out of the check valves at each end of the bulb. New clamps are included with a new squeeze bulb. If the fuel line has been exposed to considerable sunlight, it may have become hardened, causing difficulty in working it over the check valve. To remedy this situation, simply immerse the ends of the hose in boiling water for a few minutes to soften the rubber and the hose will then slip onto the check valve without further problems. After the lines on both sides have been installed, snap the camps in place to secure the line. Check a second time to be sure the arrow is pointing in the fuel flow direction, **TOWARDS** the engine.

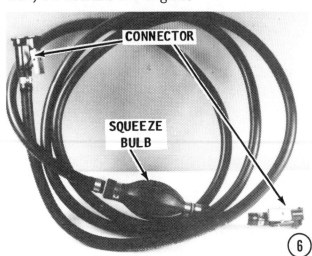

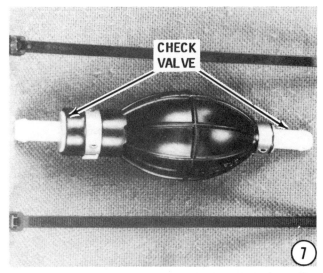

8- Use two ice picks or similar tool, and push down the center plunger of the connector and work the O-ring out of the hole.

9- Apply just a drop of oil into the hole of the connector. Apply a thin coating of oil to the surface of the O-ring. Pinch the O-ring together and work it into the hole while simultaneously using a punch to depress the plunger inside the connector.

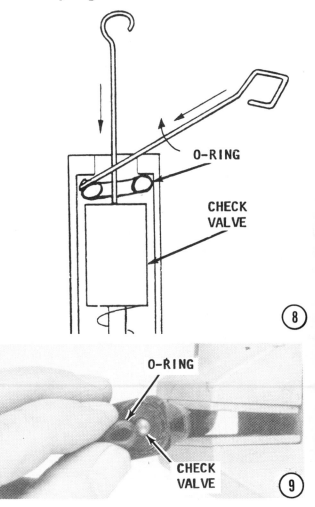

5
IGNITION

5-1 INTRODUCTION

The less an outboard engine is operated, the more care it needs. Allowing an outboard engine to remain idle will do more harm than if it is used regularly. To maintain the engine in top shape and always ready for efficient operation at any time, the engine should be operating every 3 to 4 weeks throughout the year.

The carburetion and ignition principles of two-cycle engine operation **MUST** be understood in order to perform a proper tune-up on an outboard motor.

If you have any doubts concerning your understanding of two-cyle engine operation, it would be best to study the operation theory section in the first portion of Chapter 3, before tackling any work on the ignition system.

The **ONLY** ignition system used on all Johnson/Evinrude engines covered in this manual is a flywheel magneto system. The first sections of this chapter will be devoted to an explanation of the system and its theory of operation. The latter sections will provide troubleshooting and repair instructions. For synchronizing procedures, see Section 5-8.

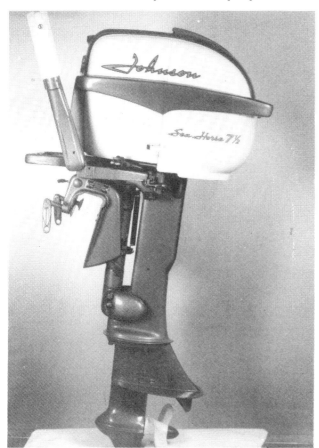

Outboard engine performance can be affected by engine non-use. The engine should be started and run periodically.

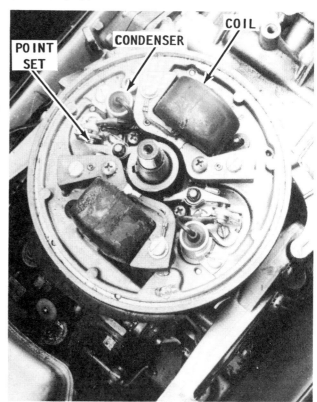

View of the magneto installed on engines covered in this manual. A one-cylinder engine will have only half the parts.

5-2 SPARK PLUG EVALUATION

Removal: Remove the spark plug wires by pulling and twisting on only the molded cap. **NEVER** pull on the wire or the connection inside the cap may become separated or the boot damaged. Remove the spark plugs and keep them in order. **TAKE CARE** not to tilt the socket as you remove the plug or the insulator may be cracked.

Examine: Line the plugs in order of removal and carefully examine them to determine the firing conditions in each cylinder. If the side electrode is bent down onto the center electrode, the piston is traveling too far upward in the cylinder and striking the spark plug. Such damage indicates the wrist pin or the rod bearing is worn excessively. In all cases, an engine overhaul is required to correct the condition. To verify the cause of the problem, turn the engine over by hand. As the piston moves to the full up position, push on the piston crown with a screwdriver inserted through the spark plug hole, and at the same time rock the flywheel back-and-forth. If any play in the piston is detected, the engine must be rebuilt.

Correct Color: A proper firing plug should be dry and powdery. Hard deposits inside the shell indicate too much oil is being mixed with the fuel. The most important evidence is the light gray color of the porcelain, which is an indication this plug has been running at the correct temperature. This means the plug is one with the correct heat range and also that the air-fuel mixture is correct.

Rich Mixture: A black, sooty condition

*Damaged spark plugs. Notice the broken electrode on the left plug. The broken part **MUST** be found and removed before returning the engine to service.*

on both the spark plug shell and the porcelain is caused by an excessively rich air-fuel mixture, both at low and high speeds. The rich mixture lowers the combustion temperature so the spark plug does not run hot enough to burn off the deposits.

Deposits formed only on the shell is an indication the low-speed air-fuel mixture is too rich. At high speeds with the correct mixture, the temperature in the combustion chamber is high enough to burn off the deposits on the insulator.

Too Cool: A dark insulator, with very few deposits, indicates the plug is running too cool. This condition can be caused by low compression or by using a spark plug of an incorrect heat range. If this condition shows on only one plug it is most usually caused by low compression in that cylinder. If all of the plugs have this appearance, then it is probably due to the plugs having a too-low heat range.

This spark plug is foul from operating with an over-rich condition, possibly an improper carburetor adjustment.

This spark plug has been operating too-cool, because it is rated with a too-low heat range for the engine.

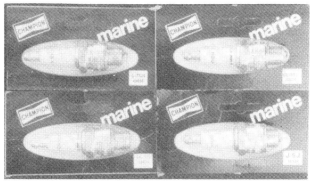

Today, numerous type spark plugs are available for service. ALWAYS check with your local marine dealer to be sure you are purchasing the proper plug for the engine being serviced.

Fouled: A fouled spark plug may be caused by the wet oily deposits on the insulator shorting the high-tension current to ground inside the shell. The condition may also be caused by ignition problems which prevent a high-tension pulse to be delivered to the spark plug.

Carbon Deposits: Heavy carbon-like deposits are an indication of excessive oil in the fuel. This condition may be the result of poor oil grade, (automotive-type instead of a marine-type); improper oil-fuel mixture in the fuel tank; or by worn piston rings.

Overheating: A dead white or gray insulator, which is generally blistered, is an indication of overheating and pre-ignition. The electrode gap wear rate will be more than normal and in the case of pre-ignition, will actually cause the electrodes to melt as shown in this illustration. Overheating and pre-ignition are usually caused by improper point gap adjustment; detonation from using too-low an octane rating fuel; an excessively lean air-fuel mixture; or problems in the cooling system.

Electrode Wear: Electrode wear results in a wide gap and if the electrode becomes carbonized it will form a high-resistance path for the spark to jump across. Such a condition will cause the engine to misfire during acceleration. If all plugs are in this condition, it can cause an increase in fuel consumption and very poor performance during high-speed operation. The solution is to replace the spark plugs with a rating in the proper heat range and gapped to specification.

Red rust-colored deposits on the entire firing end of a spark plug can be caused by water in the cylinder combustion chamber. This can be the first evidence of water entering the cylinders through the exhaust manifold because of an accumulation of scale or defective exhaust shutter. This condition **MUST** be corrected at the first opportunity. Refer to Chapter 3, Engine Service.

5-3 POLARITY CHECK

Coil polarity is extremely important for proper battery ignition system operation. If a coil is connected with reverse polarity, the spark plugs may demand from 30 to 40 percent more voltage to fire. Under such demand conditions, in a very short time the coil would be unable to supply enough voltage to fire the plugs. Any one of the following three methods may be used to quickly determine coil polarity.

1- The polarity of the coil can be checked using an ordinary D.C. voltmeter. Connect the positive lead to a good ground. With the engine running, momentarily touch the negative lead to a spark plug terminal.

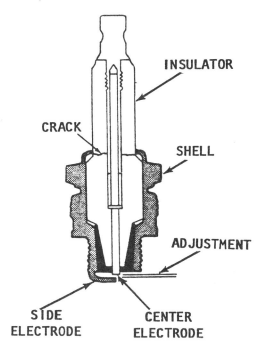

Cut-a-way drawing showing major spark plug parts.

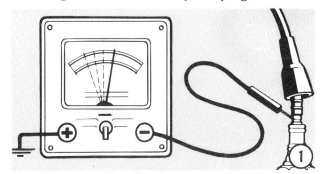

5-4 IGNITION

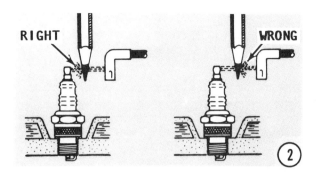

The needle should swing upscale. If the needle swings downscale, the polarity is reversed.

2- If a voltmeter is not available, a pencil may be used in the following manner: Disconnect a spark plug wire and hold the metal connector at the end of the cable about 1/4" from the spark plug terminal. Now, insert an ordinary pencil tip between the terminal and the connector. Crank the engine with the ignition switch ON. If the spark feathers on the plug side and has a slight orange tinge, the polarity is correct. If the spark feathers on the cable connector side, the polarity is reversed.

3- The firing end of a used spark plug can give a clue to coil polarity. If the ground electrode is "dished", it may mean polarity is reversed.

5-4 WIRING HARNESS

CRITICAL WORDS: These next two paragraphs may well be the most important words in this chapter. Misuse of the wiring harness is the most single cause of electrical problems with outboard power plants.

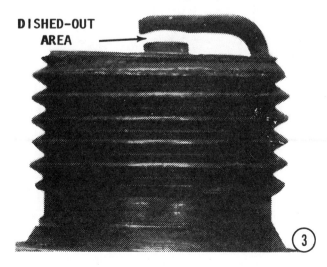

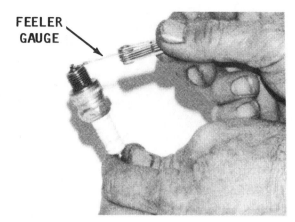

The spark plug gap should always be checked before installing new or used spark plugs.

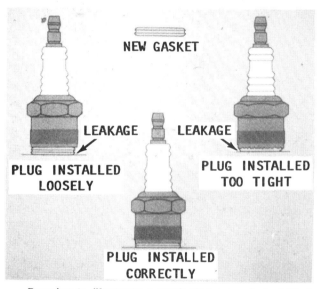

Drawing to illustrate a spark plug properly installed, center, and other plugs, left and right, improperly installed.

Damaged wiring harness lead. Such damage cannot be repaired. The lead or harness must be replaced.

A wiring harness is used between the key switch and the engine. This harness seldom contains wire of sufficient size to allow connecting accessories. Therefore, anytime a new accessory is installed, **NEW** wiring should be used between the battery and the accessory. A separate fuse panel **MUST** be installed on the dash. To connect the fuse panel, use one red and one black No. 10 gauge wire from the battery. If a small amount of 12-volt current should be accidently attached to the magneto system, the coil may be damaged or **DESTROYED**. Such a mistake in wiring can easily happen if the source for the 12-volt accessory is taken from the key switch. Therefore, again let it be said, **NEVER** connect accessories through the key switch.

5-5 FLYWHEEL MAGNETO IGNITION

Description

READ AND BELIEVE. A battery installed to crank the engine **DOES NOT** mean the engine is equipped with a battery-type ignition system. A magneto system uses the battery only to crank the engine. Once the engine is running, the battery has absolutely no affect on engine operation. Therefore, if

*The battery **MUST** be located near the engine in a well-ventilated area. It must be secured in such a manner that absolutely no movement is possible in any direction under the most violent actions of the boat.*

the battery is low and fails to crank the engine properly for starting, the engine may be cranked manually, started, and operated. Under these conditions, the key switch must be turned to the **ON** position or the engine will not start by hand cranking.

A magneto system is a self-contained unit. The unit does not require assistance from an outside source for starting or continued operation. Therefore, as previously mentioned, if the battery is dead, the engine

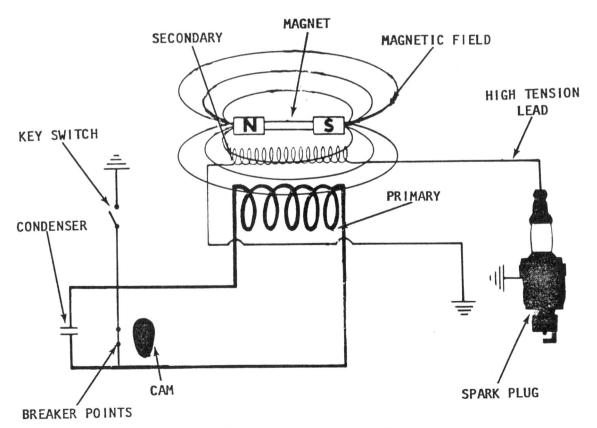

Schematic diagram of a magneto ignition system.

5-6 IGNITION

may be cranked manually and the engine started.

The flywheel-type magneto unit consists of an armature plate, and a permanent magnet built into the flywheel. The ignition coil, condenser and breaker points are mounted on the armature plate.

As the pole pieces of the magnet pass over the heels of the coil, a magnetic field is built up about the coil, causing a current to flow through the primary winding.

Now, at the proper time, the breaker points are separated by action of a cam, and the primary circuit is broken. When the circuit is broken, the flow of primary current stops and causes the magnetic field about the coil to break down instantly. At this precise moment, an electrical current of extremely high voltage is induced in the fine secondary windings of the coil. This high voltage is conducted to the spark plug where it jumps the gap between the points of the plug to ignite the compressed charge of air-fuel mixture in the cylinder.

5-6 TROUBLESHOOTING

Always attempt to proceed with the troubleshooting in an orderly manner. The shotgun approach will only result in wasted time, incorrect diagnosis, replacement of unneccessary parts, and frustration.

Begin the ignition system troubleshooting with the spark plug/s and continue through the system until the source of trouble is located.

Remember, a magneto system is a self-contained unit. Therefore, if the engine has a key switch and wire harness, remove them from the engine and then make a test for

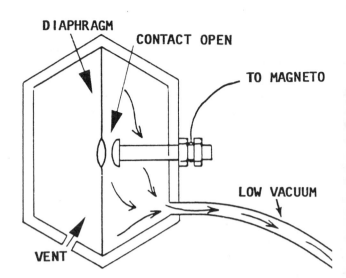

Schematic diagram of a vacuum cutout switch used on early model small horsepower engine. This illustration depicts the position of the diaphragm in relation to the ground contact when operating at normal manifold pressure. Spring omitted for clarity.

spark. If a good spark is obtained with these two items disconnected, but no spark is available at the plug when they are connected, then the trouble is in the harness or the key switch. If a test is made for spark at the plug with the harness and switch connected, check to be sure the key switch is turned to the **ON** position.

Vacuum Cutout Switch

On some smaller engine models, a cutout vacuum switch is installed. This switch is connected to one of the cylinders in the

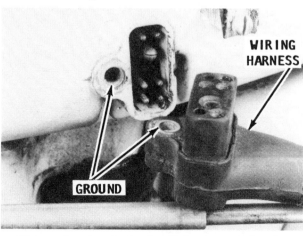

The engine terminals should be inspected and any corrosion removed to ensure a proper connection with wiring harness.

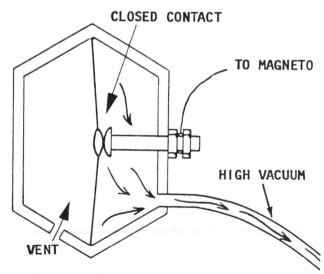

Schematic diagram of the cutout switch to depict the diaphragm making contact with the ground. This condition results from abnormally high manifold suction on the instant of rapidly throttling down from high to slow idle speed with the engine in neutral. Spring not shown for clarity.

ignition system. The switch is actuated by vacuum from the cylinder. When a high vacuum pull is exerted against the switch, during engine operation in gear without the lower unit in the water, the switch is closed and the engine is shut down. This feature is a safeguard against the engine "running away" while operating with a no-load condition on the propeller. A two-cycle engine will continue to increase rpm under a no-load condition and attempts to shut it down will fail, resulting in serious damage or destruction of the unit.

The vacuum switch also serves as a safety feature when the boat is operating in the water. If the propeller is released from the shaft, because of an accident, striking an underwater object, whatever, the engine would then be operating under a no-load condition. The vacuum switch will shut down the engine and prevent extensive damage, resulting from a "run-a-way" condition.

This cutout switch arrangement was installed on all 28, 33, and 35 hp engines, 1958-59, and all 40 hp engines covered in this manual.

Therefore, if spark is not present at the spark plug, disconnect the wires from the vacuum switch and again test for spark at the spark plug. If spark is present with the vacuum switch disconnected, the switch is defective and must be replaced.

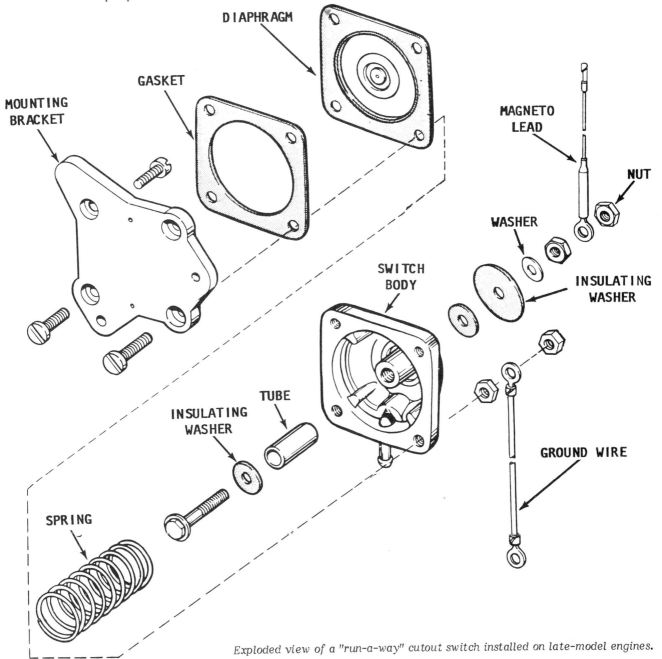

Exploded view of a "run-a-way" cutout switch installed on late-model engines.

5-8 IGNITION

WIRING HARNESS

CRITICAL WORDS: These next two paragraphs may well be the most important words in this chapter. Misuse of the wiring harness is the most single cause of electrical problems with outboard power plants.

A wiring harness is used between the key switch and the engine. This harness seldom contains wire of sufficient size to allow connecting accessories. Therefore, anytime a new accessory is installed, **NEW** wiring should be used between the battery and the accessory. A separate fuse panel **MUST** be installed on the dash. To connect the fuse panel, use one red and one black No. 10 gauge wires from the battery. If a small amount of 12-volt current should be accidently attached to the magneto system, the coil will be damaged or **DESTROYED**. Such a mistake in wiring can easily happen if the source for the 12-volt accessory is taken from the key switch. Therefore, again let it be said, **NEVER** connect accessories through the key switch.

The wiring harness installed on the 35 hp, 1959, and the 40 hp, 1960 model units, was connected to the the side of the engine through an electrical plug utilized "male" and "female" connectors. This particular type connector has been a contributing factor to a number of problems in the ignition system due to the suceptibility of the connector to corrosion. The plug is exposed and subject to moisture which is especially des-

A coil DESTROYED when 12-volts was connected into the magneto wiring system. Mechanics report in 85% of the cases, the damage occurs when an accessory is connected through the key switch.

tructive in a salt water atmosphere. Therefore, during the troubleshooting work on these engines, always disconnect this plug and make a careful check for any sign of corrosion.

Key Switch

A magneto key switch operates in **REVERSE** of any other type key switch. When the key is moved to the **OFF** position, the circuit is **CLOSED** between the magneto and ground. In some cases, when the key is turned to the **OFF** position the points are grounded. For this reason, an automotive-type switch **MUST NEVER** be used, because the circuit would be opened and closed in reverse, and if 12-volts should reach the coil, the coil will be **DESTROYED**.

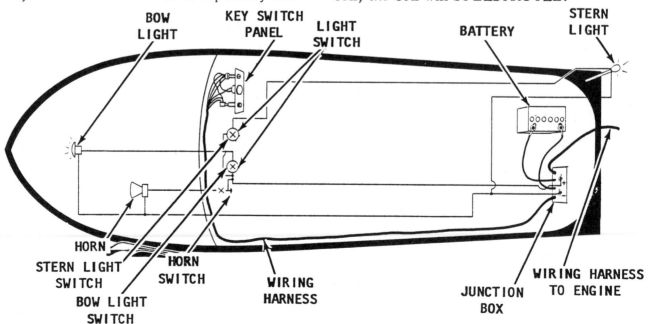

Functional diagram to illustrate proper hookup of accessories through a junction box. If a junction box is not installed on the boat, connect accessories directly to the battery. NEVER connect accessories through the key switch.

TROUBLESHOOTING 5-9

Spark Plugs

1- Check the plug wires to be sure they are properly connected. Check the entire length of the wire/s from the plug/s to the magneto under the armature plate. If the wire is to be removed from the spark plug, **ALWAYS** use a pulling and twisting motion as a precaution against damaging the connection.

2- Attempt to remove the spark plug/s by hand. This is a rough test to determine if the plug is tightened properly. You should not be able to remove the plug without using the proper socket size tool. Remove the spark plug/s and keep them in order. Examine each plug and evaluate its condition as described in Section 5-2.

If the spark plugs have been removed and the problem cannot be determined, but the plug appears to be in satisfactory condition, electrodes, etc., then replace the plugs in the spark plug openings.

A conclusive spark plug test should always been performed with the spark plugs installed. A plug may indicate satisfactory spark when it is removed and tested, but under a compression condition may fail. An example would be the possibility of a person being able to jump a given distance on the ground, but if a strong wind is blowing, his distance may be reduced by half. The same is true with the spark plug. Under good compression in the cylinder, the spark may be too weak to ignite the fuel properly.

Therefore, to test the spark plug under compression, replace it in the engine and tighten it to the proper torque value. Another reason for testing for spark with the

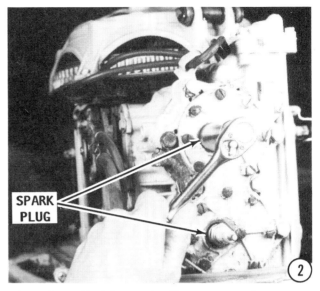

plugs installed is to duplicate actual operating conditions regarding flywheel speed. If the flywheel is rotated with the pull cord with the plugs removed, the flywheel will rotate must faster because of the no-compression condition in the cylinder, giving the **FALSE** indication of satisfactory spark.

A spark tester capable of testing for spark while cranking and also while the engine is operating, can be purchased from Better Way Tool Company. See the inside back cover of this manual for "How to Order" instructions.

3- Use a spark tester and check for spark at each cylinder. If a spark tester is

5-10 IGNITION

not available, hold the plug wire about 1/4-inch from the engine. Turn the flywheel with a pull starter or electrical starter and check for spark. A strong spark over a wide gap must be observed when testing in this manner, because under compression a strong spark is necessary in order to ignite the air-fuel mixture in the cylinder. This means it is possible to think you have a strong spark, when in reality the spark will be too weak when the plug is installed. If there is no spark, or if the spark is weak, the trouble is most likely under the flywheel in the magneto.

ONE MORE WORD: Each cylinder has its own ignition system in a flywheel-type ignition system. This means if a strong spark is observed on any one cylinder and not at another, only the weak system is at fault. However, it is always a good idea to check and service all systems while the flywheel is removed.

Compression

A compression check is extremely important, because an engine with low or uneven compression between cylinders **CANNOT** be tuned to operate satisfactorily. Therefore, it is essential that any compression problem be corrected before proceeding with the tune-up procedure. See Chapter 3.

If the powerhead shows any indication of overheating, such as discolored or scorched paint, especially in the area of the top (No. 1) cylinder, inspect the cylinders visually thru the transfer ports for possible scoring. A more thorough inspection can be made if the head is removed. It is possible for a cylinder with satisfactory compression to be scored slightly. Also, check the water pump. The overheating condition may be caused by a faulty water pump.

An overheating condition may also be caused by running the engine out of the water. For unknown reasons, many operators have formed a bad habit of running a small engine without the lower unit being submerged. Such a practice will result in a overheated condition in a matter of seconds. It is interesting to note, the same operator would never operate or allow anyone else to run a large horsepower engine without water circulating through the lower unit for cooling. Bear-in-mind, the laws governing operation and damage to a large unit **ALL** apply equally as well to the small engine.

Checking the rings and cylinder walls through the opening on the exhaust side of the engine to be sure the walls are not scored and the rings are not stuck in the piston (fail to expand properly).

The preferred method of checking the cylinder walls and rings is to pull the head and make an inspection. This method will also reveal the piston condition in each cylinder.

Checking Compression

4- Remove the spark plug wires. **ALWAYS** grasp the molded cap and pull it loose with a twisting motion to prevent damage to the connection. Remove the spark plugs and keep them in **ORDER** by cylinder for evaluation later. Ground the spark plug leads to the engine to render the ignition system inoperative while performing the compression check.

Insert a compression gauge into the No. 1, top, spark plug opening. Crank the engine with the starter, or pull on the starter cord, through at least 4 complete piston strokes with the throttle at the wide-open position, or until the highest possible reading is observed on the gauge. Record the reading.

Repeat the test and record the compression for each cylinder. A variation between cylinders is far more important than the actual readings. A variation of more than 5 psi between cylinders indicates the lower compression cylinder may be defective. The problem may be worn, broken, or sticking piston rings, scored pistons or worn cylinders. These problems may only be determined after the head has been removed. Removing the head on an outboard engine is not that big a deal, and may save many hours of frustration and the cost of purchasing unnecessary parts to correct a faulty condition.

Condenser

In simple terms, a condenser is composed of two sheets of tin or aluminum foil laid

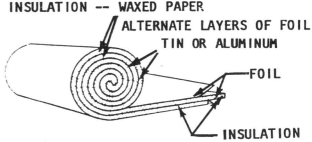

Rough sketch to illustrate how the waxed paper, aluminum foil, and insulation are rolled in a typical condenser.

one on top of the other, but separated by a sheet of insulating material such as waxed paper, etc. The sheets are rolled into a cylinder to conserve space and then inserted into a metal case for protection and to permit easy assembly.

The purpose of the condenser is to absorb or store the secondary current built up in the primary winding at the instant the breaker points are separated. By absorbing or storing this current, the condenser prevents excessive arcing and the useful life of the breaker points is extended. The condenser also gives added force to the charge produced in the secondary winding as the condenser discharges.

Modern condensers seldom cause problems, therefore, it is not necessary to install a new one each time the points are replaced. However, if the points show evidence of arcing, the condenser may be at fault and should be replaced. A faulty condenser may not be detected without the use of special test equipment. The modest cost of a new condenser justifies its purchase and installation to eliminate this item as a source of trouble.

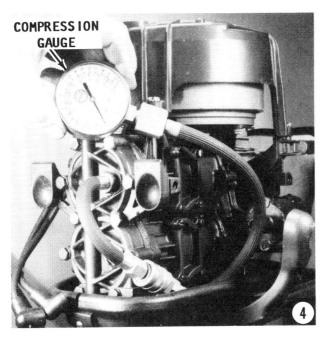

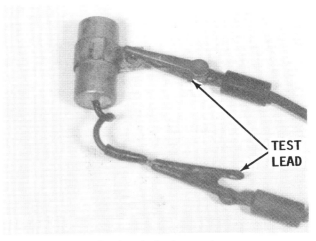

Proper hookup to test a condenser.

5-12 IGNITION

Worn and corroded breaker points unfit for further service.

Breaker Points

The breaker points in an outboard motor are an extremely important part of the ignition system. A set of points may appear to be in good condition, but they may be the source of hard starting, misfiring, or poor engine performance. The rules and knowledge gained from association with 4-cycle engines does not necessarily apply to a 2-cycle engine. The points should be replaced every 100 hours of operation or at least once a year. **REMEMBER**, the less an outboard engine is operated, the more care it needs. Allowing an outboard engine to remain idle will do more harm than if it is used regularly.

A breaker point set consists of two points. One is attached to a stationary bracket and does not move. The other point is attached to a movable mount. A spring is used to keep the points in contact with each other, except when they are separated by the action of a cam built into the flywheel or machined on the crankshaft. Both points are constructed with a steel base and a tungsten cap fused to the base.

To properly diagnose magneto (spark) problems, the theory of electricity flow must be understood. The flow of electricity through a wire may be compared with the flow of water through a pipe. Consider the voltage in the wire as the water pressure in the pipe and the amperes as the volume of water. Now, if the water pipe is broken, the water does not reach the end of the pipe. In a similar manner if the wire is broken the flow of electricity is broken. If the pipe springs a leak, the amount of water reaching the end of the pipe is reduced. Same with the wire. If the installation is defective or the wire becomes grounded, the amount of electricity (amperes) reaching the end of the wire is reduced.

Check the wiring carefully, inspect the points closely and adjust them accurately. The point setting for **ALL** engines covered in this manual is 0.020". An added item of useful information simplifying purchase of new points is that **ALL** point sets for the engines covered in this manual have the same part number, 580148.

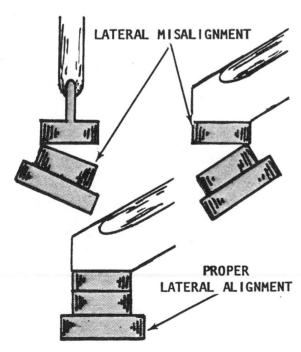

Drawing to illustrate proper point alignment, bottom set, compared with exaggerated misalignment of the other two.

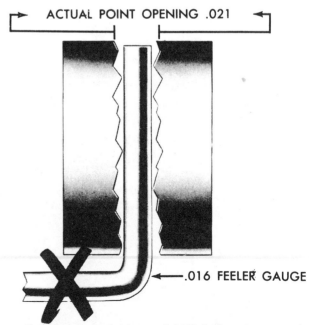

Drawing to depict how a 0.016" feeler gauge may be inserted between a badly set of worn points and the actual opening is 0.021". The point set must be in good conditon to obtain an accurate adjustment.

SERVICING 5-13

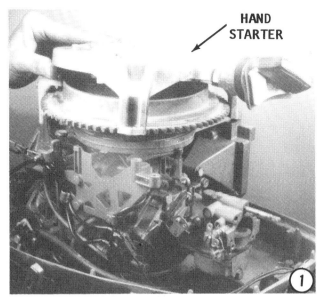

5-7 SERVICING FLYWHEEL MAGNETO IGNITION SYSTEM

General Information

Magnetos installed on outboard engines will usually operate over extremely long periods of time without requiring adjustment or repair. However, if ignition system problems are encountered, and the usual corrective actions such as replacement of spark plugs does not correct the problem, the magneto output should be checked to determine if the unit is functioning properly.

Magneto overhaul procedures may differ slightly on various outboard models, but the following general basic instructions will apply to all Johnson/Evinrude high speed flywheel-type magnetos.

REMOVAL

1- Remove the hood or enough of the engine cover to expose the flywheel. Disconnect the battery connections from the battery terminals, if a battery is used to crank the engine. If a hand starter is installed, remove the attaching hardware from the legs of the starter assembly and lift the starter free.

2- On hand started models, a round ratchet plate is attached to the flywheel to allow the hand starter to engage in the ratchet and thus turn the flywheel. This plate must be removed before the flywheel nut is removed.

3- Remove the nut securing the flywheel to the crankshaft. It may be necessary to use some type of flywheel strap to prevent the flywheel from turning as the nut is loosened.

4- Install the proper flywheel puller using the same screw holes in the flywheel

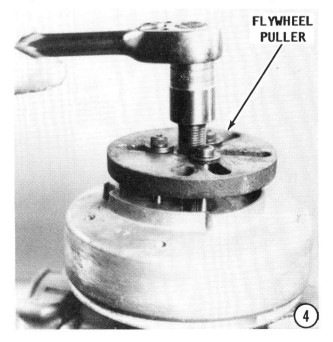

5-14 IGNITION

that are used to secure the ratchet plate removed in Step 2. **NEVER** attempt to use a puller which pulls on the outside edge of the flywheel or the flywheel may be damaged. After the puller is installed, tighten the center screw onto the end of the crankshaft. Continue tightening the screw until the flywheel is released from the crankshaft. Remove the flywheel. **DO NOT** strike the puller center bolt with a hammer in an attempt to dislodge the flywheel. Such action could seriously damage the lower seal and/or lower bearing.

5- **STOP**, and carefully observe the magneto and associated wiring layout. Study how the magneto is assembled. **TAKE TIME** to make notes on the wire routing. Observe how the heels of the laminated core, with the coil attached, is flush with the boss on the armature plate. These items must be replaced in their proper positions. You may elect to follow the practice of many professional mechanics by taking a series of photographs of the engine with the flywheel removed: one from the top, and a couple from the sides showing the wiring and arrangement of parts.

Breaker Points/Condenser Service

The armature plate does not have to be removed to service the magneto. If it is necessary to remove the plate for other

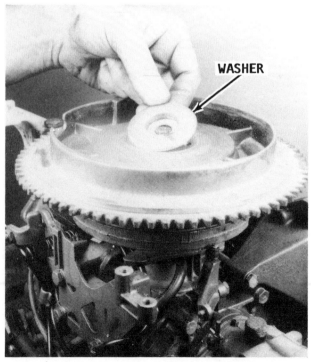

This particular engine differs from the text procedures because a washer is installed under the flywheel nut.

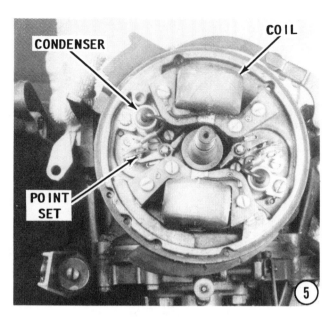

service work, such as to replace the coil or to replace the top seal, see Step 7.

For simplicity and clarity, the following procedures and accompanying illustrations cover a one-cylinder ignition system. If larger than one-cylinder is being serviced, repeat the procedures for each coil and breaker point assembly.

6- Remove the screw attaching the wires from the coil and condenser to one set of points. On engines equipped with a key switch, "kill" button, or "run-a-way" switch, a ground wire is also connected to this screw.

7- Using a pair of needle-nose pliers remove the wire clip from the post protruding through the center of the points.

8- Again, with the needle-nose pliers, remove the flat retainer holding the set of points together.

Plate installed on some engines covering the inspection hole in the flywheel. The plate is clearly marked to indicate which side is to face upward during installation.

SERVICING 5-15

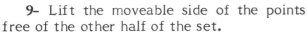

9- Lift the moveable side of the points free of the other half of the set.

10- Remove the hold-down screw securing the non-moveable half of the point set to the armature plate.

11- Remove the hold down screw securing the condenser to the armature plate. Observe how the condenser sets into a recess in the armature plate.

Repeat the procedure for the other set of points.

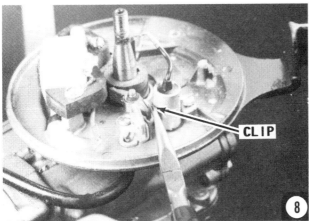
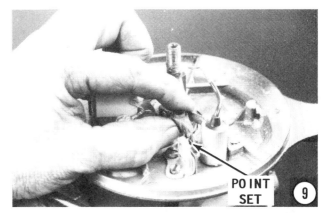
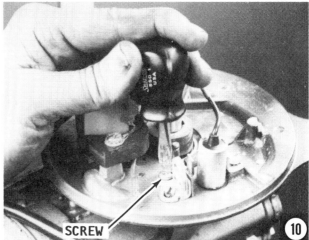

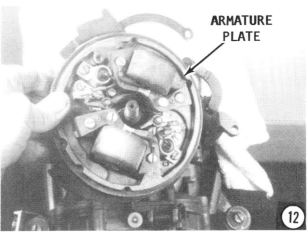

5-16 IGNITION

Armature Plate Removal

First, These Words: It is not necessary to remove the armature plate unless the top seal or the coil is to be replaced.

12- Disconnect the advance arm connecting the armature plate with the power shaft on the side of the engine. Next, remove the wires connecting the underside of the armature plate with the "kill" switch, or to the wiring harness plug. The wires of most units have a quick-disconnect fitting. Remove the wires from the vacuum (run-a-way) switch, if one is installed.

13- Observe the four screws, in a square pattern, through the armature plate. Two of these screws pass through the laminated core and the armature plate into the powerhead retainer. The other two pass just through the plate. Loosen these four screws. After the screws are loose, lift the armature plate up the crankshaft and clear of the engine. If any oil is present on top of the armature plate, or on the points, the top seal **MUST** be replaced.

Top Seal Replacement

Replacement of the top seal on a Johnson/Evinrude engine is **NOT** a difficult task, with the proper tools: a seal remover and seal installer. **NEVER** attempt to remove the seal with screwdrivers, punch, pick, or other similar tool. Such action will most likely damage the collars in the powerhead. The special tools described in the text are almost a necessity to do a proper job. The tools are usually available at the local Johnson/Evinrude dealer at reasonable cost.

14- To remove the seal, first, work the point cam up and free of the driveshaft.

Next, remove the Woodruff key from the crankshaft. A pair of side-cutters is a handy tool for this job. Grasp the Woodruff key with the side-cutters and use the leverage of the pliers against the crankshaft to remove the key.

15- Work the special tool into the seal. Observe how the special tool is tapered and has threads. Continue working and turning the tool until it has a firm grip on the inside of the seal. Now, tighten the center screw of the puller against the end of the crankshaft and the seal will begin to lift from the collars. Continue turning this center screw until the seal can be raised manually from the crankshaft.

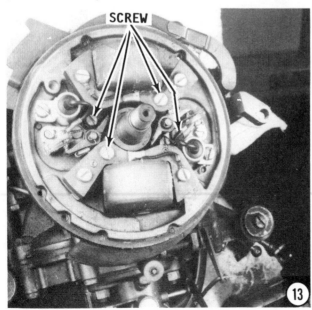

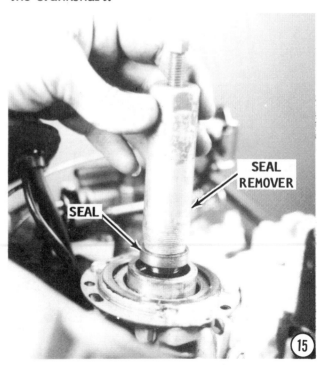

SERVICING 5-17

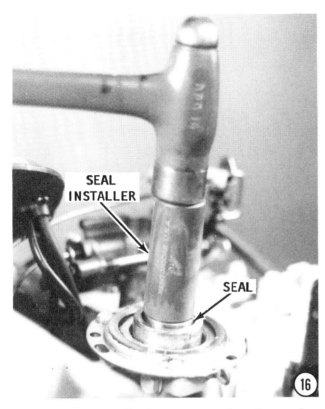

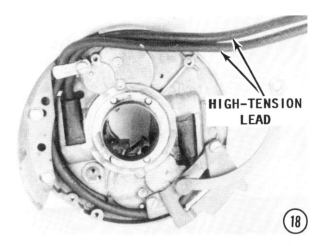

16- To install the new seal: Coat the inside diameter of the seal with a thin layer of oil. Apply OMC sealer to the outside diameter of the seal. Slide the seal down the crankshaft and start it into the recess of the powerhead. Use the special tool and work the seal completely into place in the recess.

17- Install the Woodruff key into the crankshaft. On some models, a pin was used to locate the cam for the points. If the pin was used, install it at this time. Oberve the difference to the sides of the cam. On almost all cams, the word **TOP** is stamped on one side. Also, on some cams, the groove does not go all the way through. Therefore, it is very difficult to install the cam incorrectly, with the wrong side up.

Slide the cam down the crankshaft with the word **TOP** facing upward. Continue working the cam down the crankshaft until it is in place over the Woodruff key, or pin.

If the coil is **not** to be removed, proceed directly to Step 25. To remove the coil, perform the procedures in the following section.

Coil Removal from the Armature Plate

The armature plate must be removed as described earlier in this section, Step 12 and Step 13. Notice how the coil has a laminated core. The coil cannot be separated, that is, the laminations from the core.

18- Turn the armature plate over and notice how the high-tension leads are installed on the plate in a recess. The routing of the wires is misleading. The wire to the No. 1 spark plug is **NOT** connected to the No. 1 coil as might be expected.

19- Remove the three screws attaching the coils to the armature plate.

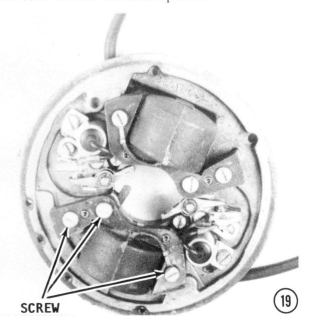

5-18 IGNITION

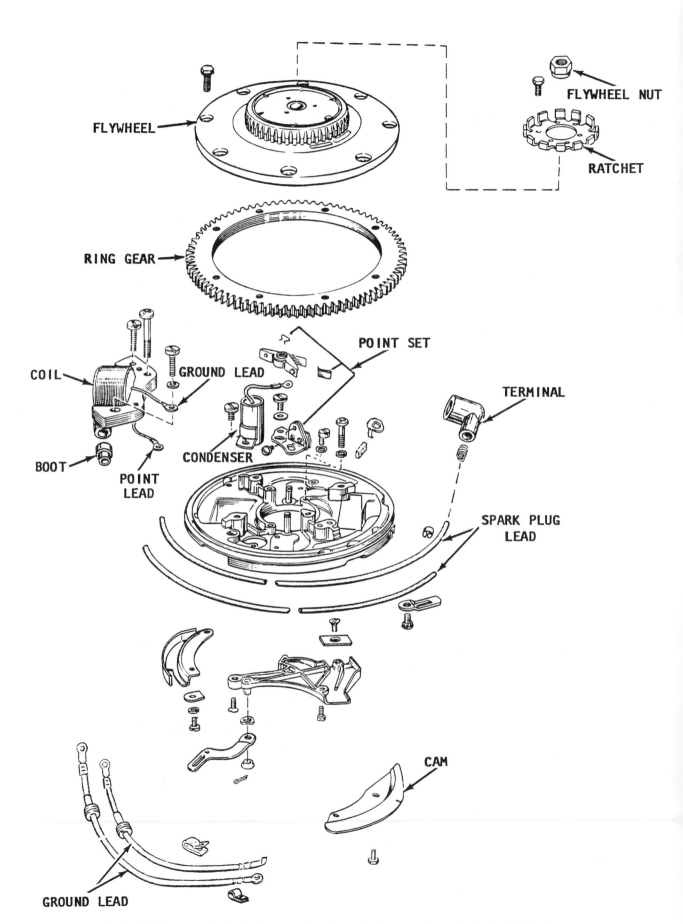

Exploded drawing of a typical magneto system. Only one coil and set of points is shown.

CLEANING AND INSPECTING 5-19

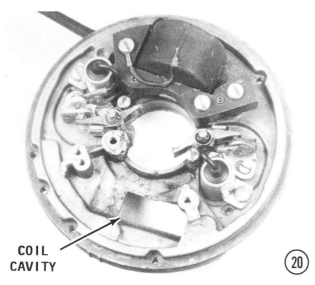

20- Hold the armature plate and separate the coils from the plate. As the coil is separated from the plate, observe the high-tension lead to the spark plug inside the coil. Work the small boot, if used, and the high-tension lead from the coil.

CLEANING AND INSPECTING

Inspect the flywheel for cracks or other damage, especially around the inside of the center hub. Check to be sure metal parts have not become attached to the magnets. Verify each magnet has good magnetism by using a screwdriver or other tool.

Throughly clean the inside taper of the flywheel and the taper on the crankshaft to prevent the flywheel from "walking" on the crankshaft while the engine is running.

Check the top seal around the crankshaft to be sure no oil has been leaking onto the armature plate. If there is **ANY** evidence the seal has been leaking, it **MUST** be replaced, as outlined earlier in this section.

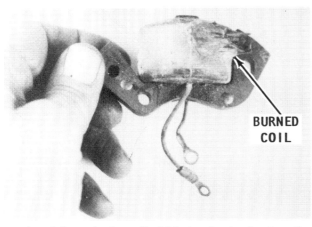

A coil burned where the high-tension lead enters the coil on the bottom side. Arcing caused the damage.

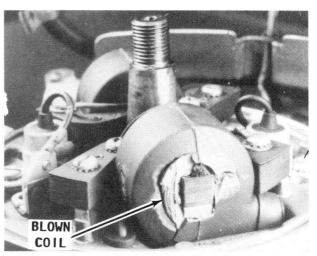

A coil destroyed when the side blew out. This damage was caused when 12-volts was connected to the magneto circuit at the key switch.

Test the armature plate to verify it is not loose. Attempt to lift each side of the plate. There should be little or no evidence of movement.

Clean the surface of the armature plate where the points and condenser attach. Install a new condenser into the recess and secure it with the hold-down screw.

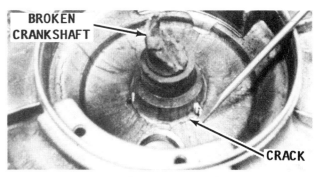

A broken crankshaft and cracked flywheel damaged when the engine was operated at a high rpm with a flush attachment and garden hose connected to the lower unit.

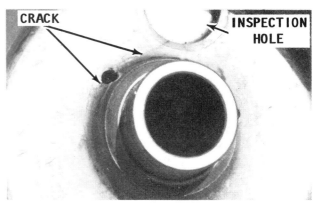

Cracks in the flywheel hub caused by metal fatigue due to flywheel construction and the inspection hole. This hole is no longer incorporated in late-model flywheels.

IGNITION

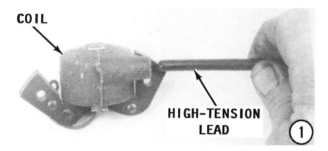

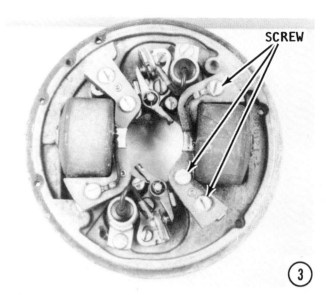

ASSEMBLING

Coil to Armature Plate

1- To install a new coil, first turn the armature plate over, and loosen the spark plug lead wires, and push them through the armature plate. Now, work the leads into the coil.

2- After the leads are into the coil, work the small boot up onto the coil. Apply a coating of rubber seal material underneath the boot, if a boot is used.

3- Start the three screws through the laminated core into the armature plate, but **DO NOT** tighten them. If the engine being serviced has a second coil, install the other coil in the same manner.

4- Check to be sure the spark plug leads are properly positioned in the coil and are securely attached to the bottom side of the armature plate.

5- To adjust the coils: A special ring tool is required that fits down over the armature plate. This tool will properly locate the coil in relation to the flywheel. Install this special tool over the armature plate. Push outward on the coil and secure the two outer screws.

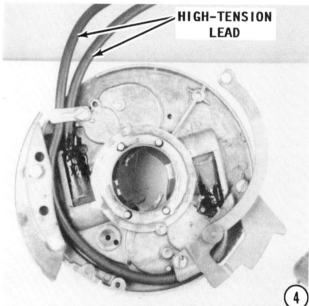

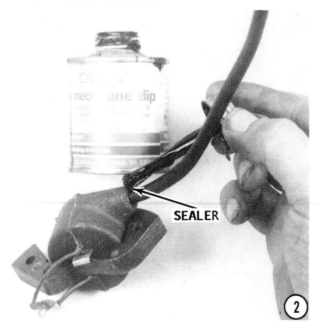

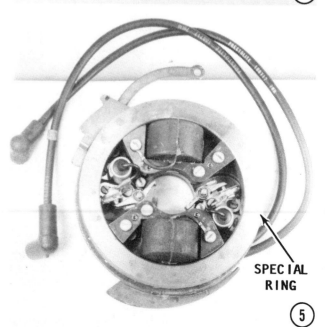

ASSEMBLING 5-21

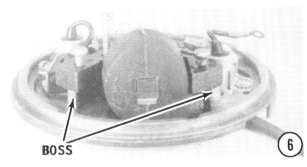

6- If a special ring tool is not available, and in a emergency, hold a straight edge against the boss on the armature plate and bring the heel of the laminated core out square against the edge of the boss on the armature plate. The ground wire for the coil should be attached under the head of the top screw passing through the laminated core.

Wick Replacement

7- The wick, mounted in a bracket under the coil, can be replaced without removing the armature plate. The wick **SHOULD** be replaced each and every time the breaker points are replaced. To replace the wick, simply loosen all three coil retaining screws and remove the one screw through the wick holder. Lift the coil slightly and remove the wick and wick holder. Slide the new wick into the holder; install the holder and wick under the coil; and secure it in place with the retaining screw. Adjust the coil as described earlier in this section, Steps 5 and 6, and tighten the three screws.

Armature Plate Installation

8- Slide the armature plate down over the crankshaft and onto the engine. Align the screw holes in the armature plate with the holes in the powerhead retainer. After the armature plate is in place, install and

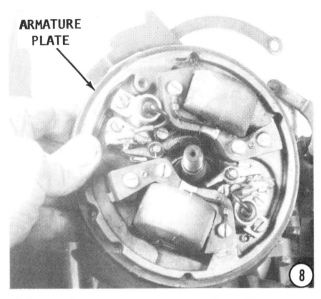

tighten the two screws securing the armature plate to the retainer. Now, take up on the two screws through the laminated core closest to the crankshaft, but **DO NOT** tighten them at this time. Attach the advance arm from the magneto to the tower shaft arm.

GOOD WORDS

All engines covered in this manual use the same set of points (Part No. 580148). The points **MUST** be assembled as they are installed. One side of each point set has has the base and is non-moveable. The other side of the set has a moveable arm. A small wire clip and a flat retainer are included in each point set package.

9- Hold the base side of the points and the flat retainer. Notice how the base has a bar at right angle to the points. Observe the hole in the bar. Observe the flat retainer. Notice that one side has a slight indentation. When the points are installed, this indentation will slip into the hole in the base bar.

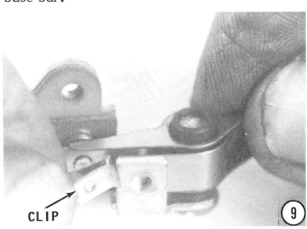

5-22 IGNITION

Point Set

10- Install the condensers and secure them in place with their hold-down screws.

11- Hold the base side of the points and slide it down over the anchor pin onto the armature plate. Install the wavy washer and hold-down screw to secure the point base to the armature plate. Tighten the hold-down screw securely.

12- Hold the moveable arm and slide the points down over post, and at the same time, hold back on the points and work the spring arm to the inside of the post of the base points. Continue to work the points on down into the base.

13- Observe the points. The points should be together and the spring part of the moveable arm on the inside of the flat post.

14- Install the flat retainer onto the flat bar of the base points. Check to be sure the flat spring from the other side of the points is on the inside of the retainer. Push the retainer inward until the indentation slips into the hole in the base. The retainer **MUST** be horizontal with the armature plate.

15- Install the wire clip into the groove of the post.

Repeat Steps No. 10 thru 15 for the second set of points.

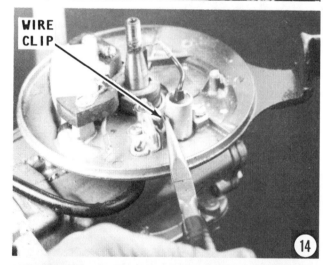

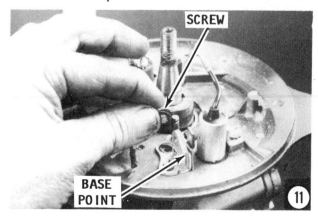

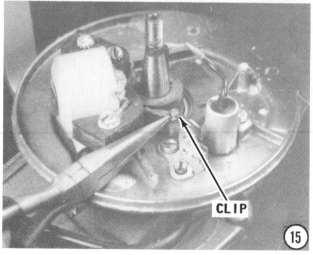

ASSEMBLING 5-23

CRITICAL WORDS

As the coil, condenser, and "kill" switch wire are being attached to the point set, take the following precautions and adjustments:

a- The wire between the coil and the points should be tucked back under the coil and as far away from the crankshaft as possible.

b- The condenser wire leaving the top of the condenser and connected to the point set, should be bent downward to prevent the flywheel from making contact with the wire. A countless number of installations have been made only to have the flywheel rub against the condenser wire and cause failure of the ignition system.

c- Check to be sure all wires connected to the point set are bent downward toward the armature plate. The wires **MUST NOT** touch the plate. If any of the wires make contact with the armature plate, the ignition system will be grounded and the engine will fail to start.

16- Connect the wire leads to the set of points.

Repeat all of these **Critcal Words** for the second set of points.

GOOD WORDS

The point spring tension is predetermined at the factory and does not require adjustment. Once the point set is properly installed, all should be well. In most cases, breaker contact and alignment will not be necessary. If a slight alignment adjustment should be required, **CAREFULLY** bend the insulated part of the point set.

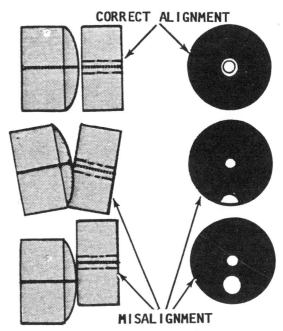

*Before setting the breaker point gap, the points must be properly aligned (top). **ALWAYS** bend the stationary point, **NEVER** the breaker lever. Attempting to adjust an old worn set of points is not practical because oxidation and pitting of the points will always give a false reading.*

Point Adjustment

17- Install the flywheel nut onto the end of the crankshaft. Now, turn the crankshaft clockwise and at the same time observe the cam on the crankshaft. Continue turning the crankshaft until the rubbing block of the point set is at the high point of the cam. At this position, use a wire gauge or feeler gauge and set the points at 0.020" for all models covered in this manual. A wire gauge will always give a more accurate adjustment than a feeler gauge. Work the gauge between the points and, at the same time, turn the eccentric on the armature plate until the proper adjustment (0.020") is obtained. Rotate the crankshaft a complete

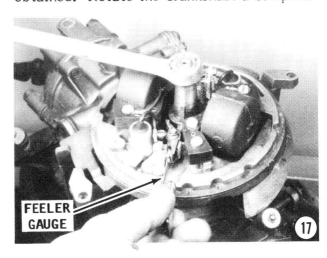

5-24 IGNITION

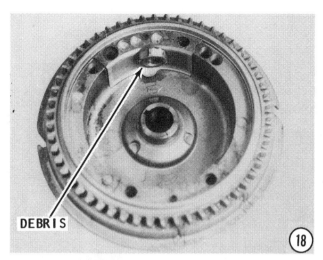

revolution and again check the gap adjustment. After the crankshaft has been turned and the points are on the high point of the cam, check to be sure the hold-down screw is tight against the base. There is enough clearance to allow the eccentric on the base points to turn. If the hold-down screw is tightened **AFTER** the point adjustment has been made, it is very likely the adjustment will be changed. Follow the same procedure and adjust the other set of points. Remove the nut from the crankshaft.

Flywheel Installation

18- Check to be sure the flywheel magnets are free of any metal parts.

19- Place the key in the crankshaft keyway. Check to be sure the inside taper of the flywheel and the taper on the crankshaft are clean of dirt or oil, to prevent the flywheel from "walking" on the crankshaft while the engine is operating. Slide the flywheel down over the crankshaft with the keyway in the flywheel aligned with the key on the crankshaft.

20- Rotate the flywheel clockwise and check to be sure the flywheel does not contact any part of the magneto or the wiring.

21- Thread the flywheel nut onto the crankshaft and tighten it to the torque value given in the Appendix.

22- Place the ratchet for the starter on top of the flywheel and install the three 7/16" screws On some model engines, a plate retainer covers these screws. On other model engines a very wide plate is used with a definite **UP** side and a **DOWN**

ASSEMBLING 5-25

Plate installed on some engines covering the inspection hole in the flywheel. The plate is clearly marked to indicate which side is to face upward during installation.

side. When installing this plate, check to be sure the word **UP** is facing upward.

23- After the ratchet plate has been installed, install the hand starter over the flywheel. Check to be sure the ratchet engages the flywheel properly.

24- Set the gap on each spark plug at 0.030".

25- Install the spark plugs and tighten them to the torque value given in the following table.

Connect the battery leads to the battery terminals, if a battery is used with a starter motor to crank the engine.

This particular engine differs from the text procedures because a washer is installed under the flywheel nut.

HP	YEAR	SPARK PLUG TORQUE (INCH-LBS)
1.5	1968-70	210-246
3.0	1956-68	240-246
4.0	1969-70	210-246
5.0	1965-68	210-246
5.5	1956-64	240-246
6.0	1965-70	210-246
7.5	1956-58	240-246
9.5	1965-70	210-246
10	1956-63	240-246
15	1956	240-246
18	1957-70	210-246
20	1966-70	210-246
25	1969-70	210-246
28	1962-64	240-246
30	1956	240-246
33	1965-70	240-246
35	1957-59	240-246
40	1960-70	240-246

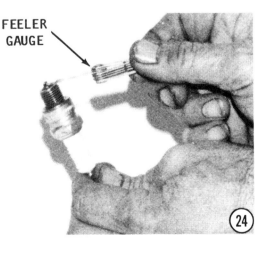

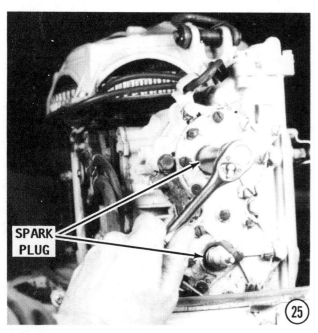

5-8 SYNCHRONIZATION
FUEL AND IGNITION SYSTEMS

Timing is **NOT** adjustable with a flywheel magneto system. The timing is controlled by the point setting. The correct point setting for **ALL** engines covered in this manual is 0.020". The fuel and ignition systems **MUST** be carefully synchronized to achieve maximum performance from the engine. In simple terms, synchronization is timing the carburetion to the ignition. This means, as the throttle is advanced to increase engine rpm, the carburetor and ignition systems are both advanced equally and at the same rate.

Therefore, any time the fuel system or the ignition system is serviced to replace a faulty part, or any adjustments are made for any reason, the engine synchronization must be carefully checked and verified.

Before making any adjustments with the synchronization, the ignition system should be thoroughly checked according to the procedures outlined in this chapter and the fuel checked according to the procedures outlined in Chapter 4.

PRIMARY PICKUP ADJUSTMENTS AND LOCATIONS

To properly adjust the synchronization, numbered primary adjustment and locations are used. These numbers are referenced in the Appendix under Tune-up Specifications. The adjustments are numberd from 1 thru 4 and the locations are numbered 5 thru 9. The number to be used is taken from the table by following across from the first column for the engine being serviced to the column titled Primary Pickup Location and the column titles Primary P/U Adjustment Note. Therefore, from the Appendix two numbers will be obtained: One for the location of the primary adjustment and the other indicating the method of making the adjustment.

GOOD WORDS

When making the synchronization adjustment, it is well to know and understand exactly what to look for and why. The critical time when the throttle shaft in the carburetor begins to move is of the utmost importance. First, realize that the time the cam follower makes contact with the cam is not the time the throttle shaft starts to move. Instead, the critical time is when the follower hits the designated position (as described in the next five paragraphs) and the throttle shaft **AT THE CARBURETOR** begins to move.

A considerable amount of play exists between the follower at the top of the carburetor through the linkage to the actual throttle shaft. Therefore, the most important consideration is to watch for movement of the **THROTTLE SHAFT**, and not the follower. Movement of the shaft can be exaggerated by attaching a short piece of stiff wire to an alligator clip; grinding down the teeth on one side of the clip; and then attaching the clip to the throttle shaft, as shown. The wire jiggling will instantly indicate movement of the shaft.

Almost all of the photographs were taken with the flywheel removed for clarity. Normally the synchronization is set with the flywheel installed.

EXAMPLE

If the engine being serviced is a 25 hp, 1970, then the primary pickup location number obtained from the Appendix is **8** and the primary pickup adjustment note is **4**. In this case:

Perform No. 8 Location procedures.
Method of adjustment is No. 4.

The following paragraphs describe the location in detail and given specific instructions as to exactly how the synchronization is to be made.

SYNCHRONIZATION 5-27

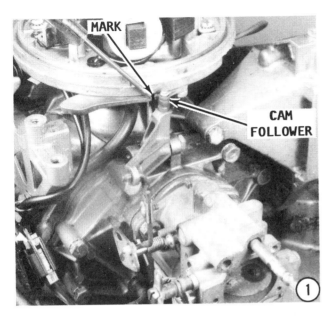

The No. 5 Location

1- The pickup location for the No. 5 location is **PORT** side of the mark. This means that the pickup on the carburetor arm should be just to the **PORT** side of the mark on the cam. To obtain this position, the cam, or the cam follower, is to be adjusted until pickup is at the proper location and the throttle shaft just begins to move.

2- Movement of the shaft can be exaggerated by attaching a short piece of stiff wire to an alligator clip; grinding down the teeth on one side of the clip; and then attaching the clip to the throttle shaft, as shown. The wire jiggling will instantly indicate movement of the shaft. The actual adjustment is accomplished by **ONLY ONE** method, depending on the engine being serviced. Check the Appendix for the adjustment to be performed. The reference numbers are listed in the Appendix under Primary Pickup Adjustment Note.

3- Per Note 1 in the Appendix, loosen the two screws under the armature plate and move the primary pickup inward or outward to meet the follower.

The No. 6 Location

4- Engines referenced in the Appendix to the No. 6 Primary Pickup Location are to be adusted to the **STARBOARD** side of the mark. To obtain this position, the cam, or the cam follower, (depending the type of engine being serviced), is to be adjusted until pickup is at the proper location and the throttle shaft just begins to move. Movement of the shaft can be exaggerated by attaching a short piece of stiff wire to an alligator clip; grinding down the teeth on

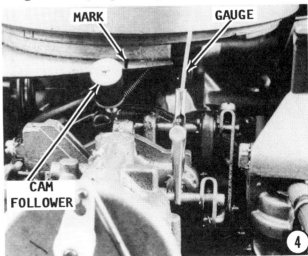

5-28 IGNITION

one side of the clip; and then attaching the clip to the throttle shaft, as shown. The wire jiggling will instantly indicate movement of the shaft. The actual adjustment is accomplished by **ONLY ONE** method, regardless of the engine being serviced for the No. 6 location. Check the Appendix for the adjustment to be performed. The reference numbers are listed in the Appendix under Primary Pickup Adjustment Note.

5- Per Note 1 in the Appendix, loosen the two screws under the armature plate and move the primary pickup inward or outward to meet the follower.

The No. 7 Location

6- Engines referenced to No. 7 are to be adjusted to the **CENTER** of the mark. To obtain this position, the cam, or the cam follower, (depending the type of engine being serviced), is to be adjusted until pickup is at the proper location and the throttle shaft just begins to move.

7- Movement of the shaft can be exaggerated by attaching a short piece of stiff wire to an alligator clip; grinding down the

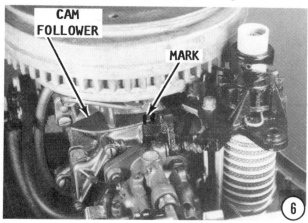

teeth on one side of the clip; and then attaching the clip to the throttle shaft, as shown. The wire jiggling will instantly indicate movement of the shaft. The actual adjustment is accomplished by **ONLY ONE** of two means, depending on the engine being serviced. Check the Appendix for the adjustment to be performed. The reference numbers are listed in the Appendix under Primary Pickup Adjustment Note.

8- Per Note 1 in the Appendix, loosen the two screws under the armature plate and move the primary pickup inward or outward to meet the follower.

9- Per Note 2 in the Appendix, loosen the center screw on the throttle lever and move the lever inward or outward to match the line on the cam. This applies only to the 9.5 hp models.

SYNCHRONIZATION 5-29

The No. 8 Location

10- Engines referenced to the No. 8 location are to be adjusted with the cam follower midway between the two marks on the cam, at the moment the throttle shaft at the carburetor begins to move. The marks on the cam are about 1/4" apart.

11- Movement of the shaft can be exaggerated by attaching a short piece of stiff wire to an alligator clip; grinding down the teeth on one side of the clip; and then attaching the clip to the throttle shaft, as shown. The wire jiggling will instantly indicate movement of the shaft. The actual adjustment is accomplished by **ONLY ONE** of two means, depending on the engine being serviced. Check the Appendix for the adjustment to be performed. The reference numbers are listed in the Appendix under Primary Pickup Adjustment Note.

12- Per Note 1 in the Appendix, loosen the two screws under the armature plate and move the primary pickup inward or outward to meet the follower.

13- Per Note 4 in the Appendix, loosen the clamp on the throttle shaft and move the roller to meet the armature cam.

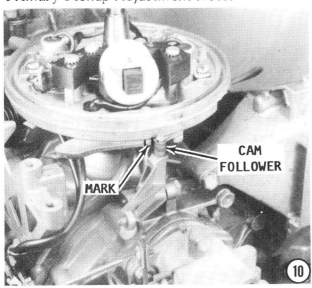

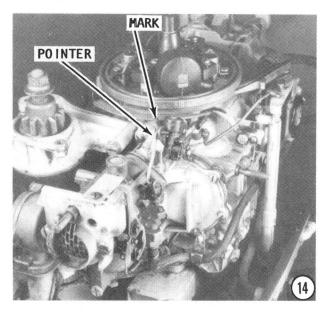

The No. 9 Location

14- Engines referenced to the No. 9 location have a pointer attached to the intake manifold. Synchronization is made by advancing the magneto until the mark on the cam is aligned with the pointer on the intake manifold. At this point the throttle shaft should just begin to move.

15- Movement of the shaft can be exaggerated by attaching a short piece of stiff wire to an alligator clip; grinding down the teeth on one side of the clip; and then attaching the clip to the throttle shaft, as shown. The wire jiggling will instantly indicate movement of the shaft. The actual adjustment is accomplished by **ONLY ONE** of three means, depending on the engine being serviced. Check the Appendix for the adjustment to be performed. The reference

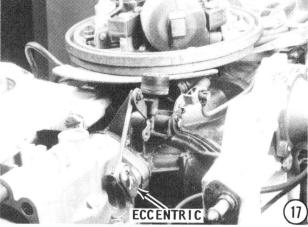

numbers are listed in the Appendix under Primary Pickup Adjustment Note.

16- Per Note 1 in the Appendix, loosen the two screws under the armature plate and move the primary pickup inward or outward to meet the follower.

17- Per Note 3 in the Appendix, loosen the eccentric lock screw on the throttle shaft and turn the eccentric to move the roller to meet the armature cam.

18- Per Note 4 in the Appendix, loosen the clamp on the throttle shaft and move the roller to meet the armature cam.

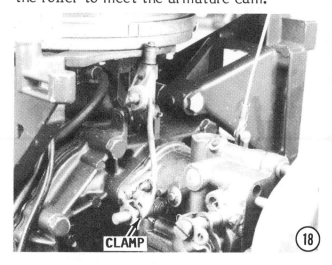

6
ELECTRICAL

6-1 INTRODUCTION

The battery, gauges and horns, charging system, and the cranking system are all considered subsystems of the electrical system. Each of these units or subsystems will be covered in detail in this chapter beginning with the battery.

All engines covered in this manual use the magneto ignition system and a battery is not required to operate the engine. Most of the larger horsepower units use a cranking motor for starting and the battery is only used to supply power for this motor.

The starting circuit consists of a cranking motor and a starter-engaging mechanism. A solenoid is used as a heavy-duty switch to carry the heavy current from the battery to the starter motor. The solenoid is actuated by turning the ignition key to the **START** position. On some models, a pushbutton is used to actuate the solenoid.

These engines are also equipped with a hand starter for use when the electric starter motor system is inoperative.

6-2 BATTERIES

The battery is one of the most important parts of the electrical system. In addition to providing electrical power to start the engine, it also provides power for operation of the the running lights, radio, electrical accessories, and possibly the pump for a bait tank.

Because of its job and the consequences, (failure to perform in an emergency) the best advice is to purchase a well-known brand, with an extended warranty period, from a reputable dealer.

The usual warranty covers a prorated replacement policy, which means you would be entitled to a consideration for the time left on the warranty period if the battery should prove defective before its time.

Do not consider a battery of less than 70-ampere hour capacity. If in doubt as to how large your boat requires, make a liberal estimate and then purchase the one with the next higher ampere rating.

MARINE BATTERIES

Because marine batteries are required to perform under much more rigorous conditions than automotive batteries, they are constructed much differently than those used in automobiles or trucks. Therefore, a marine battery should always be the No. 1 unit for the boat and other types of batteries used only in an emergency.

Marine batteries have a much heavier exterior case to withstand the violent pounding and shocks imposed on it as the boat moves through rough water and in extremely tight turns.

The plates in marine batteries are thicker than in automotive batteries and each

A fully charged battery, filled to the proper level with electrolyte, is the heart of the ignition system. Engine starting and efficient performance can never be obtained if the battery is below a fully charged rating.

6-2 ELECTRICAL

plate is securely anchored within the battery case to ensure extended life.

The caps of marine batteries are "spill proof" to prevent acid from spilling into the bilges when the boat heels to one side in a tight turn, or is moving through rough water.

Because of these features, the marine battery will recover from a low charge condition and give satisfactory service over a much longer period of time than any type of automotive-type unit.

BATTERY CONSTRUCTION

A battery consists of a number of positive and negative plates immersed in a solution of diluted sulfuric acid. The plates contain dissimilar active materials and are kept apart by separators. The plates are grouped into what are termed elements. Plate straps on top of each element connect all of the positive plates and all of the negative plates into groups. The battery is divided into cells which hold a number of the elements apart from the others. The entire arrangement is contained within a hard-rubber case. The top is a one-piece cover and contains the filler caps for each cell. The terminal posts protrude through the top where the battery connections for the boat are made. Each of the cells is connected to its neighbor in a positive-to-negative manner with a heavy strap called the cell connector.

The battery MUST be located near the engine in a well-ventilated area. It must be secured in such a manner that absolutely no movement is possible in any direction under the most violent actions of the boat.

BATTERY RATINGS

Two ratings are used to classify batteries: one is a 20-hour rating at $80°F$ and the other is a cold rating at $0°F$. This second figure indicates the cranking load capacity and is referred to as the Peak Watt Rating of a battery. This Peak Watt Rating (PWR) has been developed to measure the cold-cranking ability of the battery. The numerical rating is embossed on each battery case at the base and is determined by multiplying the maximum current by the maximum voltage.

The ampere-hour rating of a battery is its capacity to furnish a given amount of amperes over a period of time at a cell voltage of 1.5. Therefore, a battery with a capacity of maintaining 3 amperes for 20 hours at 1.5 volts would be classified as a 60-ampere hour battery.

Do not confuse the ampere-hour rating with the PWR, because they are two unrelated figures used for different purposes.

A replacement battery should have a power rating equal or as close to the old unit as possible.

BATTERY LOCATION

Every battery installed in a boat must be secured in a well-protected ventilated area. If the battery area is not well ventilated, hydrogen gas which is given off during charging could become very explosive if the gas is concentrated and confined. Because of its size, weight, and acid content, the battery must be well-secured. If the battery should break loose during rough boat maneuvers, considerable damage could be done, including damage to the hull.

BATTERY SERVICE

The battery requires periodic servicing and a definite maintenance program to ensure extended life. If the battery should test satisfactorily, but still fails to perform properly, one of four problems could be the cause.

1- An accessory might have accidently been left on overnight or for a long period during the day. Such an oversight would result in a discharged battery.

2- Slow speed engine operation for long periods of time resulting in an undercharged condition.

3- Using more electrical power than the generator or alternator can replace resulting in an undercharged condition.

4- A defect in the charging system. A faulty generator or alternator system, a defective regulator, or high resistance somewhere in the system, could cause the battery to become undercharged.

5- Failure to maintain the battery in good order. This might include a low level of electrolyte in the cells; loose or dirty cable connections at the battery terminals; or possibly an excessively dirty battery top.

Electrolyte Level

The most common practice of checking the electrolyte level in a battery is to remove the cell cap and visually observe the level in the vent well. The bottom of each vent well has a split vent which will cause the surface of the electrolyte to appear distorted when it makes contact. When the distortion first appears at the bottom of the split vent, the electrolyte level is correct.

Some late-model batteries have an electrolyte-level indicator installed which operates in the following manner:

A transparent rod extends through the center of one of the cell caps. The lower tip of the rod is immersed in the electrolyte when the level is correct. If the level should drop below normal, the lower tip of the rod is exposed and the upper end glows as a warning to add water. Such a device is only necessary on one cell cap because if the electrolyte is low in one cell it is also low in the other cells. **BE SURE** to replace the cap with the indicator onto the second cell from the positive terminal.

During hot weather and periods of heavy use, the electrolyte level should be checked more often than during normal operation. Add colorless, odorless, drinking water to bring the level of electrolyte in each cell to the proper level. **TAKE CARE** not to overfill, because adding an excessive amount of water will cause loss of electrolyte and any loss will result in poor performance, short battery life, and will contribute quickly to corrosion. **NEVER** add electrolyte from another battery. Use only clean pure water.

Cleaning

Dirt and corrosion should be cleaned from the battery just as soon as it is discovered. Any accumulation of acid film or dirt will permit current to flow between the terminals. Such a current flow will drain the battery over a period of time.

Clean the exterior of the battery with a solution of diluted ammonia or a soda solution to neutralize any acid which may be present. Flush the cleaning solution off with clean water. **TAKE CARE** to prevent any of the neutralizing solution from entering the cells, by keeping the caps tight.

A poor contact at the terminals will add

One of the most effective means of cleaning the battery terminals is to use a wire brush designed for this specific purpose.

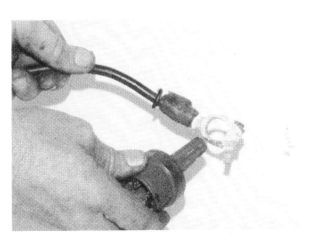

An inexpensive brush can be purchased and used to clean battery lead connectors to ensure a proper connection.

resistance to the charging circuit. This resistance will cause the voltage regulator to register a fully charged battery, and thus cut down on the alternator output adding to the low battery charge problem.

Scrape the battery posts clean with a suitable tool or with a stiff wire brush. Clean the inside of the cable clamps to be sure they do not cause any resistance in the circuit.

Battery Testing

A hydrometer is a device to measure the percentage of sulfuric acid in the battery electrolyte in terms of specific gravity. When the condition of the battery drops from fully charged to discharged, the acid leaves the solution and enters the plates, causing the specific gravity of the electrolyte to drop.

The following six points should be observed when using a hydrometer.

1- NEVER attempt to take a reading immediately after adding water to the battery. Allow at least 1/4 hour of charging at a high rate to thoroughly mix the electrolyte with the new water and to cause vigorous gassing.

2- ALWAYS be sure the hydrometer is clean inside and out as a precaution against contaminating the electrolyte.

3- If a thermometer is an integral part of the hydrometer, draw liquid into it several times to ensure the correct temperature before taking a reading.

4- BE SURE to hold the hydrometer vertically and suck up liquid only until the float is free and floating.

5- ALWAYS hold the hydrometer at eye level and take the reading at the surface of the liquid with the float free and floating.

Disregard the light curvature appearing where the liquid rises against the float stem. This phenomenon is due to surface tension.

6- DO NOT drop any of the battery fluid on the boat or on your clothing, it is extremely caustic. Use water and baking soda to neutralize any battery liquid that does accidently drop.

After withdrawing electrolyte from the battery cell until the float is barely free, note the level of the liquid inside the hydrometer. If the level is within the green band range, the condition of the battery is satisfactory. If the level is within the white band, the battery is in fair condition, and if the level is in the red band, it needs charging badly or is dead and should be replaced. If the level fails to rise above the red band after charging, the only answer is to replace the battery.

A check of the electrolyte in the battery should be on the maintenance schedule for any boat. A hydrometer reading of 1.300 or in the green band, indicates the battery is in satisfactory condition. If the reading is 1.150 or in the red band, the battery needs to be charged. Observe the six safety points given in the text when using a hydrometer.

A pair of pliers should be used to tighten the wingnuts, when they are used. Securing the wingnuts by hand is not adequate, the connections will vibrate loose.

JUMPER CABLES

If booster batteries are used for starting an engine the jumper cables must be connected correctly and in the proper sequence to prevent damage to either battery, or to the alternator diodes.

ALWAYS connect a cable from the positive terminal of the dead battery to the positive terminal of the good battery **FIRST**. **NEXT**, connect one end of the other cable to the negative terminal of the good battery and the other end to the **ENGINE** for a good ground. By making the ground connection on the engine, if there is an arc when you make the connection it will not be near the battery. An arc near the battery could cause an explosion, destroying the battery and causing serious personal **INJURY**.

DISCONNECT the battery ground cable before replacing an alternator or before connecting any type of meter to the alternator.

If it is necessary to use a fast-charger on a dead battery, **ALWAYS** disconnect one of the boat cables from the battery **FIRST**, to prevent burning out the diodes in the rectifier.

NEVER use a fast-charger as a booster to start the engine because the voltage regulator may be **DAMAGED**.

STORAGE

If the boat is to be laid up for the winter or for more than a few weeks, special attention must be given to the battery to prevent complete discharge or possible damage to the terminals and wiring. Before putting the boat in storage, disconnect and remove the batteries. Clean them thoroughly of any dirt or corrosion, and then charge them to full specific gravity reading. After they are fully charged, store them in a clean cool dry place where they will not be damaged or knocked over.

NEVER store the battery with anything on top of it or cover the battery in such a manner as to prevent air from circulating around the fillercaps. All batteries, both new and old, will discharge during periods of storage, more so if they are hot than if they remain cool. Therefore, the electrolyte level and the specific gravity should be checked at regular intervals. A drop in the specific gravity reading is cause to charge them back to a full reading.

In cold climates, care should be exercised in selecting the battery storage area. A fully-charged battery will freeze at about 60 degrees below zero. A discharged battery, almost dead, will have ice forming at about 19 degrees above zero.

DUAL BATTERY INSTALLATION

Three methods are available for utilizing a dual-battery hook-up.

Corroded battery terminals such as these result in high resistance at the connections. Such corrosion places a strain on all electrically operated devices on the boat and causes hard engine starting.

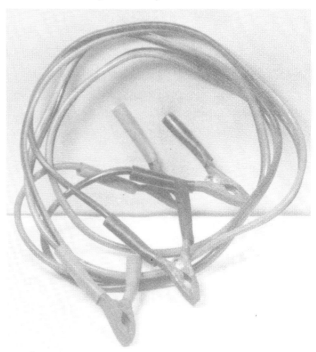

A common set of heavy-duty jumper cables. Observe the safety precautions given in the text when using jumper cables.

6-6 ELECTRICAL

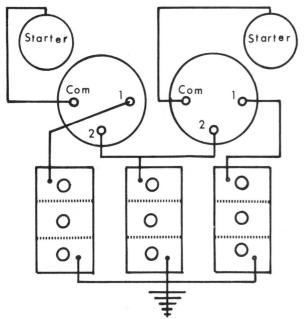

Schematic drawing of a three battery, two engine hookup.

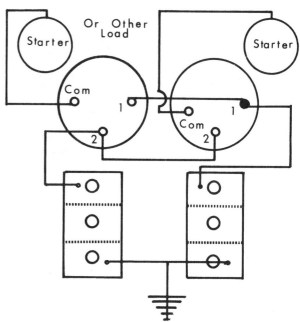

Schematic drawing for a two battery, two engine hookup.

1- A high-capacity switch can be used to connect the two batteries. The accompanying illustration details the connections for installation of such a switch. This type of switch installation has the advantage of being simple, inexpensive, and easy to mount and hookup. However, if the switch is accidently left in the closed position, it will cause the convenience loads to run down both batteries and the advantage of the dual installation is lost. The switch may be closed intentionally to take advantage of the extra capacity of the two batteries, or it may be temporarily closed to help start the engine under adverse conditions.

2- A relay, can be connected into the ignition circuit to enable both batteries to be automatically put in parallel for charging or to isolate them for ignition use during engine cranking and start. By connecting the relay coil to the ignition terminal of the ignition-starting switch, the relay will close during the start to aid the starting battery. If the second battery is allowed to run down,

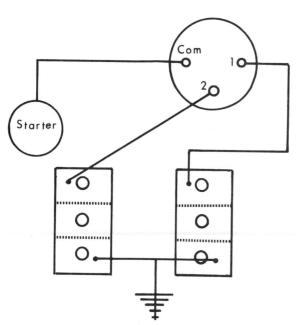

Schematic drawing for a two battery, one engine hookup.

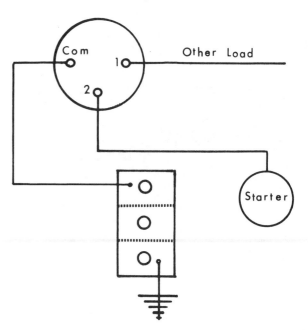

Schematic drawing for a single battery, one engine hookup.

this arrangement can be a disadvantage since it will draw a load from the starting battery while cranking the engine. One way to avoid such a condition is to connect the relay coil to the ignition switch accessory terminal. When connected in this manner, while the engine is being cranked, the relay is open. But when the engine is running with the ignition switch in the normal position, the relay is closed, and the second battery is being charged at the same time as the starting battery.

3- A heavy duty switch installed as close to the batteries as possible can be connected between them. If such an arrangement is used, it must meet the standards of the American Boat and Yacht Council, Inc. or the Fire Protection Standard for Motor Craft, N.F.P.A. No. 302.

6-3 GAUGES AND HORNS

Gauges or lights are installed to warn the operator of a condition in the cooling and lubrication systems that may need attention. The fuel gauge gives an indication of the amount of fuel in the tank. If the engine overheats, a warning light will come on or a horn sound advising the operator to shut down the engine and check the cause of the warning before serious damage is done.

CONSTANT-VOLTAGE SYSTEM

In order for gauges to register properly, they must be supplied with a steady voltage. The voltage variations produced by the engine charging system would cause erratic gauge operation, too high when the generator or alternator voltage is high, and too low when the generator or alternator is not charging. To remedy this problem, a constant-voltage system is used to reduce the 12-14 volts of the electrical system to an average of 5 volts. This steady 5 volts ensures the gauges will read accurately under varying conditions from the electrical system.

6-4 SERVICE PROCEDURES

Systems utilizing warning lights do not require a constant-voltage system, therefore, this service is not needed.
Service procedures for checking the gauges and their sending units is detailed in the following sections.

TEMPERATURE GAUGES

The body of temperature gauges must be grounded and they must be supplied with 12 volts. Many gauges have a terminal on the mounting bracket for attaching a ground wire. A tang from the mounting bracket makes contact with the gauge. **CHECK** to be sure the tang does make good contact with the gauge.

Ground the wire to the sending unit and the needle of the gauge should move to the full right position indicating the gauge is in serviceable condition.

See Chapter 3, to test the sender unit.

WARNING LIGHTS

If a problem arises on a boat equipped with water and temperature lights, the first area to check is the light assembly for loose wires or burned-out bulbs.

When the ignition key is turned on, the light assembly is supplied with 12 volts and grounded through the sending unit mounted on the engine. When the sending unit makes contact because the water temperature is too hot, the circuit to ground is completed and the lamp should light.

Check The Bulb: Turn the ignition switch on. Disconnect the wire at the engine sending unit, and then ground the wire. The

The gauges and controls on the dashboard should be kept clean and protected from water spray, especially when operating in a salt water atmosphere.

6-8 ELECTRICAL

lamp on the dash should light. If it does not light, check for a burned-out bulb or a break in the wiring to the light.

THERMOMELT STICKS

Thermomelt sticks are an easy method of determining if the engine is running at the proper temperature. Thermomelt sticks are not expensive and are available at your local marine dealer.

Start the engine with the propeller in the water and run it for about 5 minutes at roughly 3000 rpm.

CAUTION: Water must circulate through the lower unit to the engine any time the engine is run to prevent damage to the water pump in the lower unit. Just five seconds without water will damage the water pump.

The 140 degree stick should melt when you touch it to the lower thermostat housing or on the top cylinder. If it does not melt, the thermostat is stuck in the open position and the engine temperature is too low.

Touch the 170 degree stick to the same spot on the lower thermostat housing or on the top cylinder. The stick should not melt. If it does, the thermostat is stuck in the closed position or the water pump is not operating properly because the engine is running too hot. For service procedures on the cooling system, see Chapter 8.

A thermomelt stick is a quick, simple, inexpensive, and fairly accurate method to determine engine running temperature.

6-5 FUEL SYSTEM

FUEL GAUGE

The fuel gauge is intended to indicate the quantity of fuel in the tank. As the experienced boatman has learned, the gauge reading is seldom an accurate report of the fuel available in the tank. The main reason for this false reading is because the boat is rarely on an even keel. A considerable difference in fuel quantity will be indicated by the gauge if the bow or stern is heavy, or if the boat has a list to port or starbaord.

Therefore, the reading is usually low. The amount of fuel drawn from the tank is dependent on the location of the fuel pickup tube in the tank. The engine may cutout while cruising because the pickup tube is above of the fuel level. Instead of assuming the tank is empty, shift weight in the boat to change the trim and the problem may be solved until you are able to take on more fuel.

FUEL GAUGE HOOKUP

The Boating Industry Association recommends the following color coding be used on all fuel gauge installations:

Black -- for all grounded current-carrying conductors.

Pink -- insulated wire for the fuel gauge sending unit to the gauge.

Red -- insulated wire for a connection from the positive side of the battery to any electrical equipment.

Connect one end of a pink insulated wire to the terminal on the gauge marked **TANK** and the other end to the terminal on top of the tank unit.

Connect one end of a black wire to the terminal on the fuel gauge marked **IGN** and the other end to the ignition switch.

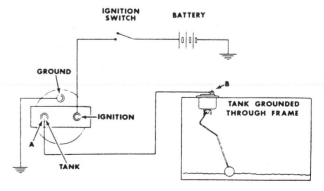

Schematic for a safe fuel tank gauge hookup.

FUEL SYSTEM 6-9

Connect one end of a second black wire to the fuel gauge terminal marked **GRD** and the other end to a good ground. It is important for the fuel gauge case to have a good common ground with the tank unit. Aboard an all-metal boat, this ground wire is not necessary. However, if the dashboard is insulated, or made of wood or plastic, a wire **MUST** be run from the gauge ground terminal to one of the bolts securing the sending unit in the fuel tank, and then from there to the **NEGATIVE** side of the battery.

FUEL GAUGE TROUBLESHOOTING

In order for the fuel gauge to operate properly the sending unit and the receiving unit must be of the same type and preferably of the same make.

The following symptoms and possible corrective actions will be helpful in restoring a faulty fuel gauge circuit to proper operation.

If you suspect the gauge is not operating properly, the first area to check is all electrical connections from one end to the other. Be sure they are clean and tight.

Next, check the common ground wire between the negative side of the battery, the fuel tank, and the gauge on the dash.

If all wires and connections in the circuit are in good condition, remove the sending unit from the tank. Run a wire from the gauge mounting flange on the tank to the flange of the sending unit. Now, move the float up-and-down to determine if the receiving unit operates. If the sending unit does not appear to operate, move the float to the midway point of its travel and see if the receiving unit indicates half full.

If the pointer does not move from the **EMPTY** position one of four faults could be to blame:

1- The dash receiving unit is not properly grounded.

2- No voltage at the dash receiving unit.

3- Negative meter connections are on a positive grounded system.

4- Positive meter connections are on a negative grounded system.

If the pointer fails to move from the **FULL** position, the problem could be one of three faults.

1- The tank sending unit is not properly grounded.

2- Improper connection between the tank sending unit and the receiving unit on the dash.

3- The wire from the gauge to the ignition switch is connected at the wrong terminal.

If the pointer remains at the 3/4 full mark, it indicates a six-volt gauge is installed in a 12-volt system.

If the pointer remains at about 3/8 full, it indicates a 12-volt gauge is installed in a six-volt system.

Preliminary Inspection

Inspect all of the wiring in the circuit for possible damage to the insulation or conductor. Carefully check:

1- Ground connections at the receiving unit on the dash.

2- Harness connector to the dash unit.

3- Body harness connector to the chassis harness.

4- Ground connection from the fuel tank to the tank floor pan.

5- Feed wire connection at the tank sending unit.

GAUGE ALWAYS READS FULL when the ignition switch is **ON**:

1- Check the electrical connections at the receiving unit on the dash; the body harness connector to chassis harness connector; and the tank unit connector in the tank.

2- Make a continuity check of the ground wire from the tank to the tank floor pan.

3- Connect a known good tank unit to the tank feed wire and the ground lead. Raise and lower the float and observe the receiving unit on the dash. If the dash unit follows the arm movement, replace the tank sending unit.

GAUGE ALWAYS READS EMPTY when the ignition switch is **ON**:

Disconnect the tank unit feed wire and do not allow the wire terminal to ground. The gauge on the dash should read **FULL**.

If Gauge Reads Empty:

1- Connect a spare dash unit into the dash unit harness connector and ground the unit. If the spare unit reads **FULL**, the original unit is shorted and must be replaced.

2- A reading of **EMPTY** indicate a short in the harness between the tank sending unit and the gauge on the dash.

6-10 ELECTRICAL

If Gauge Reads Full:

1- Connect a known good tank sending unit to the tank feed wire and the ground lead.

2- Raise and lower the float while observing the dash gauge. If dash gauge follows movement of the float, replace the tank sending unit.

GAUGE NEVER INDICATES FULL

This test requires shop test equipment.

1- Disconnect the feed wire to the tank unit and connect the wire to a good ground through a variable resistor or through a spare tank unit.

2- Observe the dash gauge reading. The reading should be **FULL** when resistance is increased to about 90 ohms. This resistance would simulate a full tank.

3- If the check indicates the dash gauge is operating properly, the trouble is either in the tank sending unit rheostat being shorted, or the float is binding. The arm could be bent, or the tank may be deformed. Inspect and correct the problem.

6-6 TACHOMETER

An accurate tachometer can be installed on any engine. Such an instrument provides an indication of engine speed in revolutions per minute (rpm). This is accomplished by measuring the number of electrical pulses per minute generated in the primary circuit of the ignition system.

The meter readings range from 0 to 6,000 rpm, in increments of 100. Tachometers have solid-state electronic circuits which eliminates the need for relays or batteries and contributes to their accuracy. The electronic parts of the tachometer susceptible to moisture are coated to prolong their life.

6-7 HORNS

The only reason for servicing a horn is because it fails to operate properly or because it is out of tune. In most cases, the problem can be traced to an open circuit in the wiring or to a defective relay.

Cleaning:

Crocus cloth and carbon tetrachloride should be used to clean the contact points. **NEVER** force the contacts apart or you will bend the contact spring and change the operating tension.

Check Relay and Wiring:

Connect a wire from the battery to the horn terminal. If the horn operates, the problem is in the relay or in the horn wiring. If both of these appear satisfactory, the horn is defective and needs to be replaced.

Before replacing the horn however, connect a second jumper wire from the horn frame to ground to check the ground connection.

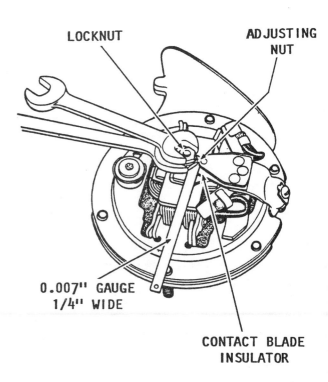

*The tone of a horn can be adjusted with a 0.007" feeler gauge, as described in the text. **TAKE CARE** to prevent the feeler gauge from making contact with the case, or the circuit will be shorted out.*

Maximum engine performance can only be obtained through proper tuning using a tachometer.

Test the winding for an open circuit, faulty insulation, or poor ground. Check the resistor with an ohmmeter, or test the condenser for capacity, ground, and leakage. Inspect the diaphragm for cracks.

Adjust Horn Tone

Loosen the locknut, and then rotate the adjusting screw until the desired tone is reached. On a dual horn installation, disconnect one horn and adjust each, one-at-a-time. The contact point adjustment is made by inserting a 0.007" feeler gauge blade between the adjusting nut and the contact blade insulator. **TAKE CARE** not to allow the feeler gauge to touch the metallic parts of the contact points because it would short them out. Now, loosen the locknut and turn the adjusting nut down until the horn fails to sound. Loosen the adjusting nut slowly until the horn barely sounds. The locknut **MUST** be tightened after each test. When the feeler gauge is withdrawn the horn will operate properly and the current draw will be satisfactory.

Small horsepower engines use a hand starter instead of an electric starter motor. The hand starter may be mounted on top of the engine or on the side, depending on the model.

6-8 ELECTRICAL SYSTEM GENERAL INFORMATION

Probably 75-80% of all Johnson/Evinrude engines covered in this manual are started by pulling on a rope. As the manufacturer increased the size and horsepower of the engines, it was necessary to incorporate some form of power cranking system in addition to the rope starter method. Today, most small engines are still started only by pulling on a rope.

On the larger hp engines, an electric starter motor coupled with a mechanical gear mesh between the cranking motor and the engine flywheel, similar to the method used to crank an automobile engine, was added. This system provided an alternate method to the hand starter rope arrangement. If the electric cranking system is inoperative for any reason, including a dead or weak battery, the engine may still be cranked and started by hand.

Since the starting motor requires a large amount of electrical current, it is necessary to have a fully charged battery available for the starting system. If the boat is equipped with several electrical accessories, such as bait tank with circulating pump, radio, a number of running and accessory lights etc., the charging system must be performing properly to keep the battery charged.

Charging Circuit

The charging circuit consists of a generator driven by a belt connected to the flywheel. The flywheel is equipped with a pulley arrangement to transfer flywheel rotation to the generator puller through the belt.

Choke Circuit

The choke is activated by a solenoid. This solenoid attracts a plunger to close the choke valves. The solenoid is energized when the ignition key is turned to the **START** position and the choke button is depressed. When using the electric choke, the manual choke **MUST** be in the **NEUTRAL** position.

Only the electric choke is covered in this chapter. Service procedures for the heat/electric choke, and the water choke, are presented in Chapter 4.

Starting Circuit

The starting circuit consists of a cranking motor and a starter-engaging mechanism. A solenoid is used as a heavy-duty

6-12 ELECTRICAL

switch to carry the heavy current from the battery to the starter motor. The solenoid is actuated by turning the ignition key to the **START** position. On some models, a pushbutton is used to actuate the solenoid. See Section 6-11 for detailed service procedures on the starter motor circuit.

6-9 CHARGING CIRCUIT SERVICE

The generator has two terminals on the lower end. One terminal is larger than the other and the wires connected have different size connectors to ensure the proper wire is connected to the correct terminal.

If several electrical accessories are used and the engine is operating at idle speed, or below 1500 rpm for extended periods of time, the battery will not recieve adequate current to remain in serviceable condition.

The rated capacity of the generator is 10-amps. Therefore, the electrical accessory load should not exceed 10-amps or current will be drawn from the battery at a greater rate than the generator is able to produce. Such a negative draw on the battery will result in a run-down condition and failure of the battery to provide the required current to the starter for cranking the engine.

To calculate the amperage draw of an accessory the following simple formula may be used: Amps equals watts divided by volts. $\text{Amps} = \dfrac{\text{Watts}}{\text{Volts}}$

Front view of a Johnson outboard with the generator mounted on the port side.

The volts will always be 12. Accessories will usually be given in watts. If the obsolete measurement of candlepower is used, then one candle power is equal to approximately one watt. Example: A boat has running lights requiring 8-watts; auxiliary lights use 10 watts; and a radio rated at 30-watts.

$\text{Amps} = 48 \text{ Watts}/12 \text{ volts} = 4 \text{ Amps}.$

In this case, if all the lights are on and the radio is being used, the total draw on the battery would be 4 amps. If the engine is running at 1500 rpm or higher, and the generator circuit is performing properly by charging the battery with 10 amps, then a net postive gain of 6 amps is being recieved by the battery.

The battery can be externally charged or the engine can be equipped with a generator to charge the battery while the engine is operating.

A voltage regulator, mounted in a junction box on the rear of the engine, is connected between the generator and the battery to prevent overcharging of the battery while the engine is operating. The junction box also houses a fuse to protect the charging circuit.

The generator circuit requires at least 1500 rpm engine speed to effectively charge the battery. At this speed, the ampere meter on the dash will indicate a positive charge to the battery.

If the boat has a twin engine installation, the usual practice is to use only one battery for cranking both units. With such a twin installation, only one engine generator should be used to charge and maintain the battery at its full ampereage rating.

Most mechanics have discovered if both generators of a twin installation are connected to charge the battery, one seems to "fight" the other. Instead of having an improved system, this type of hook-up causes many serious electrical problems that are unexplainable. See Section 6-9 for detailed service procedures on the generator circuit.

TROUBLESHOOTING

One of three areas causing problems in the generating circuit and failure of the system to provide sufficient current to maintain the battery as a satisfactory charge. Remember, the generator will only produce approximately 10 amps of current.

CHARGING CIRCUIT 6-13

Most amp-meters have a 20-amp scale. Therefore, it is only necessary for the scale to register in the 10-amp area, while the engine is operating above 1500 rpm, to indicate satisfactory performance

a- The 4-amp or 20-amp fuse in the junction box may have burned, opening the circuit. If the fuse requires replacement, a check should be made immediately, to determine why the fuse burned protecting the circuit.

b- The voltage regulator may be defective. If the regulator has failed, a thorough check of the circuit to determine the cause. Simply replacing the regulator usually will not solve the problem and the new regulator may be damaged when the engine and generator are operating.

c- The generator may be defective and fail to produce the current necessary to maintain the battery. The problem may simply be worn brushes. Replacement with new brushes may solve the problem. However, if the brushes are in good condition, and testing reveals the generator must be replaced, the conservative mechanic will install a new voltage regulator at the same time.

CRITICAL WORDS

The engine must be operated, in gear, at speeds in excess of 1500 rpm to test the generator circuit. Therefore, the engine **MUST** be mounted in a body of water to prevent a **RUNAWAY** condition and serious damage to internal parts, or destruction of the unit. **NEVER** attempt to operate the engine above idle speed with a flush attachment connected to the lower unit or with the engine mounted in a small test tank, such as a fifty-gallon drum.

CAUTION: Water must circulate through the lower unit to the engine any time the engine is run to prevent damage to the water pump in the lower unit. Just five seconds without water will damage the water pump.

1- Check to be sure all electrical connections in the circuit are secure and free of corrosion. Double check the battery connections and terminals. If the terminals are badly corroded, there is no way on this green earth for the current produced by the generator to reach the battery cells. A special wire brush can be purchased at very modest cost to clean the inside of the wire connectors. A common wire brush may be used to clean the battery terminals. Baking

soda and water is a good cleaning agent for the battery surface.

2- Check the wiring in the circuit for broken insulation, or an actual break in the line. Disconnect the **POSITIVE** electrical lead from the battery as a precaution against an accidental short causing damage to the voltage regulator. Remove the junction box cover. Check the condition of the

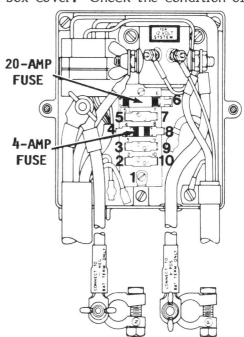

6-14 ELECTRICAL

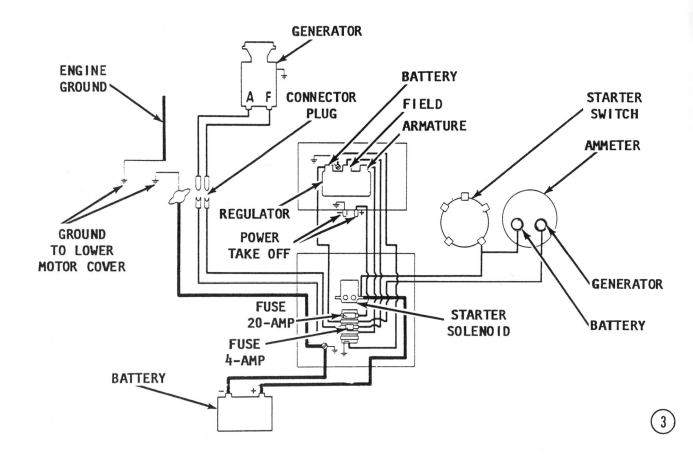

4-amp or 20-amp fuse with a continuity light, or install a new fuse and check the charging circuit again with the engine operating. The 4-amp and 20-amp fuses can easily be "popped" out their retainers in the panel of the junction box base and replaced after the cover has been removed.

3- Ground the field of the generator very **QUICKLY** and only **MOMENTARILY** with a jumper wire while the engine is operating at approximately 2000 rpm. By **MOMENTARILY** grounding the field, the voltage regulator is actually bypassed and the generator will "run wild". If the amp meter registers a high reading while the generator field is grounded, the circuit has a broken wire, or the voltage regulator is defective. If the amp meter reading does not change while the generator field is grounded, then the indication is a faulty generator.

Voltage Regulator

1- Disconnect the positive lead from the battery terminal an accidental short causing damage in the circuit. Loosen the two wingnuts on both sides of the junction box cover. These wingnuts are "captive" with the cover and cannot be completely removed, only released from the junction box. (This arrangement prevents loss of the wingnuts). Disconnect the wires between the generator and the terminal board in the junction box at the board, if the regulator is to be replaced.

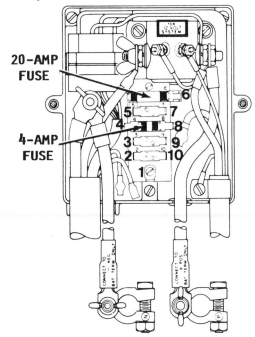

CHARGING CIRCUIT 6-15

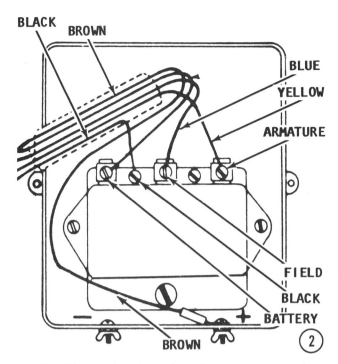

2- Place the junction box cover on its back and remove the five leads to the voltage regulator. Notice how the leads are color-coded, as an assist in connecting them correctly during installation. Remove the attaching hardware securing the voltage regulator to the cover. Remove the regulator. Place the new regulator in position in the junction box cover and secure it with the attaching hardware. Connect the four color-coded wires to the regulator: Yellow, from the generator armature; blue, from the generator field; brown, from the battery; and the black is the ground wire. A second brown wire is connected to the bottom of the junction box to allow additional electrical accessories to be connected. This second wire would not have been disconnected to remove the regulator.

CRITICAL WORDS

When a new voltage regulator is installed, the generator must be "polarized" BEFORE the cover is installed.

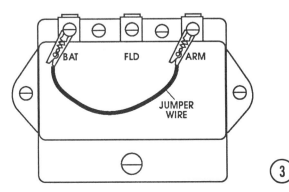

3- "Polarize" the new regulator by first connecting the positive lead to the battery, and then using a small jumper wire to make a **MOMENTARY** connection between the battery terminal and the armature terminal of the regulator. The generator is now properly "polarized" with the new regulator for service. **TAKE CARE** not to touch the field terminal when making the connections.

4- Now, disconnect the positive battery lead again, before installing the junction box cover. The few moments involved in disconnecting and connecting the positive lead at the battery is well spent. This small task will prevent any possible short from causing damage to the circuit when working with the wires.

5- Install the junction box cover to the box base. As the cover is moved into place, work the wires alongside the regulator. Secure the cover in place with the two "captive" wingnuts.

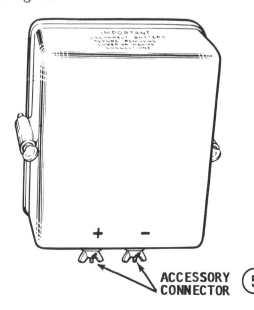

6-16 ELECTRICAL

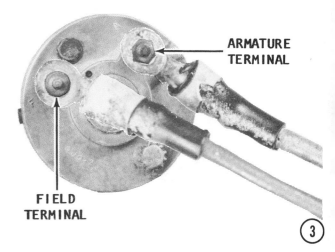

GENERATOR SERVICE

1- Disconnect the positive lead at the battery terminal. Remove the hood from the engine. If the engine has a hand starter, remove the retaining bolts and lift off the hand starter. Prevent the shaft from turning by engaging an open-end wrench with the flats on the generator shaft underneath the pulley. Now, while continuing to hold the wrench on the shaft, remove the nut from the top of the generator pulley. Remove the cover from the inside of the generator pulley.

2- Remove the generator belt from the pulley. Use a screwdriver or other similar tool and pry the puller up and free of the generator shaft. Hold the generator, and at the same time remove the nuts from the top of the generator support bracket, and the generator is free.

SPECIAL WORDS

In some cases, it is just as easy to remove the 7/16" bolts securing the bracket to the engine, and then to remove the bracket and generator together. The bracket is then removed from the generator.

3- Remove the two wires on the bottom side of the generator. Notice how one generator stud is smaller than the other and the electrical connectors are different sizes to match the studs.

4- Remove the two nuts from the wire terminals at the bottom of the generator. Work out the two white insulators from around the studs. Remove the two thru-bolts.

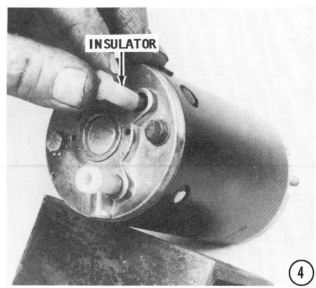

CHARGING CIRCUIT 6-17

5- Remove the end cap from the generator. After the cap has been removed, take notice of the small dowel in the end of the generator. This dowel ensures the cap will be installed correctly when the dowel is indexed in a matching hole in the cap. Pull the armature and upper cap out of the frame.

ARMATURE TESTING

Testing for a Short

6- Position the armature on a growler, then hold a hacksaw blade over the armature core. Turn the growler switch to the ON position. Slowly rotate the armature. If the hacksaw blade vibrates, the armature or commutator has a short. Clean the grooves

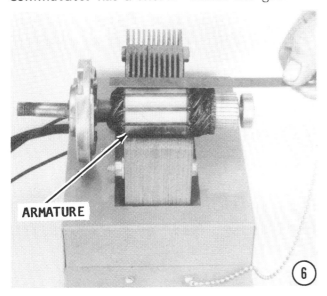

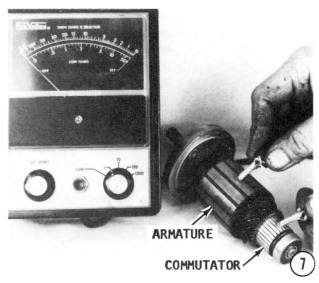

between the commutator bars on the armature. Perform the test again. If the hacksaw blade still vibrates during the test, the armature has a short and **MUST** be replaced.

Testing for a Ground

7- Obtain a test lamp or continuity meter. Make contact with one probe lead on the armature core and the other probe lead on the commutator bar. If the lamp lights, or the meter indicates continuity, the armature is grounded and **MUST** be replaced.

Checking the Commutator Bar

8- Check between or check bar-to-bar as shown in the accompanying illustration. The test light should light, or the meter should indicate continuity. If the commutator fails the test, the armature **MUST** be replaced.

Field Coil Test for Ground

9- Check to be sure the free end of the field wire is not grounded to the frame and

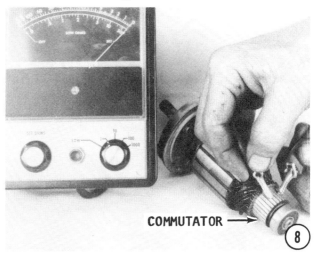

ELECTRICAL

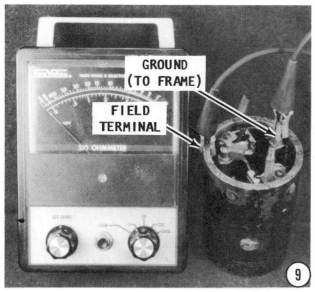

the field insulation is not broken. Using a test lamp or ohmmeter, make contact with one probe lead to the ground of the generator frame. Make contact with the other lead to the field terminal. If the lamp lights or the ohmmeter indicates continuity, the field coils are grounded. If the location of the ground in the field coils cannot be determined, or repaired, the coils **MUST** be replaced.

Armature Terminal Test for Ground

10- Check to be sure the loose end of the armature ternimal lead of the generator is **NOT** grounded to the frame. Using a test lamp or ohmmeter, make contact with one probe lead to the armature terminal of the generator. Make contact with the other probe lead to a good ground on the generator frame. If the test lamp lights or the ohmmeter indicates continuity, the positive terminal insulation through the generator frame is broken down and **MUST** be replaced.

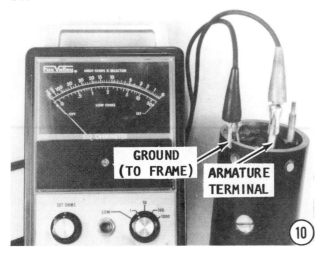

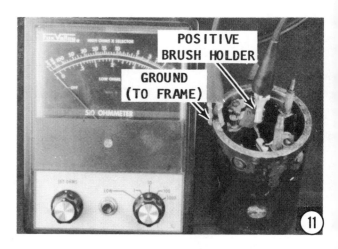

Positive Brush Test for Ground

11- Using a test lamp or ohmmeter, make contact with one probe lead to the positive or insulated brush holder. Make contact with the other probe lead to a good ground on the generator frame. If the lamp lights, or the ohmmeter indicates continuity, the brush holder is grounded due to defective insulation at the frame.

Field Test

12- Using a test lamp or ohmmeter, make contact with one probe lead to the armature stud. Make contact with the other probe lead to the armature brush. The lamp should light or the ohmmeter indicate continuity. If this test is not successful, check for a poor connection between the stud and the brush.

CLEANING AND INSPECTING

Check the ball bearing at the end of the commutator bar. Verify that the bearing turns free with no sign of "rough spots" or binding. Hold the armature in one hand and

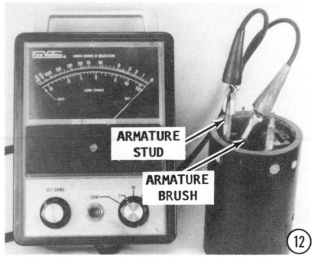

CHARGING CIRCUIT 6-19

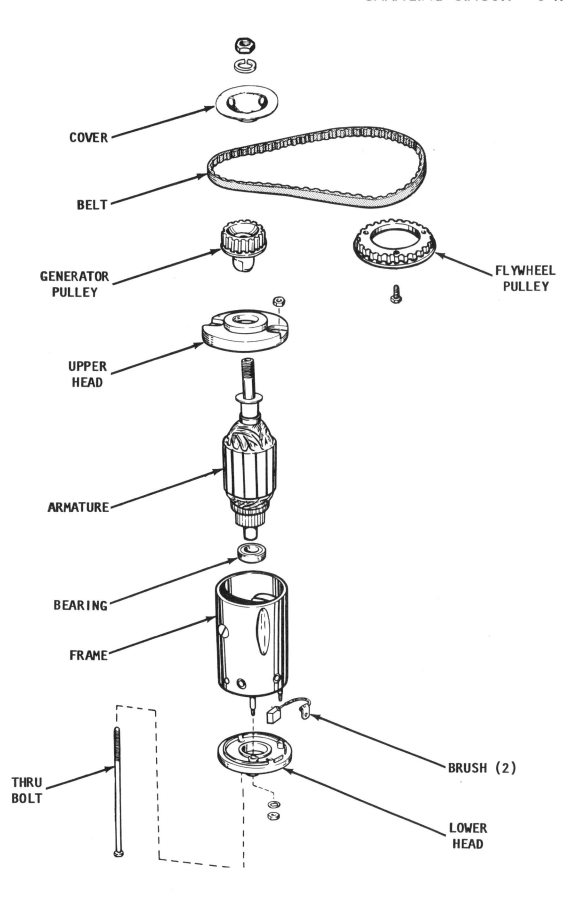

Exploded view of a typical generator showing arrangement of major parts.

6-20 ELECTRICAL

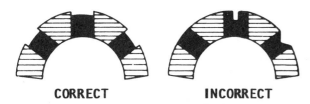

Armature segments properly cleaned (left) and improperly cleaned (right).

the turn the upper cap on the shaft with the other hand. The cap and shaft should turn freely with no sign of binding. If either of these tests are not successful, the bearing **MUST** be replaced.

* Check the amount of brush wear. If the brush is worn more than 50% of its original size, or to within 1/4" of the base, it should be replaced. Replacement of the brushes is a simple task. First, remove the brush retaining screw, and then remove the old brush and install a new brush. Secure the new brush in place with the retaining screw.

If the armature commutator requires turning, it should be turned in a lathe to ensure accuracy. The local generator shop can perform this task, usually for a very reasonable fee. If the turning is accomplished by other than generator shop personnel, the following words are necessary. After the turning, an undercut should be made. The insulation between the commutator bars should be 1-3/4". This undercut must be the full width of the insulation and flat at the bottom. A triangular groove is **NOT** satisfactory. After the undercut work is completed the slot should be thoroughly cleaned to remove any foreign material, dirt or copper dust. Sand the commutator **LIGHTLY** with "00" sandpaper to remove any slight burrs left from the undercutting. After all work has been completed, test the unit again, on the growler.

ASSEMBLING THE GENERATOR

1- Slide the armature into the frame and align the top armature cap with the dowel in the frame. Proper alignment is achieved when the dowel in the frame indexes into a matching hole in the cap. As the armature is moved into place, pull back on the brushes, and work them around the commutator bar.

2- Install the end cap down over the studs of the field and armature. Check to be sure the dowel in the frame has indexed with the hole in the cap.

3- Install the two thru-bolts and secure the complete assembly with the nuts.

4- Place the two bushings over the terminal studs of the armature and field. Secure the bushings in place with the washers and proper nuts (one terminal is larger than the other).

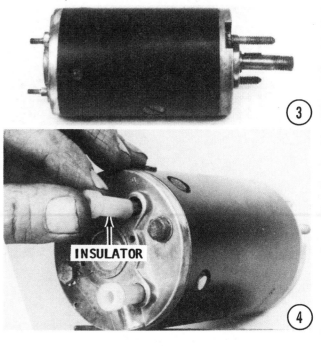

GENERATOR SERVICE

Testing by Rotating the Armature

Performing this test will also "polarize" the new or rebuilt generator. If this test is not performed, the new or rebuilt generator **MUST** still be "polarized" following installation. "Polarization" at that time is accomplished by first connecting the battery to the system in the normal manner, and then connecting a jumper lead to the **BAT** terminal of the voltage regulator. Next, **MOMENTARILY** making contact with the other end of the jumper lead to the **ARM** terminal of the regulator. The generator is now "polarized" for service.

Now, returning to bench testing after rebuilding the generator:

CAUTION: The armature will turn rapidly during this test. Therefore, the generator **MUST** be well **SECURED** before making the test to prevent personal **INJURY** or damage to the generator.

1- Connect a jumper wire between the field terminal and a good ground on the case. Connect a second jumper wire between the positive battery terminal and the armature stud. **MOMENTARILY** make contact with the negative lead from the battery to any good ground on the generator. The generator should rotate

rapidly. If the generator fails to rotate, the generator must be disassembled again and the service work carefully checked. Sorry about that, but some phase of the rebuild task was not performed properly.

2- Install the holding bracket to the generator, or if the bracket remained on the engine, install the generator into the bracket and secure it in place with the attaching hardware, but **DO NOT** tighten the nuts on the generator thru-bolts at his time. Check to be sure the Woodruff key is in place in the generator shaft. Slide the pulley onto the armature shaft with the slot in the

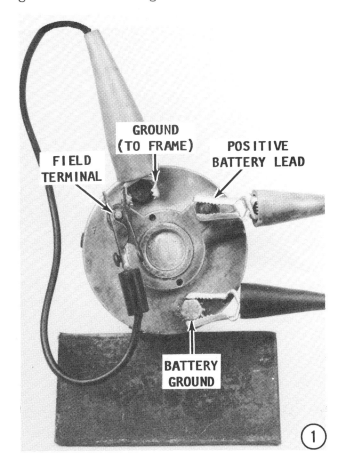

pulley indexed with the Woodruff key. Install the drive belt around the pulley on the flywheel and the generator pulley.

3- Install the pulley cap, lockwasher, and nut, to secure the generator pulley in place. Hold the generator shaft from turning with an open-wrench on the flats of the shaft underneath the pulley. Adjust tension on the generator pulley by pulling the generator away from the engine, and then tightening the thru-bolt nuts securing the generator in the bracket. The pulley is properly adjusted when it may be depressed approximately 1/4" at a point mid-way between the two pulleys.

6-10 CHOKE CIRCUIT SERVICE

This short section provides instructions to test the choke circuit. If the system fails the test, the attaching hardware can be removed and the choke assembly replaced.

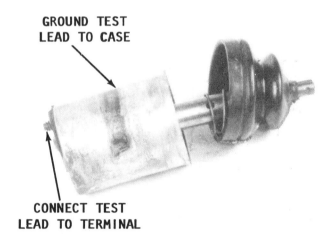

Hookup points to test an electric choke.

Choke Circuit Testing

The choke circuit may be quickly tested to determine if it is functioning properly as follows:

a- Obtain an ohmmeter.
b- Connect the black meter lead to an unpainted portion of the engine block for a good ground.
c- Connect the red meter lead to the choke terminal.
d- Test the circuit using the Rx1 scale of the ohmmeter. A satisfactory reading is approximately 3 ohms.
e- After the test is completed, check to be sure the choke plunger is pulled into the choke solenoid.

6-11 STARTER MOTOR CIRCUIT SERVICE

CIRCUIT DESCRIPTION

As the name implies, the sole purpose of the starter motor circuit is to control operation of the starter motor to crank the engine until the engine is operating. The circuit includes a solenoid or magnetic switch to connect or disconnect the starter from the battery. The operator controls the switch with a pushbutton or key switch.

A cutout switch is installed in the system to prevent starting the engine if the throttle is advanced too far, beyond idle speed. When the throttle is advanced, the starter solenoid is not grounded and the starter motor will not rotate. On electric shift models, a cut-out switch is installed in the circuit to permit operation of the starter motor **ONLY** if the shift control lever is in **NEUTRAL**. This switch is a safety device to prevent accidental engine start when the engine is in gear.

STARTER MOTOR DESCRIPTION

Delco-Remy, Autolite, and Prestolite starter motors are used on the Johnson/Evinrude engines covered in this manual. Any of the three may be installed on a 35 hp or 40 hp engine. The early model starters (especially the Delco-Remey) were shorter and just a bit less powerful than the later models. If a replacement unit must be purchased, any one of the three may be obtained and installed. The recommendation is to spend a few dollars more for the longer, more powerful unit. One more word: If a long starter motor is replacing the short model, an additional bracket **MUST** be bought and installed on the lower end of the starter.

Marine starter motors are very similar in construction and operation to the units used in the automotive industry. All marine starter motors use the inertia-type drive assembly. This type assembly is mounted on an armature shaft with external spiral splines which mate with the internal splines of the drive assembly.

The starter motor is a series wound electric motor which draws a heavy current from the battery. It is designed to be used only for short periods of time to crank the engine for starting. To prevent overheating the motor, cranking should not be continued for more than 30-seconds without allowing the motor to cool for at least three minutes. Actually, this time can be spent in making preliminary checks to determine why the engine fails to start.

Most starter motors operate in much the same manner and the service work involved in restoring a defective unit to service is almost identical. Therefore, the information in this chapter is grouped together for the major components of the starter under separate headings. Differences, where they occur, between the various manufacturers, are clearly indicated.

Theory of Operation

Power is transmitted from the starter motor to the engine flywheel through a Bendix drive. This drive has a pinion gear mounted on screw threads. When the motor is operated, the pinion gear moves up into mesh with the teeth on the flywheel ring gear.

When the engine starts, the pinion gear is driven faster than the shaft, and as a result, it screws out of mesh with the flywheel. A rubber cushion is built into the Bendix drive to absorb the shock when the pinion meshes with the flywheel ring gear. The parts of the drive **MUST** be properly assembled for efficient operation. If the drive is removed for cleaning, **TAKE CARE** to assemble the parts as shown in the accompanying illustration. If the screw shaft assembly is reversed, it will strike the splines and the rubber cushion will not absorb the shock.

The sound of the motor during cranking is a good indication of whether the starter motor is operating properly or not. Naturally, temperature conditions will affect the speed at which the starter motor is able to crank the engine. The speed of cranking a cold engine will be much slower than when cranking a warm engine. An experienced operator will learn to recognize the favorable sounds of the cranking engine under various conditions.

Faulty Symptoms

If the starter spins, but fails to crank the engine, the cause is usually a corroded or gummy Bendix drive. The drive should be removed, cleaned, and given an inspection.

If the starter motor cranks the engine to slowly, the following are possible causes and the corrective actions that may be taken:

a- Battery charge is low. Charge the battery to full capacity.
b- High resistance connections at the battery, solenoid, or motor. Clean and tighten all connections.
c- Undersize battery cables. Replace cables with sufficient size.
d- Battery cables too long. Relocate the battery to shorten the run to the starter solenoid.

Maintenance

The starter motor does not require periodic maintenance or lubrication **EXCEPT** just a drop of light-weight oil on the starter shaft to ease movement of the Bendix drive. If the motor fails to perform properly, the

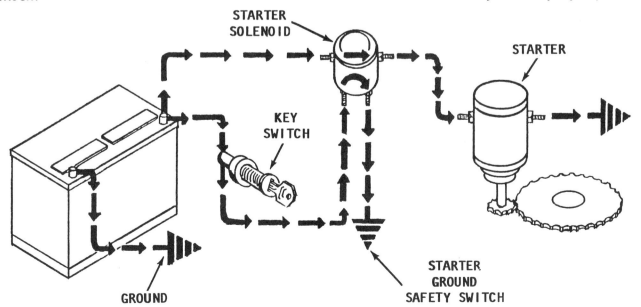

Functional diagram to show current flow when the key switch is turned to the START position.

Two handy instruments for use in checking the generating circuit (left) and starter motor circuit (right). These meters do not require any wire connections. A reading will be obtained by simply placing the meter on the line.

checks outlined in the previous paragraph should be performed.

The frequency of starts governs how often the motor should be removed and reconditioned. The manufacturer recommends removal and reconditioning every 1000 hours.

Naturally, the motor will have to be removed if the corrective actions outlined under **Faulty Symptoms** above, does not restore the motor to satisfactory operation.

STARTER MOTOR TROUBLESHOOTING

Before wasting too much time troubleshooting the starter circuit, the following checks should be made. Many times, the problem will be corrected.

a- Battery fully charged.

b- Throttle advanced too far (beyond fast idle speed).

Typical Bendix spring arrangement on a starter motor. A small amount of oil on the shaft in the spring area will prolong satisfactory operation.

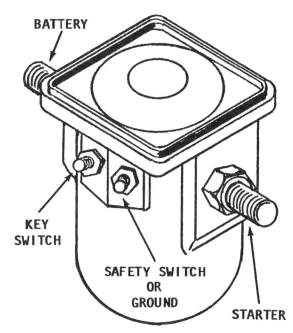

*Functional diagram of a starter motor solenoid. Notice the separate terminal for a ground wire. This solenoid is **NOT** grounded through the mounting bracket.*

c- Shift control lever not in **NEUTRAL** (electric shift models only).

d- All electrical connections clean and tight.

e- Wiring in good condition, insulation not worn or frayed.

f- One of the cut-out switches may be defective.

Two more areas may cause the engine to turn over slowly even though the starter motor circuit is in excellent condition: A tight or "frozen" engine; and water in the lower unit causing the bearings to tighten up. The following troubleshooting procedures are presented in a logical sequence,

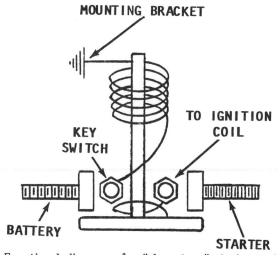

*Functional diagram of a "slave-type" starter motor solenoid used on four-cycle engine installation. This solenoid **CANNOT** be used on a two-cycle engine.*

STARTER MOTOR 6-25

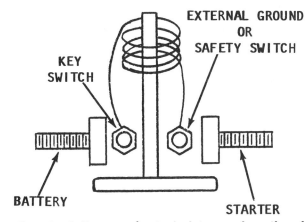

*Functional diagram of a typical two-cycle outboard engine starter motor solenoid. Notice that the right-hand small terminal is connected to ground or a safety switch. The unit is **NOT** grounded through the mounting bracket.*

with the most common and easily corrected areas listed first in each problem area. The connection number refers to the numbered positions in the accompanying illustrations.

Perform the following quick checks and corrective actions for following problems:

TESTING

FIRST THESE WORDS

The starter solenoid is actually nothing more than a switch between the battery and the starter motor. Several types of solenoids are used and many appear similar. **NEVER** attempt to use an automotive-type solenoid in a marine installation. Such practice will lead to more problems than can be imagined. An automotive-type solenoid has a completely different internal wiring circuit. If such a solenoid is connected into the starter system, and the system is activated, current will be directed to ground. The wires will be burned and the cut-out switch will be burned and rendered useless. Therefore, when installing replacement parts in the starter or other circuits on a marine installation, always take time to obtain parts from a **MARINE** outlet to ensure proper service and to prevent damage to other expensive components.

SAFETY WORD

Before making any test of the cranking system, disconnect the spark plug leads at the spark plugs to prevent the engine from possibly starting during the test and causing personal injury.

The following tests are to be performed according to the faulty condition described. The numbers referenced in the steps are correlated with numbers on the accompany circuit diagram on Page 26, to identify exactly where the connection or test is to made.

Starter Motor Turns Slowly

a- Battery charge is low. Charge the battery to full capacity.

b- Electrical connections corroded or loose. Clean and tighten.

c- Defective starter motor. Perform an amp draw test. Lay an amp draw-gauge on the cable leading to the starter motor No. 5. Turn the key to the **START** position and attempt to crank the engine. If the gauge indicates an excessive amperage draw, the starter motor **MUST** be replaced or rebuilt.

Starter Motor Fails To Turn Over
Voltage Check

a- Check the voltage at No. 2, the battery and ground.

b- If satisfactory voltage is indicated at the battery, check the voltage at No. 3, the positive side of the starter solenoid. Weak, or no voltage at this point indicates corroded battery terminals, poor connection at the solenoid, or defective wiring between the battery and the solenoid.

c- Test the voltage at No. 4, the key. A full 12-volt reading should be registered at the key. Weak or no voltage at the key indicates a poor connection at the solenoid, or a broken wire between the starter solenoid and the key.

d- If satisfactory voltage is indicated during Steps a, b, and c, connect a volt meter at No. 5 and ground, and then turn the key switch to the **START** position. If 12-volts is registered at No. 5 and the starter still fails to operate, the starter is defective and requires service. If voltage is **NOT** present at No. 5, proceed to the next section, Testing Starter Solenoid.

Testing Starter Solenoid

a- Remove the heavy starter cable at No. 5, at the starter. This cable **MUST** be disconnected prior to performing this test to prevent the starter motor from turning and cranking the engine. Connect a voltmeter to No. 6 (the starter solenoid), and ground. Turn the key to the **START** position. The meter should indicate 12-volts. If voltage is not present at No. 6, the key switch is defective, or the wire is broken between the key switch and the starter solenoid.

6-26 ELECTRICAL

b- If voltage is present at No. 6, connect a voltmeter at No. 3 and to No. 7. Connect one end of a jumper wire to No. 2, the positive terminal of the battery and **MOMENTARILY** make contact with the other end at No. 6, the starter solenoid. If voltage is indicated through the starter solenoid, the solenoid is satisfactory and the problem has been corrected while making the tests. Sometimes, when working with electrical circuits, corrective action has been taken almost accidently, a bad connection has been made good, etc. If the solenoid test failed, it does not necessarily mean the solenoid is defective. The solenoid may not be properly grounded through the cutout switch. Therefore, the cutout switch may be defective and should be checked as outlined later in this section.

c- With the voltmeter still connected at No. 3 and No. 7, connect one end of a jumper wire at No. 8, the starter solenoid, and the other lead to a good ground. Connect a second jumper wire at No. 2, the positive terminal of the battery, to No. 6, the starter solenoid. The voltmeter should indicate voltage is present. If voltage is not present, the starter solenoid is defective and **MUST** be replaced.

Testing Throttle Advance Cut-out Switch

a- Remove the existing wire from No. 1 the switch terminal. Connect one probe lead of a ohmmeter to the terminal. Connect the other test probe lead to a good ground. Depress the switch button and the ohmmeter should indicate continuity. If continuity is not indicated, the switch is defective and **MUST** be replaced. Connect the heavy cable at No. 5, the starter motor.

6-12 STARTER DRIVE GEAR SERVICE

STARTER REMOVAL

Before beginning any work on the starter motor, disconnect the positive (+) lead from the battery terminal. Remove the hood. Disconnect the red cable at the starter motor terminal.

Remove the 1/2" bolt securing the starter bracket. This bolt is located on the starboard side of the engine just above the

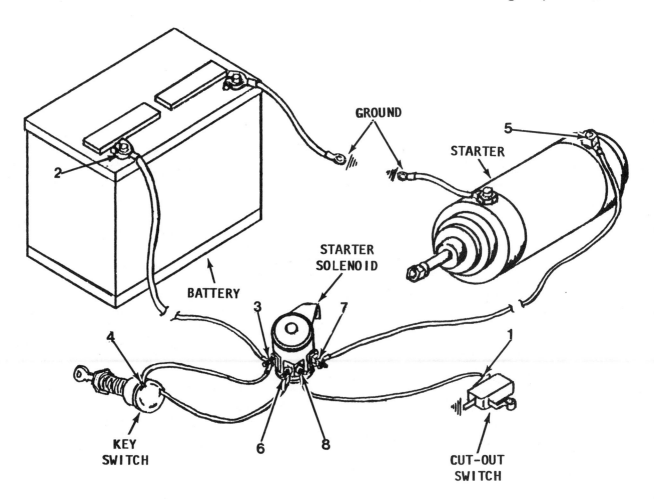

STARTER DRIVE GEAR   6-27

Typical starter motor installation.

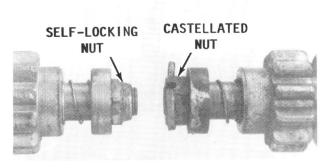

Two different type drive nuts used on outboard starter motors. The nut on the left is a self-locking type and the right one is a castellated nut with a cotter pin.

carburetor. Remove the three 7/16" bolts (or nuts in some cases) securing the starter motor bracket to the engine. Remove the starter motor and bracket together.

DRIVE GEAR DISASSEMBLING

Two types of drive gear arrangements are used on Johnson/Evinrude outboard starter motors covered in this manual. Both units use a drive gear with a rubber cushion. However one has a castillated nut with a cotter pin or possibly a self-locking nut to secure the drive gear on the armature shaft. The arrangement of parts behind the nut is, of course, unique. This drive gear is referred to as Type I. The other is very similar except for the arrangement of parts on the armature shaft. This second unit is referred to as Type II.

To determine which starter motor drive type is being serviced observe the unit and make a comparison with the two exploded illustrations in this section, especially the pinion gear and the screw shaft.

Type I Drive Gear with Rubber Cushion

These starter motors have the drive secured to the shaft by means of a castillated nut. Begin diassembling by removing the cotter pin from the armature shaft end, and then unscrew the castillated nut. Some models use a self-locking nut without the cotter pin. The drive mechanism is now ready to be removed in the following sequence: pinion stop, anti-drift spring, pinion washer, anti-drift spring sleeve, pinion gear, screw shaft, thrust washer, cushion cup, cushion, cushion spacer, and thrust washer. The exploded drawing accompanying this section will be helpful in assembling the starter motor in the proper sequence.

CLEANING AND INSPECTING

Inspect the drive gear teeth for chips, cracks, or a broken tooth. Check the spline inside the drive gear for burrs and to be sure the drive gear moves freely on the armature shaft. Check to be sure the return spring is flexible and has not become distorted. In-

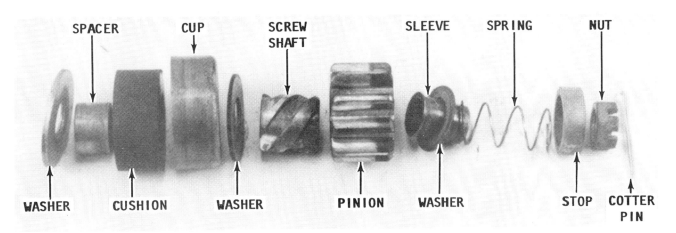

Arrangement of parts for a typical starter motor drive gear.

6-28 ELECTRICAL

flexible and has not become distorted. Inspect the rubber cushion for cracks and for signs of oil on the cushion. Clean the armature shaft with crocus cloth.

ASSEMBLING
Type I Starter Drive

Begin by assembling the following parts in the order given. The accompanying illustration will be most helpful in assembling the parts in the proper sequence. First, slide the thrust washer onto the armature shaft. Next, and in the sequence given, install the cushion spacer; cushion; cushion cup with the open end over the cushion; the cup; and then the thrust washer.

Install the screw shaft with the splined end facing **UP**. Lubricate the underside with Multipurpose Lubricant, or equivalent. Install the pinion gear with the screw end facing **DOWN**.

Install the anti-drift spring washer, anti-drift spring, and the pinion gear stop with the recessed end facing **DOWN**.

Secure the assembly with the castillated nut tightened to a torque value of 200 in.-lbs. Now, turn the nut slightly more until the cotter pin can be inserted through the nut and the hole in the shaft. Some models use a self-locking nut without the cotter key.

To test the complete starter motor, proceed directly to Section 6-16.

To install the starter motor onto the engine, if no further work is to be performed, proceed directly to Section 6-17.

DISASSEMBLING

Type II Drive Gear with Rubber Cushion

First, remove the nut from the end of the armature shaft. Scribe a mark on the top of the screw shaft and one of the top of the pinion as an aid during assembling. These marks will identify the top of both parts. Next, remove the following parts in the sequence given: The pinion stop; anti-drift spring; sleeve; screw shaft cup; screw shaft; pinion; thrust washer; cushion cap; cushion; and the cushion retainer. The exploded drawing accompanying this section will be helpful in assembling the starter motor in the proper sequence.

CLEANING AND INSPECTING

Inspect the drive gear teeth for chips, cracks, or a broken tooth. Check the spline inside the drive gear for burrs and to be sure the drive gear moves freely on the armature shaft. Check to be sure the return spring is flexible and has not become distorted. Inspect the rubber cushion for cracks and for signs of oil on the cushion. Clean the armature shaft with crocus cloth.

ASSEMBLING

Type II Drive Gear with Rubber Cushion

Begin by assembling the following parts in the sequence given: The accompanying

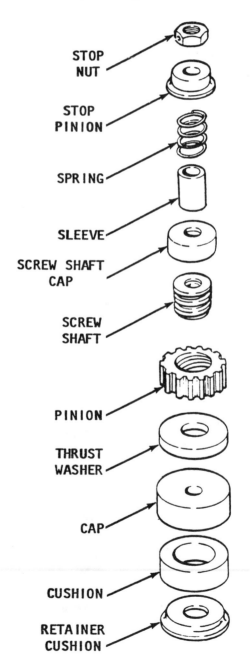

Exploded drawing of the Type II drive gear.

illustration will be most helpful in assembling the parts in the proper sequence.

First, slide the cushion retainer down onto the drive end cap, with the shoulder facing **UPWARD**. Next, slide the cushion down the armature shaft and seat it over the shoulder of the retainer. Install the cushion cap over the cushion. Slide thrust washer down the armature shaft onto the top of the cap. Rotate the screw shaft clockwise into the the pinion, and then slide the pinion and screw shaft down the armature shaft onto the thrust washer. Install the cap over the end of the screw shaft.

Slide the following parts onto the armature shaft in the order given: The sleeve; spring; pinion stop washer; and finally thread the nut onto the end of the shaft. Tighten the nut securely.

To test the complete starter motor, proceed directly to Section 6-16.

To install the starter motor onto the engine, if no further work is to be performed, proceed directly to Section 6-17.

6-13 DELCO-REMY SERVICE

REMOVAL

1- Before beginning any work on the starter motor, disconnect the positive (+) lead from the battery terminal. Remove the engine hood. Disconnect the red cable at the starter motor terminal. Remove the 1/2" bolt securing the starter bracket. This bolt is located on the starboard side of the engine just above the carburetor. Remove the three 7/16" bolts (or nuts in some cases) securing the starter motor bracket to the engine. Remove the starter motor and bracket together.

GOOD NEWS

If the only motor repair necessary is replacement of the brushes, the drive gear does not have to be removed. All starter motors have thru-bolts securing the upper and lower cap to the field frame assembly. In all cases both caps have some type of mark or boss. These marks are used to properly align the caps with the field frame assembly.

DISASSEMBLING

FIRST THESE WORDS

Two models of Delco-Remy starter motors are installed, a short model and longer, more powerful model. The short model has one positive and one negative brush. The longer model has two positive and two negative brushes. The negative brushes can always be identified as the brushes with the lead connected to the frame. The other brush or brushes are connected to the field coil.

2- Observe the caps and find the identifying mark or boss on each. If the marks are not visible, make an identifying mark prior to removing the thru-bolts as an essential aid during asembling. Remove the thru-bolts from the bracket and the starter motor. Use a small hammer and **CAREFULLY** tap the lower cap free of the starter motor. On the Delco-Remy starter motor, the brushes are mounted in the frame assembly. Pull on the armature shaft from the drive gear end and remove it from the field frame assembly. Remove the brushes from their holders, and then remove the brush springs.

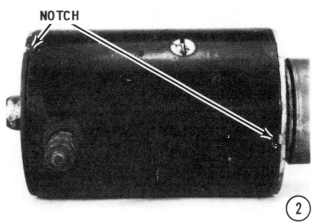

6-30 ELECTRICAL

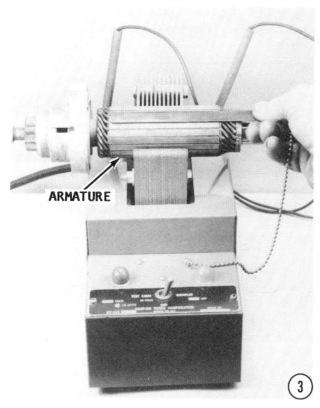

ARMATURE TESTING

Testing for a Short

3- Position the armature on a growler, then hold a hacksaw blade over the armature core. Turn the growler switch to the **ON** position. Slowly rotate the armature. If the hacksaw blade vibrates, the armature or commutator has a short. Clean the grooves between the commutator bars on the armature. Perform the test again. If the hacksaw blade still vibrates during the test, the armature has a short and **MUST** be replaced.

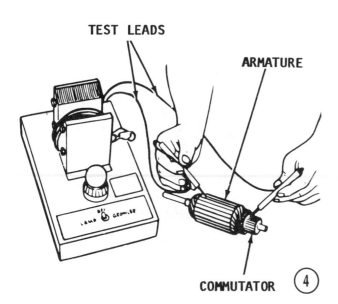

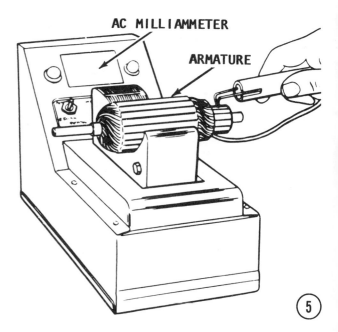

Testing for a Ground

4- Obtain a test lamp or continuity meter. Make contact with one probe lead on the armature core and the other probe lead on the commutator bar. If the lamp lights, or the meter indicates continuity, the armature is grounded and **MUST** be replaced.

Checking the Commutator Bar

5- Check between or check bar-to-bar as shown in the accompanying illustration. The test light should light, or the meter should indicate continuity. If the commutator fails the test, the armature **MUST** be replaced.

Turning the Commutator

6- True the commutator, if necessary, in a lathe. **NEVER** undercut the mica because the brushes are harder than the insulation. Undercut the insulation between the commutator bars 1/32" to the full width of the insulation and flat at the bottom. A triangular groove is not satisfactory. After the under-cutting work is completed, clean out the slots carefully to remove dirt and copper dust. Sand the commutator lightly with No. 00 sandpaper to remove any burrs left from the undercutting. Check the armature a second time on the growler for possible short circuits.

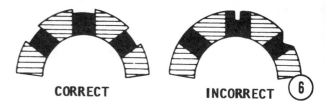

Positive Brushes

7- The positive brush or brushes can always be identified as the brushes connected to the field coil.

Obtain an ohmmeter. Connect one lead of the meter to the end of the brush and the other lead to the terminal. The ohmmeter **MUST** indicate continuity between the brush and the terminal. If the meter indicates any resistance, check the lead to the brush and the lead to the positive terminal solder connection. If the connection cannot be repaired, the brushes **MUST** be replaced. If the unit being tested has a double set of positive brushes, repeat the test for the other positive brush.

Move the test lead from the brush to a good ground on the frame. If continuity is indicated, the field coil is grounded to the case.

Negative Brushes

8- The negative brushes can always be identified because the lead is connected to the brush retainer as a ground to the frame.

Obtain an ohmmeter. Make contact with one lead on the negative brush and make contact with the other lead on the starter frame. If the meter does not indicate continuity, the brush or brush retainer is not grounded to the frame.

If the unit being tested has a double set of negative and positive brushes, move the test lead from the one negative brush to the other negative brush lead and again check for continuity. If the meter does not indicate continuity, the brush or brush retainer is not grounded to the frame.

Check to be sure none of the soldered connections are touching the frame. The

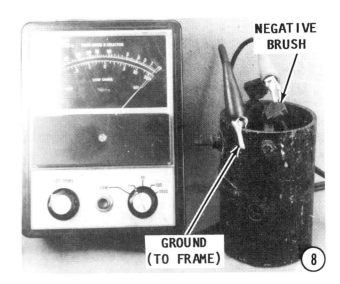

fields would be grounded if the connections make contact with the frame.

CLEANING AND INSPECTING

Clean the field coils, armature, commutator, armature shaft, brush-end plate and drive-end housing with a brush or compressed air. Wash all other parts in solvent and blow them dry with compressed air.

Inspect the insulation and the unsoldered connections of the armature windings for breaks or burns.

Perform electrical tests on any suspected defective part, according to the procedures outlined earlier in this Section.

Check the commutator for run-out. Inspect the armature shaft and both bearings for scoring.

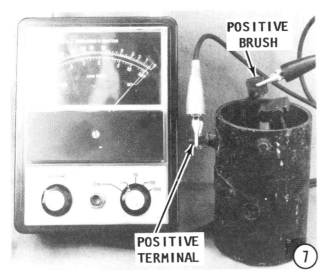

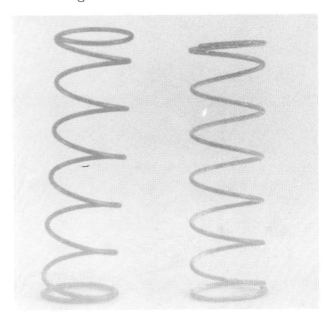

Typical brush spring. If the springs have turned blue in color, they must be replaced.

6-32 ELECTRICAL

Turn the commutator in a lathe if it is out-of-round by more than 0.005".

Check the springs in the brush holder to be sure none are broken. Check the spring tension and replace if the tension is not 32-40 ounces. Check the insulated brush holders for shorts to ground. If the brushes are worn down to 1/4" or less, they must be replaced.

Check the field brush connections and lead insulation. A brush kit and a contact kit are available at your local marine dealer, but all other assemblies must be replaced rather than repaired.

The armature, fields, and brush holders must be checked before assembling the starter motor. See the testing section in this chapter for detailed procedures to test the starter motor.

ASSEMBLING THE DELCO-REMY

Negative Brushes and Retainer

The following procedures apply to brushes mounted to the field frame assembly.

1- Remove the old set of ground brushes by cutting off the rivets with a chisel or by drilling them out. Replacement brush holder kits are available at marine outlets.

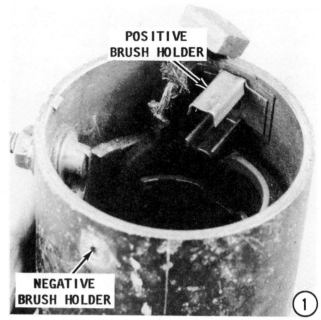

These kits are complete with screws, washers, and nuts for attachment to the frame. Replacement brush springs are also available. The brush spring is removed from the holder by compressing one side of the spring with a small screwdriver until the spring flips out of its seat. After the spring pops out, turn the spring clockwise until it is free of the holder.

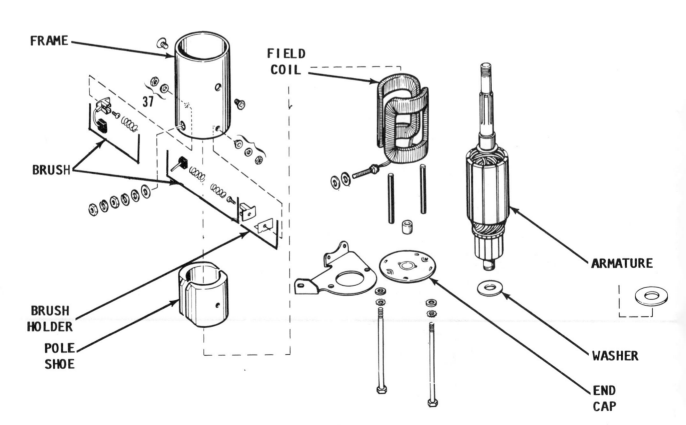

Exploded view showing arrangement of major Delco-Remy starter motor parts.

DELCO-REMY STARTER 6-33

A new negative brush and brush holder as it appears when removed from the package.

Replacement brush sets are available and usually contain the following parts:
One insulated brush, with flexible lead attached.
One ground brush holder with brush and lead attached.
Necessary attaching screws, washers, and nuts.

Positive Brushes

Cut off the old brush lead where it is attached to the field coil. Prepare the ends of the coil for soldering the new brush lead assembly. Clean the ends of the coil by filing or grinding off the old brush lead connection. Remove the varnish only as far back as necessary to enable a good soldered connection to be made.

Use rosin flux and solder the leads to the **BACK SIDES** of the coil to prevent any excess solder from rubbing against the armature. Be sure the leads are in the right position to reach the brush holders. Do not

Installing the negative brush into the frame assembly. A new bolt and washer is provided in the replacement kit.

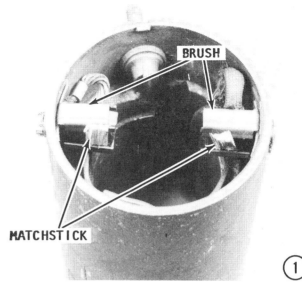

overheat the leads, because the solder will run onto the lead and the lead will loose its flexibility.

Assembling the Starter Motor

1- Clamp the drive gear in a vise equipped with soft jaws and with the drive gear down. Insert the brush springs into the brush holders, and then install the brushes in place.

TAKE TIME to whittle the end of two or four match sticks in the shape of a tiny spade. Now push one brush outward, and then wedge one of the match sticks in between the brush and the lip of the retainer. The match stick will hold the brush in the retracted position during installation of the armature. Retract each of the remaining brushes and hold them with a match stick.

2- After all the brushes have been retracted and held in place with the match sticks, lower the field frame assembly until

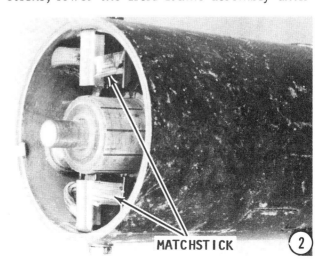

6-34 ELECTRICAL

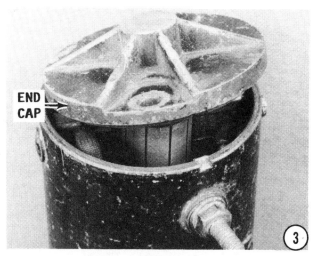

the brushes make contact with the commutator. Now, remove the match sticks. Align the mark on the upper cap with the matching mark on the field frame.

3- Place the washer onto the commutator shaft, then place the cap onto the end of the field frame assembly. Align the mark on the lower cap with the mark on the field frame.

4- Slide the rubber spacer or collar into the starter bracket, if used. Now, install the starter bracket over the starter motor. Install the thru-bolts through the end cap, the frame, and thread them into the starter bracket. Tighten the thru-bolts securely.

To test the complete starter motor, proceed directly to Section 6-16.

To install the starter motor onto the engine, proceed directly to Section 6-17.

6-14 AUTOLITE STARTER MOTOR SERVICE

REMOVAL

1- Before beginning any work on the starter motor, disconnect the positive (+) lead from the battery terminal. Remove the engine hood. Disconnect the red cable at the starter motor terminal. Remove the 1/2" bolt securing the starter bracket. This bolt is located on the starboard side of the engine just above the carburetor. Remove the three 7/16" bolts (or nuts in some cases) securing the starter motor bracket to the engine. Remove the starter motor and bracket together.

GOOD NEWS

If the only motor repair necessary is replacement of the brushes, the drive gear does not have to be removed. All starter motors have thru-bolts securing the upper and lower cap to the field frame assembly. In all cases both caps have some type of mark or boss. These marks are used to properly align the caps with the field frame assembly.

AUTOLITE STARTER

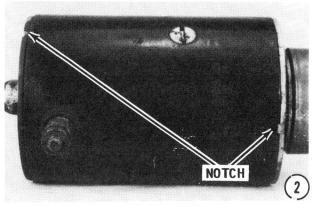

DISASSEMBLING

2- Observe the caps and find the identifying mark or boss on each. If the marks are not visible, make an identifying mark prior to removing the thru-bolts as an essential aid during asembling. Remove the thru-bolts from the bracket and the starter motor.

3- Use a small hammer and **CAREFULLY** tap the lower cap free of the starter motor. On the Autolite starter motor, the brushes are mounted in the end cap.

4- Pull on the armature shaft from the drive gear end and remove it from the field frame assembly. Remove the positive brush from its holder.

ARMATURE TESTING

Testing for a Short

5- Position the armature on a growler, then hold a hacksaw blade over the armature core. Turn the growler switch to the **ON** position. Slowly rotate the armature. If

the hacksaw blade vibrates, the armature or commutator has a short. Clean the grooves between the commutator bars on the armature. Perform the test again. If the hacksaw blade still vibrates during the test, the armature has a short and **MUST** be replaced.

Testing for a Ground

6- Obtain a test lamp or continuity meter. Make contact with one probe lead on the armature core and the other probe lead on the commutator bar. If the lamp lights,

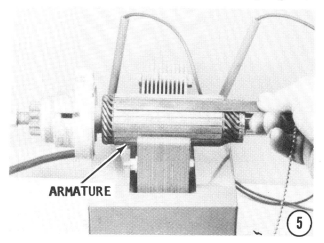

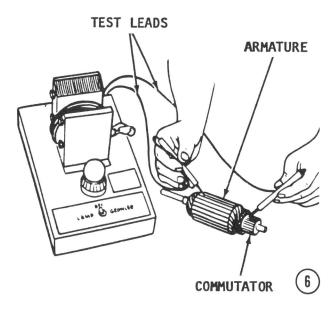

or the meter indicates continuity, the armature is grounded and **MUST** be replaced.

Checking the Commutator Bar

7- Check between or check bar-to-bar as shown in the accompanying illustration. The test light should light, or the meter should indicate continuity. If the commutator fails the test, the armature **MUST** be replaced.

Turning the Commutator

8- True the commutator, if necessary, in a lathe. **NEVER** undercut the mica because the brushes are harder than the insulation. Undercut the insulation between the commutator bars 1/32" to the full width of the insulation and flat at the bottom. A triangular groove is not satisfactory. After the under-cutting work is completed, clean out the slots carefully to remove dirt and copper dust. Sand the commutator lightly with No. 00 sandpaper to remove any burrs left from the undercutting. Check the armature a second time on the growler for possible short circuits.

Positive Brushes

9- The positive brushes can always be identified as the brush with the lead connected to the field coil.

Obtain an ohmmeter. Connect one lead of the meter to the end of the brush and the other lead to the terminal. The ohmmeter **MUST** indicate continuity between the brush and the terminal. If the meter indicates any resistance, check the lead to the brush and the lead to the positive terminal

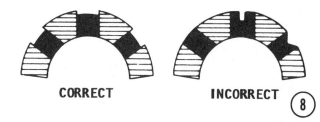

solder connection. If the connection cannot be repaired, the brush **MUST** be replaced.

Negative Brush

10- The negative brush can always be identified because the lead is connected to the starter motor end cap. Obtain an ohmmeter. Make contact with one lead on the negative brush and make contact with the other lead on the starter end cap. If the meter does not indicate continuity, the brush or brush retainer is not grounded to the end cap.

Positive Brush Installation

11- First, cut the old field coil brush free. Next, attach the lead of the new brush to the stiff wire lead on the field coil. Wrap a fine piece of copper wire around the lead and the stiff wire from the coil to hold the brush lead in place while it is soldered. If the wrapped wire becomes soldered also, no problem leave it in place. If it did not become soldered, pull it free after the soldering is complete.

Use rosin flux and solder the leads to the **BACK SIDES** of the wire to prevent any excess solder from rubbing against the armature. Be sure the leads are in the right

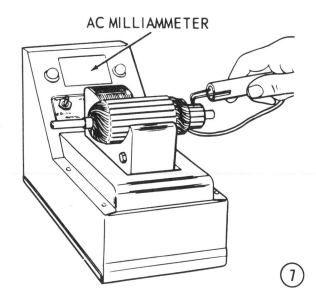

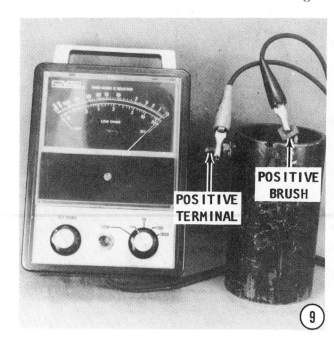

AUTOLITE STARTER 6-37

position to reach the brush holders. Do not overheat the leads, because the solder will run onto the lead and the lead will loose its flexibility.

Check to be sure none of the soldered connections are touching the frame. The fields would be grounded if the connections make contact with the frame.

Negative Brush Installation

12- Cut the old brush free from the end cap. Clean the surface thoroughly. Next, solder the lead of the new brush to the end cap. If the retainer are no longer fit for service, the entire end cap must be replaced. New brushes are **NOT** included with the end cap.

CLEANING AND INSPECTING

Clean the field coils, armature, commutator, armature shaft, brush-end plate and drive-end housing with a brush or compressed air. Wash all other parts in solvent and blow them dry with compressed air.

Inspect the insulation and the unsoldered connections of the armature windings for breaks or burns.

Perform electrical tests on any suspected defective part, according to the procedures outlined earlier in this Section.

Check the commutator for run-out. Inspect the armature shaft and both bearings for scoring.

Turn the commutator in a lathe if it is out-of-round by more than 0.005".

Check the springs in the brush holder to be sure none are broken. Check the spring

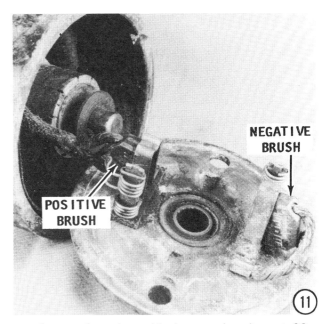

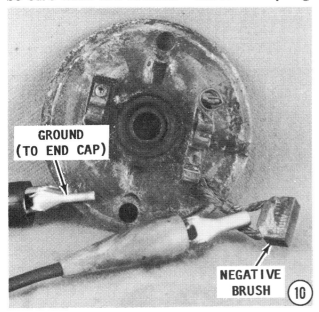

tension and replace if the tension is not 32-40 ounces. Check the insulated brush holders for shorts to ground. If the brushes are worn down to 1/4" or less, they must be replaced.

Check the field brush connections and lead insulation. A brush kit and a contact kit are available at your local marine dealer, but all other assemblies must be replaced rather than repaired.

The armature, fields, and brush holders must be checked before assembling the starter motor. See the testing section in this chapter for detailed procedures to test the starter motor.

ASSEMBLING THE AUTOLITE

1- Clamp the armature in a vise equipped with soft jaws with the drive end **DOWN**. Slide the thrust washers onto the armature shaft. Lower the field assembly down over the armature The spring action

6-38 ELECTRICAL

against the brush is built into the retainer. Therefore, a separate brush spring is not required. With one brush attached to the end cap and the other to the frame assembly, positioning the brushes properly is not the easiest task, but it can be done with patience.

Insert the negative brush into the retainer and push it in until the back part of the spring rests on the side of the brush. This force secures the brush in the retainer in a retracted position. Lower the end cap over the frame assembly. Install the positive brush into the retainer and push the brush backward until the spring part of the retainer is on the side of the brush.

2- Work the cap over the end of the armature shaft and the brushes over the commutator. Just before the cap is completely into place against the frame, use a punch and push on the back side of each brush and the brush will snap in to ride against the commutator. The spring will then be behind the brush.

3- Align the end cap notch or mark with the mark on the frame and the upper cap mark with its matching mark. Now, install

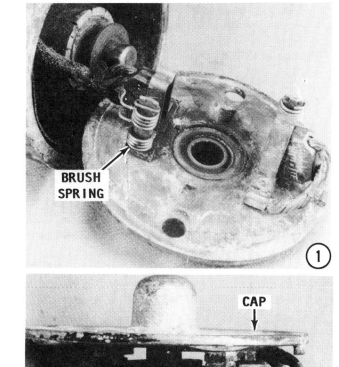

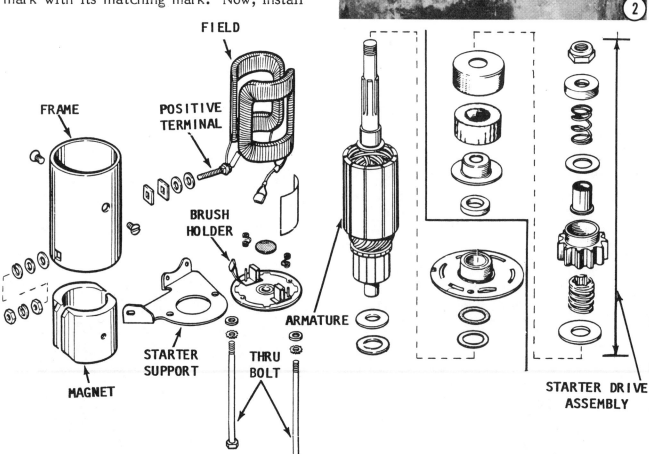

Exploded view showing arrangement of major Autolite starter motor parts.

PRESTOLITE STARTER 6-39

6-15 PRESTOLITE SERVICE

REMOVAL

1- Before beginning any work on the starter motor, disconnect the positive (+) lead from the battery terminal. Remove the engine hood. Disconnect the red cable at the starter motor terminal. Remove the 1/2" bolt securing the starter bracket. This bolt is located on the starboard side of the engine just above the carburetor. Remove the three 7/16" bolts (or nuts in some cases) securing the starter motor bracket to the engine. Remove the starter motor and bracket together.

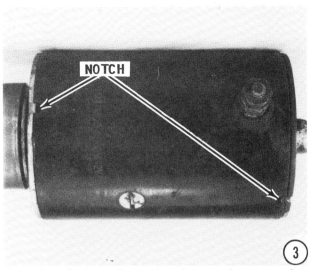

GOOD NEWS

If the only motor repair necessary is replacement of the brushes, the drive gear does not have to be removed. All starter motors have thru-bolts securing the upper and lower cap to the field frame assembly. In all cases both caps have some type of mark or boss. These marks are used to properly align the caps with the field frame assembly.

the thru-bolts through the end cap, the frame, and thread them into the upper cap.

4- Slide the rubber spacer or collar into the starter bracket, if used. Now, install the starter bracket over the starter motor. Install the thru-bolts through the end cap, the frame, and thread them into the starter bracket. Tighten the thru-bolts securely.

To test the complete starter motor, proceed directly to Section 6-16.

To install the starter motor onto the engine, proceed directly to Section 6-17.

6-40 ELECTRICAL

DISASSEMBLING

2- Observe the caps and find the identifying mark or boss on each. If the marks are not visible, make an identifying mark prior to removing the thru-bolts as an essential aid during asembling. Remove the thrubolts from the bracket and the starter motor.

3- Use a small hammer and **CAREFULLY** tap the lower cap free of the starter motor.

4- Pull on the armature shaft from the drive gear end and remove it from the field frame assembly. Remove the brushes from their holders, and then remove the brush springs. Lift the white plastic retainer free from the frame. Observe the location of the notch on the retainer in relation to the frame. The retainer must be installed in the same position.

ARMATURE TESTING

Testing for a Short

1- Position the armature on a growler, then hold a hacksaw blade over the armature core. Turn the growler switch to the

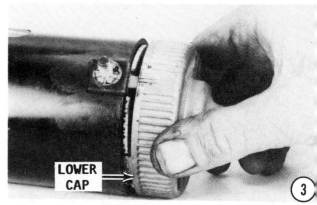

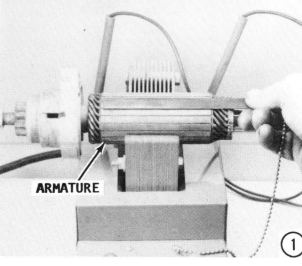

ON position. Slowly rotate the armature. If the hacksaw blade vibrates, the armature or commutator has a short. Clean the grooves between the commutator bars on the armature. Perform the test again. If the hacksaw blade still vibrates during the test, the armature has a short and **MUST** be replaced.

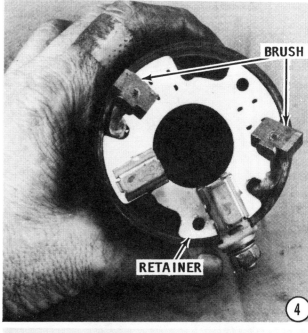

PRESTOLITE STARTER 6-41

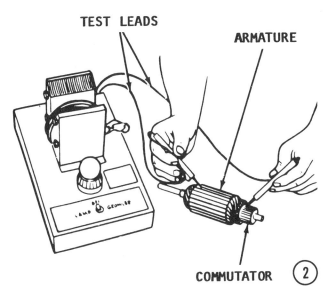

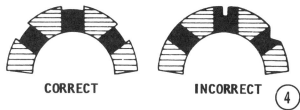

Testing for a Ground

2- Obtain a test lamp or continuity meter. Make contact with one probe lead on the armature core and the other probe lead on the commutator bar. If the lamp lights, or the meter indicates continuity, the armature is grounded and **MUST** be replaced.

Checking the Commutator Bar

3- Check between or check bar-to-bar as shown in the accompanying illustration. The test light should light, or the meter should indicate continuity. If the commutator fails the test, the armature **MUST** be replaced.

Turning the Commutator

4- True the commutator, if necessary, in a lathe. **NEVER** undercut the mica because

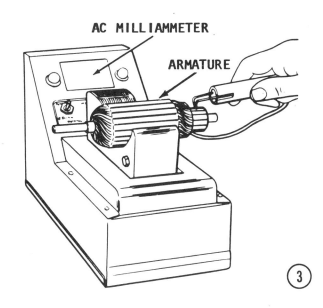

the brushes are harder than the insulation. Undercut the insulation between the commutator bars 1/32" to the full width of the insulation and flat at the bottom. A triangular groove is not satisfactory. After the under-cutting work is completed, clean out the slots carefully to remove dirt and copper dust. Sand the commutator lightly with No. 00 sandpaper to remove any burrs left from the undercutting. Check the armature a second time on the growler for possible short circuits.

Positive Brushes

5- Notice how the positive brush lead is attached to the terminal on the end of the frame. This is the same terminal to which the heavy battery cable is attached. The terminal may be removed from the frame. Pull the terminal free of the frame.

Obtain an ohmmeter. Connect one test lead of an ohmmeter to the brush and the other test lead to the terminal. Continuity should be indicated on the ohmmeter. If continuity is not indicated, the brush must be replaced. The brush and terminal are sold as an assembly, eliminating the necessity for soldering.

Negative Brushes

6- The complete terminology for Prestolite negative brushes is: Field Coil --Negative Brush.

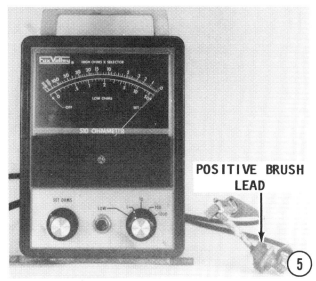

Obtain an ohmmeter. Make contact with one test lead to the negative brush and make contact with the other lead to the starter frame. If the meter does not indicate continuity, the field coils are open and **MUST** be replaced.

Check to be sure of the soldered connections are **NOT** touching the frame. The fields must not be grounded. If the connections make contact with the frame, the fields would be grounded.

CLEANING AND INSPECTING

Clean the field coils, armature, commutator, armature shaft, brush-end plate and drive-end housing with a brush or compressed air. Wash all other parts in solvent and blow them dry with compressed air.

Inspect the insulation and the unsoldered connections of the armature windings for breaks or burns.

Perform electrical tests on any suspected defective part, according to the procedures outlined earlier in this Section.

Check the commutator for run-out. Inspect the armature shaft and both bearings for scoring.

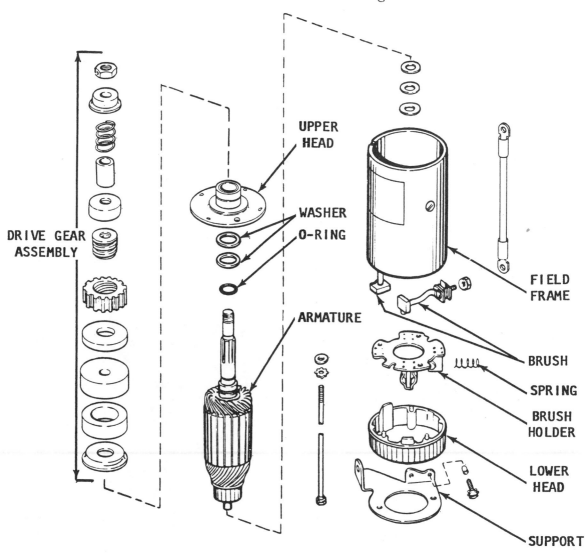

Exploded view showing arrangement of major Prestolite starter motor parts.

Turn the commutator in a lathe if it is out-of-round by more than 0.005".

Check the springs in the brush holder to be sure none are broken. Check the spring tension and replace if the tension is not 32-40 ounces. Check the insulated brush holders for shorts to ground. If the brushes are worn down to 1/4" or less, they must be replaced.

Check the field brush connections and lead insulation. A brush kit and a contact kit are available at your local marine dealer, but all other assemblies must be replaced rather than repaired.

The armature, fields, and brush holders must be checked before assembling the starter motor. See the testing section in this chapter for detailed procedures to test the starter motor.

ASSEMBLING THE PRESTOLITE

1- Slide the plastic terminal and brush lead retainer into the groove in the frame with the small protrusion on one side facing **DOWNWARD**. Continue pushing the retainer into the groove until it is fully seated.

Work the brush retainer down on top of the frame with the postive lead through the cut-a-way in the retainer plate. Check to be sure the field coil negative brush passes through the cut-a-way in the plate.

2- Install the spring into the retainer. Push the negative brush into its retainer and then, wrap a fine piece of wire around the front side of the brush and the back side of the retainer. Tighten the wire snugly. This wire will hold the brush in the retainer. Repeat the procedure for the positive brush.

Check to be sure the plate is secured onto the frame and the cut-a-way is over

the protrusion of the positive plastic terminal.

Clamp the armature in a vise equipped with soft jaws with the drive gear facing **DOWNWARD**. Install the thrust washers onto the end of the armature shaft. Lower the frame assembly down over the armature until the brushes are over the commutator.

3- After the armature is in place, cut and remove the wire wrapped around the brushes to hold them in place. The brushes should then make firm contact with the commutator.

4- Install the end cap onto the end of the starter motor. Observe three small nipples on the inside of the end cap. These nipples **MUST** index with matching dimples in the retaining plate. Align the mark on the side of the end cap with the terminal. Lower the cap onto the frame, and seat it **GENTLY**. **NEVER** tap with a hammer or other tool, because the nipples may not be indexed with the dimples and the tapping may cause damage.

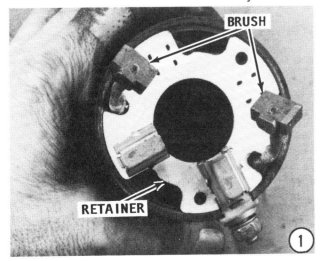

6-44 ELECTRICAL

Align the end cap notch or mark with the mark on the frame and the upper cap mark with its mark.

Slide the rubber spacer or collar into the starter bracket, if used. Now, install the starter bracket over the starter motor. Install the thru-bolts through the end cap, the frame, and thread them into the starter bracket. Tighten the thru-bolts securely.

To test the assembled starter motor, see the next section.

To install the starter motor, proceed directly to Section 6-17.

6-16 STARTER MOTOR TESTING

Clamp the starter motor in a vise equipped with soft jaws, as shown, and test its operation.

CAUTION: The armature will turn rapidly during this test. Therefore, the starter motor **MUST** be well **SECURED** before making the test to prevent personal **INJURY** or damage to the starter motor.

Firmly connect one end of a heavy-duty jumper wire to the **POSITIVE** terminal of a battery. Firmly connect the other end of the jumper lead to the starter motor terminal.

Connect a second heavy-duty jumper wire to the negative terminal of the battery. Now, **MOMENTARILY** make contact with the other end of the second jumper lead anywhere to the frame of the starter motor. **NEVER** make the momemtary contact with the positive lead to the terminal, because any arcing at the terminal may damage the terminal threads and the nut may not take to the damaged threads. The motor should turn rapidly. If the starter motor fails to rotate, the starter motor must be disassembled again and the service work carefully checked. Sorry about that, but some phase of the rebuild task was not performed properly.

6-17 STARTER MOTOR INSTALLATION

Mount the starter motor and bracket onto the engine. Align the top bolt above the carburetor, and then thread it into the block about half-way. Align the other three bolts on the starboard side or start the nuts onto the studs, depending the model engine being serviced. Tighten the three on the side evenly and alternately until they are secure.

Connect the positive red lead to the starter motor. Connect the electrical lead to the battery. Test the completed work by cranking the engine with the starter motor.

DO NOT, under any circumstances, start the engine unless it is mounted in a test tank or body of water.

CAUTION: Water must circulate through the lower unit to the engine any time the engine is run to prevent damage to the water pump in the lower unit. Just five seconds without water will damage the water pump.

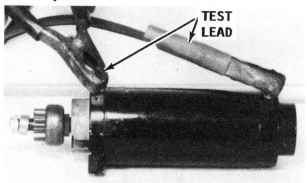

Hook-up to test an assembled starter motor.

Typical starter motor installation.

7
ACCESSORIES

7-1 INTRODUCTION

Boat accessories are seldom obtained from the original equipment manufacturer, except in the case of the electric shift unit. The electric shift box is considered a part of the new engine. Therefore, unless an owner made a change, the electric shift unit with the engine is probably original engine manufacturer equipment. Mechanical shift units are sold and installed separately.

Shift boxes, steering, bilge pumps, blowers, and other similar equipment may be added after the boat leaves the plant. Because of the wide assortment, styles, and price ranges of such accessories, the distributor, dealer, or customer has a wide selection from which to draw, when outfitting the boat.

Therefore, the procedures and suggestions in this chapter are general in nature in order to cover as many units as possible, but still specific and in enough detail to allow troubleshooting, repair, and adjustment for each of these accessories. Proper operation will do much for maximum comfort, performance, and enjoyment.

Complete procedures for removal, installation, and adjustment of four shift arrangements are covered in this Chapter: The old-style manual shift; new updated manual shift; electric shift; and the pushbutton shift mechanism. These shift boxes are all considered original Johnson/Evinrude equipment.

7-2 SHIFT BOXES DESCRIPTION

Undoubtedly, the most used accessory on any boat is the the shift control box. This unit is a remote-control device for shifting the outboard and at the same time controlling the throttle. The shift box on OMC equipped boats is considered an accessory, except in the case of the electric shift. Therefore, many installations may have other than a factory installed unit. OMC equipped boats may be equipped with one of four different type shift boxes: The old-style manual shift; new updated manual shift; electric shift; and the pushbutton shift arrangement.

The mechanical shift box units have two levers, a long lever handle and a short lever handle. The long handle controls the throttle and the short one the shift mechanism. The electric shift units, including the pushbutton models, have only one lever handle for control of the shift and throttle. The shift box installed with Johnson engines has one handle for shifting, and another at the

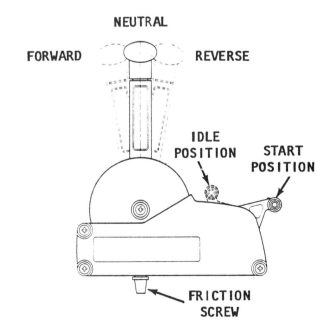

Single lever electric shift box used only on the Johnson units. These boxes incorporate a warm-up throttle lever at the rear and a friction screw on the bottom to hold the throttle position after the operator releases the handle.

rear of the box considered a "warmup" lever. This warmup lever may be adjusted for low and fast idle speeds. The pushbutton type locks out the shift lever to prevent shifting if the throttle is advanced too far while the engine is in neutral.

The old style mechanical shift boxes were very simple in design. The lower end of the shift lever handle incorporated a rocker-type arrangement with teeth and a nylon sleeve-type slider with teeth. As the handle moves, the rocker "walks" in the teeth of the slider, pulling on the inner throttle cable or the shift cable, for shifting or advancing the throttle, depending on which lever is being actuated.

Several accessories have been added to the updated mechanical shift box. This new box still uses the two lever principle for the throttle and shift. A friction feature on the throttle mechanism permits the operator to release his grip on the lever handle without the throttle changing position. An idle stop is also built into the shift box. This feature prevents the throttle from being retracted past normal idle to the point where the engine would shut down.

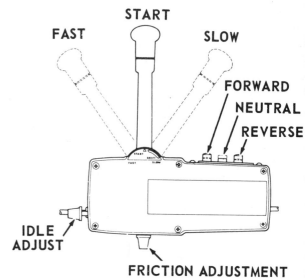

Single lever electric shift box installed with the Evinrude units. This box incorporates pushbuttons for shift control, an idle adjustment at the front, and a friction adjustment on the bottom to hold the throttle position.

Outboard models are equipped with a cut-out switch in the cranking system to open the circuit to the starter solenoid. This arrangement prevents the cranking system from operating unless the throttle is

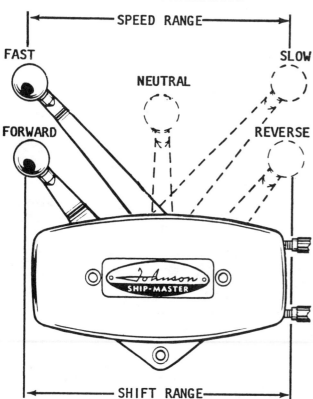

Double lever manual shift box installed with early model Johnson units. This box is no longer available. However some replacement parts may still be in stock at some dealers.

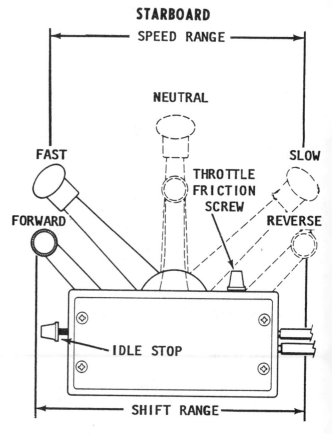

Double lever shift box installed with manual shift lower unit engines. This shift box is still in use and parts are available from OMC dealers.

DOUBLE SHIFT LEVER 7-3

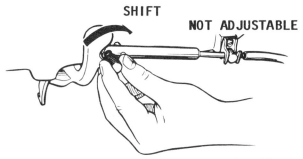

Attaching a shift cable, with non-adjustable trunnion, to the shift arm.

in the proper idle range. Stating it another way, the throttle **MUST** be in the idle position or the starter system will not operate. The position of the shift lever does not affect the starting motor circuit. All shift box models have a means of advancing the throttle without moving the shift lever into gear. This device is commonly known as the "warm-up" lever and may be adjusted for low and fast idle speeds.

7-3 OLD-STYLE DOUBLE LEVER

TROUBLESHOOTING

The following paragraphs provide a logical sequence of tests, checks, and adjustments, designed to isolate and correct a problem in the shift box operation.

The procedures and suggestions are keyed by number to matching numbered illustrations as an aid in performing the work.

The two-lever shift boxes are fairly simple in construction and operation. Seldom do they fail creating problems, requiring service more than normal lubrication.

Hard Shifting or Difficult Throttle Advance

Remove the throttle and shift control at the engine. Now, at the shift box, attempt to move the throttle or shift lever. If the lever moves smoothly, without difficulty, the problem is immediately isolated to the

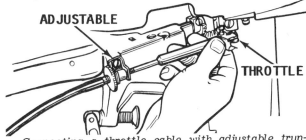

Connecting a throttle cable with adjustable trunnion.

engine. Attempt to identify if the problem is in the throttle or the shift side of the shift box. The problem may be in the tower shaft between the connector of the throttle and the armature plate or the armature plate may be "frozen", unable to move properly.

If the problem with shifting is at the engine, the first place to check is the area where the shift lever extends through the exhaust housing. The bushing may be worn, or corroded. If the bushing requires replacement, the engine powerhead must be removed. Another cause of hard shifting is water entering the lower unit. In this case the lower unit must be disassembled and the problem corrected. See Chapter 8.

If hard shifting is still encountered at the shift box when the controls are disconnected from the engine, the cables may be corroded, and require replacement, or lack of lubrication in the shift box has resulted in excessive wear, or corrosion.

Unable to Obtain Full Shift Movement or Full Throttle

Normally, this type of problem is the result of improper shift box installation. This area includes connection of the shift and throttle cables in the shift box. If the

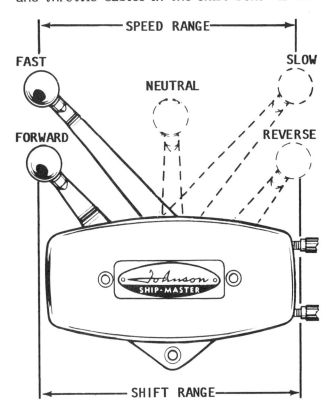

Johnson early model double lever shift box. This box is no longer available.

7-4 ACCESSORIES

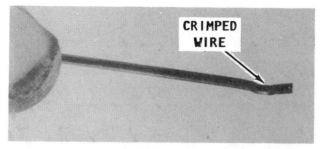

Inner shift wire after it has been removed showing the proper type crimp necessary to hold the adjustment.

stainless steel inner wire was not heated and the clamp did not hold the inner cable (wire), the wire could slip inside the sleeve thus shortening the cable. Therefore, if it is not possible to obtain full shift or full throttle, the shift box must be removed, opened, and checked for proper installation work. The inner wire could also slip at the engine end of the control, but problems at that end are very rare. Usually if improper installation work has been done at the engine end, the ability to shift at all is lost, or the throttle cannot not be actuated.

Failure to obtain full movement of the throttle or shift lever may be caused by worn or broken teeth on the slider or on the rocker of the shift handle. This type of damage results in the mechanism jumping a tooth and loosing its "timing".

DISASSEMBLING

Removing Single and Double Lever Shift Boxes

1- Remove the attaching hardware securing the shift box to the side of the boat. Once the shift box is free, the service work may be performed in the boat. The cables may remain as routed. Remove the two screws, on the side at the rear of the shift box holding the two halves together. Separate the two halves.

OBSERVE

Observe how one side accommodates the throttle and the other side the shift mechanism. Notice the plastic plate between the two halves. This plate prevents any contact between the shift parts and those for the throttle.

Throttle Half Disassembling

2- Remove the screw attaching the throttle handle to the shift box. Lift the throttle handle and rocker free of the shift box. If the handle is to be replaced, the

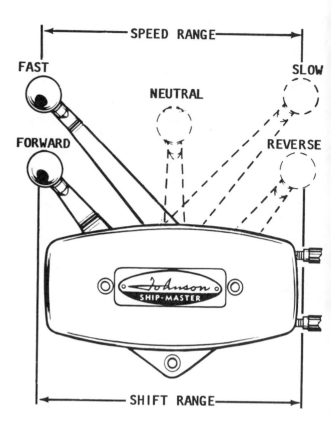

shift ball on the end of the handle must be removed and **SAVED** because a new ball is not included with a new handle or a new rocker.

3- Remove the two Allen screws, securing the ratchet to the end of the shift cable,

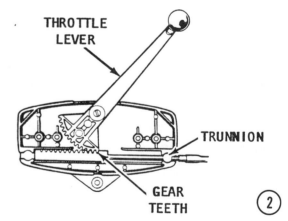

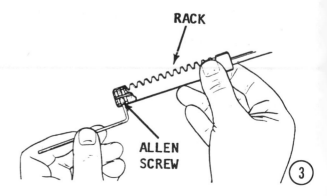

and then pull the ratchet free of the cable end. Take care not to loose the small brass sleeve.

GOOD WORDS

As the ratchet is removed from the cable end, take notice of exactly where the Allen screws made contact with the cable as an aid during installation. Remove the small sleeve from the ratchet.

Shift Half Disassembling

4- Lift the lever handle asembly free of the shift box.

5- Remove the ratchet from the end of shift cable. This is accomplished by loosening the Allen screws in the sleeve and pulling the ratchet free of the cable. Take care not to loose the small brass sleeve.

CLEANING AND INSPECTING

Check to be sure the teeth on the rocker of the shift and throttle handle are not worn or damaged. Inspect the ratchets removed from the end of the cables. The teeth should not be damaged or worn excessively.

Wash the outside and inside of the shift box halves with solvent and dry them thoroughly with a cloth or compressed air.

Throttle Cable Lubrication

6- If the throttle or shift cables are not to be replaced, now is an excellent time to lubricate the inner wires. To lubricate the inner wire, remove the casing guide from

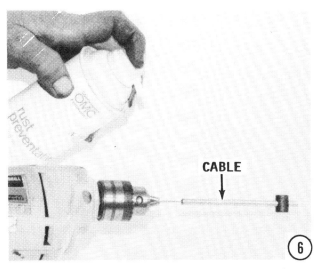

the cable at both ends. Attach an electric drill to one end of the wire. Momentarily turn the dirll on and off to rotate the wire and at the same time allow lubricant to flow into the cable, as shown.

ASSEMBLING SHIFT BOX

CRITCAL WORDS

Location of the cable end is of the upmost importance. One Allen screw must be tightened hard, until there is a definite crimp in the wire. If the Allen screw is not tightened enough, the cable will slip in the sleeve and the adjustment will be lost.

1- On some shift and throttle cables not manufactured by Johnson/Evinrude, the inner cable is made of stainless steel. When working with one of these wires, the end must be heated to remove the temper and thus allow the Allen screw to make a crimp in the wire. If the end of a stainless steel wire has a blue color, the temper has been removed. **TAKE CARE** not to overheat the stainless steel wire with a torch because it has a very low melting point.

Throttle Half Assembling

2- Insert the small sleeve onto the end of the ratchet with the two Allen screws started in the sleeve. Notice how the sleeve has a hole completely through it. The shift

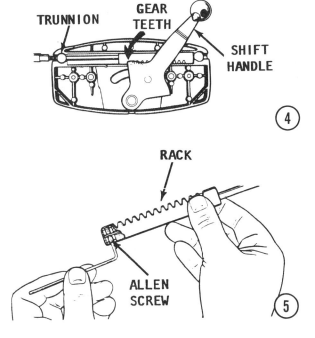

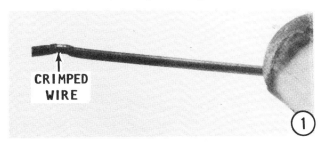

7-6 ACCESSORIES

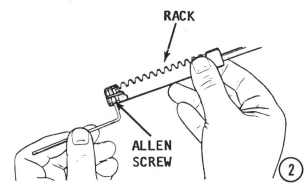

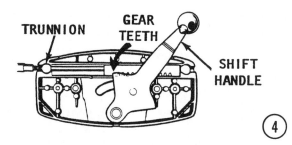

wire **MUST** pass through the hole and protrude out the end of the ratchet. Work the ratchet onto the end of the shift cable. Continue working the ratchet onto the cable until the inner cable end is completely through the ratchet, but just flush with end surface of the ratchet. Tighten one Allen screw until a definite crimp is made in the cable, then tighten the other Allen screw securely.

3- Apply a coating of lubricant around the pivot point of the throttle lever bushing. Set the lever in position inside the box half. Install the washer with the concave side of the washer on the same side as the screw. Install the screw. Move the throttle lever back-and-forth and check for freedom of movement.

Slip the ratchet and trunion into the shift box starting the last tooth on the throttle rocker engaged with the last tooth of the slider. Again, move the throttle lever back-and-forth and check for freedom of movement. Check to be sure the same number of teeth are engaging on the rocker as on the slider. The teeth indexing is extremely important and the key to a successful installation. The ratchet and rocker both have the same number of teeth. All teeth must be used on the ratchet and rocker when the lever handle is moved to maximum forward or aft.

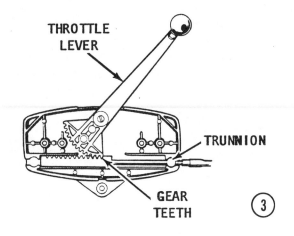

Shift Half Assembling

4- Lay the shift ratchet in position inside the shift half of the box. Place the shift lever down over the top of the ratchet, with the last tooth on the lever rocker engaged with the last tooth of the slider.

Apply a coating of light lubricant onto the surface of the gear teeth; the bottom and side walls of the rachet; and onto the full length of the gear rack. Do not use a hard grease because it will ring and groove, leaving the assembly dry.

Install the divider separating the throttle and shift halves of the box. Bring the two box halves together, and then install the retaining screws. Double check operation of both levers for smoothness and no evidence of binding.

Install the shift box onto the side of the boat and secure it in place with the attaching hardware.

7-4 NEW-STYLE SHIFT LEVER SERVICE

TROUBLESHOOTING

The following paragraphs provide a logical sequence of tests, checks, and adjustments, designed to isolate and correct a problem in the shift box operation.

The procedures and suggestions are keyed by number to matching numbered illustrations as an aid in performing the work.

The two-lever shift boxes are fairly simple in construction and operation. Seldom do they fail creating problems requiring service in addition to normal lubrication.

Hard Shifting or Difficult Throttle Advance

Checking Throttle Side

Remove the throttle and shift control at the engine. Now, at the shift box, attempt to move the throttle or shift lever. If the lever moves smoothly, without difficulty, the problem is immediatly isolated to the

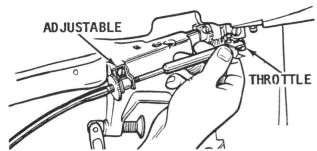

Connecting a throttle cable with adjustable trunnion.

engine. The problem may be in the tower shaft between the connector of the throttle and the armature plate. The armature plate may be "frozen", unable to move properly.

If the problem with shifting is at the engine, the first place to check is the area where the shift lever extends through the exhaust housing. The bushing may be worn, or corroded. If the bushing requires replacement, the engine powerhead must be removed, see Chapter 3. Another cause of hard shifting is water entering the lower unit. In this case the lower unit must be disassembled, see Chapter 8.

If hard shifting is still encountered at the shift box when the controls are disconnected from the engine, the cables may be corroded, and require replacment, or lack of lubrication in the shift box has resulted in excessive wear, or corrosion.

Unable to Obtain Full Shift Movement or Full Throttle

Normally, this type of problem is the result of improper shift box installation. This area includes connection of the shift and throttle cables in the shift box. If the stainless steel inner wire was not heated and the clamp did not hold the inner cable (wire), the wire could slip inside the sleeve

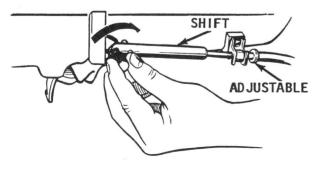

Attaching a shift cable, with non-adjustable trunnion, to the shift arm.

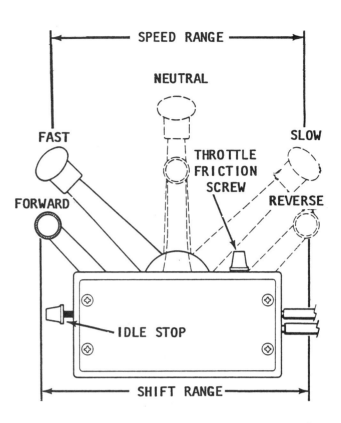

Double lever shift box installed with manual shift lower unit engines. This shift box is still in use and parts are available from OMC dealers.

and the cable would be shortened. Therefore, if it is not possible to obtain full shift or full throttle, the shift box must be removed, opened, and checked for proper installation work. The inner wire could also slip at the engine end of the control, but problems at that end are very rare. Usually if improper installation work has been done at the engine end, the ability to shift at all is lost, or the throttle cannot not be actuated.

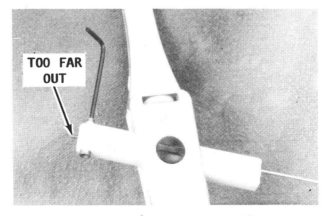

Wire extended too far through the cable connector (slider). The end of the wire should be flush with the slider surface.

ACCESSORIES

REMOVAL

Removing Double Lever Shift Box

1- Remove the attaching hardware securing the shift box to the side of the boat. Once the shift box is free, the service work may be performed in the boat. The cables may remain as routed. Remove the two screws, at the rear side of the shift box, holding the two halves together. Separate the two halves.

OBSERVE

Observe how one side accommodates the throttle and the other side the shift mechanism. Notice the metal plate between the two halves. This plate prevents any contact between the shift parts and those for the throttle. Notice the friction screw and throttle stop on the throttle side of the box.

The shift side of the box does not have any adjustments, except for the low idle stop. Observe how the shift lever pivots at the bottom and the throttle lever pivots at the top.

DISASSEMBLING

Throttle Half

2- Remove the screw from the center of the throttle lever. Notice how the washer has a concave side to allow the screw to fit flush with the washer. Loosen the screw on the top side of the shift box to relieve

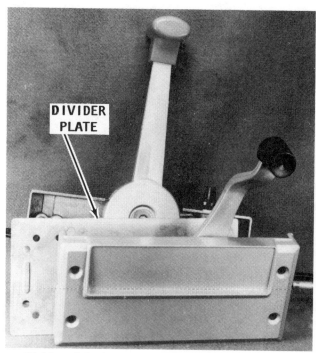

Divider plate to separate the shift half components from the throttle half parts inside the control box.

pressure on the anti-friction knob. Lift the lever and throttle cable free of the shift box. Notice how the cable on the trunnion has two small caps --one on the underside and the other on top.

3- Loosen the two screws securing the gear rack to the end of the throttle cable and remove the end of the cable from the throttle lever. Take care not to lose the

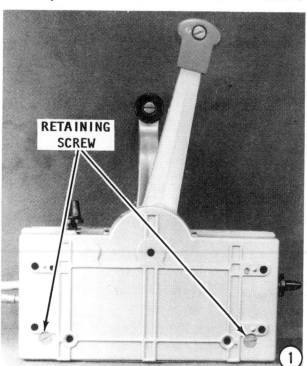

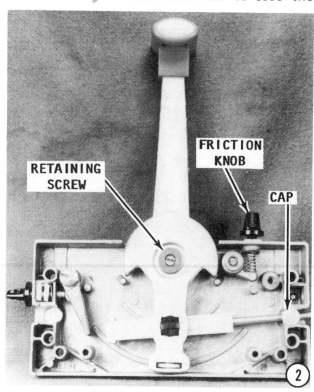

5- Loosen the two screws securing the gear rack to the end of the throttle cable and remove the end of the cable from the throttle lever. Take care not to lose the small sleeve from the end of the rack to which the screws were attached. Push out the center square button, and then remove the rack from the shift lever.

CLEANING AND INSPECTING

Check the nylon wear block on the end of the anti-friction cap. The cap has teeth which index into the inside diameter of the throttle lever. If the teeth are damaged a new block may be purchased and slipped into place. Clean the box halves thoroughly inside and out with solvent, and then dry them with compressed air. Inspect the spring on the anti-friction lever to be sure it is not distorted. Check the screw on the throttle idle stop to ensure it moves in and out freely without any sign of binding.

Throttle Cable Lubrication

If the throttle or shift cables are not to be replaced, now is an excellent time to lubricate the inner wire.

6- To lubricate the inner wire, remove the casing guide from the cable at both ends. Attach an electric drill to one end of

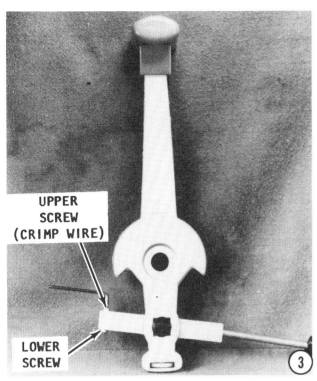

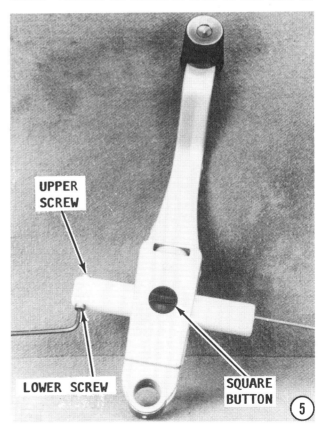

small sleeve from the end of the rack to which the screws were attached. Push out the center square button, and then remove the rack from the shift lever.

Shift Half Disassembling

4- Remove the screw and washer from the bottom of the shift lever. Notice how this washer also has a concave side to accommodate the screw. Lift the lever and shift cable free of the shift box. Notice how the cable on the trunnion has two small caps -- one on the underside and the other on top.

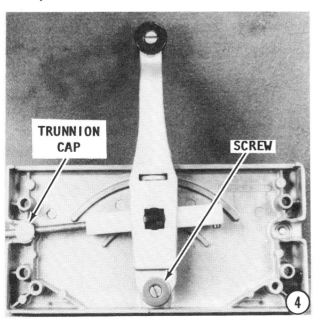

7-10 ACCESSORIES

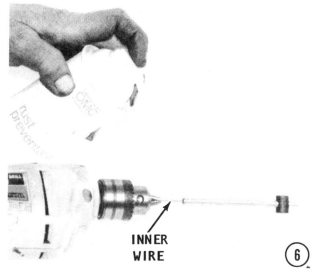

the wire. Momentarily turn the drill on and off to rotate the wire and at the same time allow lubricant to flow into the cable, as shown.

ASSEMBLING

Throttle Cable into Shift Box

1- If the slider sleeve was removed from the throttle lever, install the ratchet into the throttle lever with the hole on the end for securing the throttle cable on the opposite end of the hole that accommodates the cable. Position the center of the ratchet with the center of the throttle lever. Install the square nylon plug with the holes in the plug in a vertical position to permit the cable to slide through. Two different size screws, or possibly Allen screws, are used on each end of the sleeve. Install the short screw into the bottom of the sleeve to prevent the sleeve from rubbing on the shift box. Install the longer screw on the top part of the sleeve. Install the sleeve with the hole in the sleeve aligned with the hole for the cable.

CRITICAL WORDS

Check the end of the cable to determine if the temper has been removed. If the end has a bluish appearance, it has been heated at an earlier date and the temper removed. The temper **MUST** be removed to permit the holding screw to make a crimp in the wire to hold an adjustment. If the wire has not been tempered, heat the end, but not enough to melt the wire.

2- Slide the cable into the ratchet. Work the inner wire into the sleeve and out the end of the ratchet. Push the wire back until the end is flush with the ratchet surface. Tighten the **TOP** holding screw enough to make a definite crimp in the wire, as shown. If this screw is not tightened to make the crimp, the wire will slip during operation and the adjustment will be lost. After the top screw has been fully tightened, bring the other screw up tight against the wire. It is not necessary for this second screw to make a crimp in the wire.

3- Work the throttle lever handle down over the friction nylon block and at the same time feed the throttle cable into place in the box half. Check to be sure one of the small caps is on the bottom side of the trunnion. Install the washer and the screw with the concave side of the washer on the same side as the screw. Tighten the screw securely. Install the other trunnion cap on top of the trunnion. Check the throttle lever for ease of movement with no sign of binding.

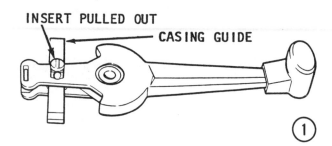

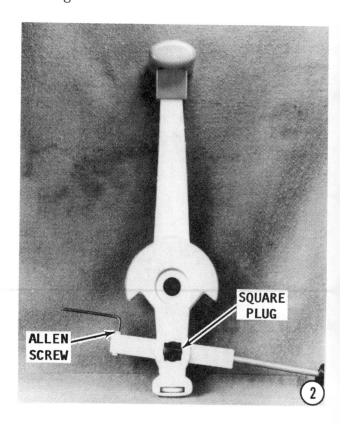

NEW SHIFT LEVER 7-11

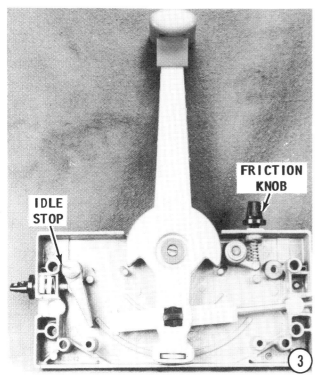

Shift Cable into Shift Box

4- If the slider sleeve was removed from the shift lever, install the slide rack into the shift lever with the hole on the end for securing the shift cable on the opposite end of the hole that will accommodate the cable. Position the center of the slide with the center of the shift lever. Install the square nylon plug with the holes in the plug in a vertical position to permit the cable to slide through. Two different size screws are used on each end of the sleeve. Install the short screw into the bottom of the sleeve to prevent the sleeve from rubbing on the shift box. Install the longer screw on the top part of the sleeve. Install the sleeve with the hole in the sleeve aligned with the hole for the cable.

CRITICAL WORDS

Check the end of the cable to determine if the temper has been removed. If the end has a bluish appearance, it has been heated at an earlier date and the temper removed. The temper **MUST** be removed to permit the holding screw to make a crimp in the wire to hold an adjustment. If the wire has not been tempered, heat the end, but not enough to melt the wire.

Slide the cable into the rack. Work the inner wire into the sleeve and out the end of the rack. Push the wire back until the end is flush with the rack surface. Tighten the **TOP** holding screw enough to make a definite crimp in the wire, as shown. If this screw is not tightened to make the crimp, the wire will slip during operation and the adjustment will be lost. After the top screw has been fully tightened, bring the other screw up tight against the wire. It is not necessary for this second screw to make a crimp in the wire.

5- Place the wavy washer and regular washer into the shift box, and then work the shift lever handle down into the shift box with one of the small caps under the shift

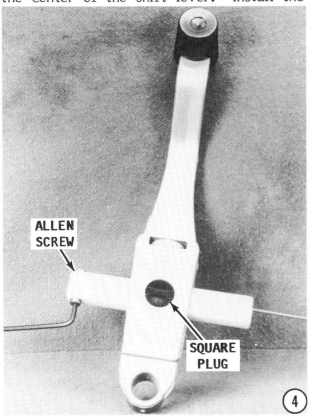

7-12 ACCESSORIES

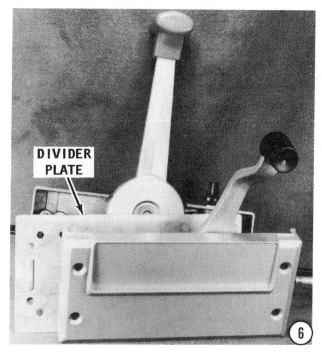

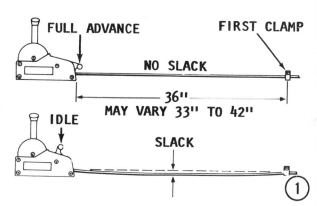

cable trunion. Install the bushing into the bottom of the shift handle. Install the washer and the screw with the concave side of the washer on the same side as the screw. Tighten the screw securely. Install the other trunion cap on top of the trunion. Check the shift lever for ease of movement with no sign of binding.

6- Place the metal separator between the two halves. Bring the two halves together and secure them with the two screws from the back side. Install the box in the boat and secure it in place with the attaching hardware. Again check the levers for ease of movement and no sign of binding.

7-5 ELECTRIC GEAR BOXES AND SINGLE LEVER CONTROL

TROUBLESHOOTING

The following paragraphs provide a logical sequence of tests, checks, and adjustments, designed to isolate and correct a problem in the Johnson single lever shift box with the warm-up lever to the rear and the Evinrude single lever pushbutton shift box operation.

The procedures and suggestions are keyed by number to matching numbered illustrations as an aid in performing the work.

1- Difficult Shift Operation

Many times this type of problem is the result of incorrect cable installation -- the cable is not the proper length or there are too many bends or kinks in the routing. Such an installation will cause the inner cable to travel much further than necessary and therefore, wear on the outer cable. Over a period of time, inner cable wear will result is difficult shifting or throttle operation.

BE SURE to cycle the shift lever to the full position in both directions, when making any test on the shift box. The shift switch may have a dead spot and will not indicate the switch is defective unless the shift lever is fully cycled for each test.

2- Amp Draw Test

Turn the ignition switch to the **ON** position and note the ammeter reading. Now, operate the shift control lever to the **FORWARD** position, and then to the **REVERSE** position. Note how much the ammeter reading increased each time the shift lever was moved. If the reading was more than 2.5 amperes for either shift positions, continue with the following checks. If the boat is not equipped with an ampere gauge, then temporarily disconnect the **GREEN** and **BROWN (or RED)** wires from the back side of the key switch and temporarily install an

ELECTRIC GEAR BOXES 7-13

amp gauge for the test. Replace the wires after the test is completed.

Disconnect the shift leads at the rear of the engine. Temporarily lay a piece of cloth or other insulating material under the wires to prevent them from shorting out during the following tests.

Again operate the shift lever and note the current loss. If the current draw is still more than 2.5 amperes, then check for a short in the control box switch or wiring. If the current draw is normal with the leads disconnected from the engine, then check for a short in the gear case coil(s) or wiring. If the coil leads are shorted to each other, both shift coils would be energized, stalling the engine or causing serious damage to the driveshaft.

3- Shift Coil Tests

Testing the shifting coils is accomplished by first disconnecting the wires at the rear of the engine. Next, connect an ohmmeter first to one shift coil lead and ground, and then to the other in the same manner. A reading of more than 4.5 to 6.5 ohms indicates a short in the coil or lead. No reading at all indicates an open circuit. If the results of this test indicate a short in the cirucit, the lower unit must be disassembled and inspected. See Chapter 8.

4- Testing Shift Switch — Forward

To test the switch for the FORWARD position, make contact with one probe of a continuity meter (or a test light) to the terminal (purple or red lead) and to the forward (green lead) terminal with the other probe. Now, move the shift lever to the FORWARD position. The meter should indicate continuity (or the test light come on),

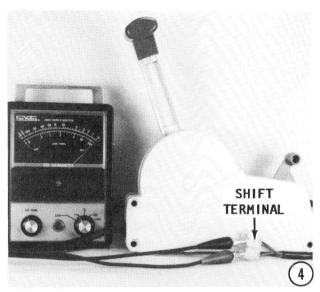

when the shift lever is in the FORWARD position.

5- Testing Shift Switch — Reverse

To test the shift switch for reverse, make contact with one probe of a continuity meter (or test light), to the terminal (purple or red lead) and to the reverse terminal (blue lead) with the other probe. Move the shift lever to the REVERSE position.

The meter should indicate continuity (or the test light come on), when the lever is in the reverse position.

6- Testing Shift Switch — Neutral

After the forward and reverse tests have been completed check for continuity with the shift lever in the NEUTRAL position. Leave the red lead connected and check the blue lead (REVERSE) and the green lead (FORWARD). Continuity should not be indicated when the shift handle is in NEUTRAL.

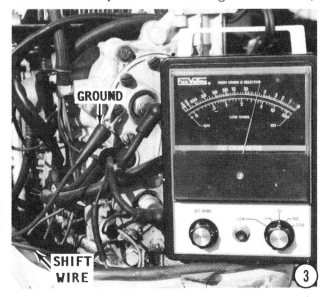

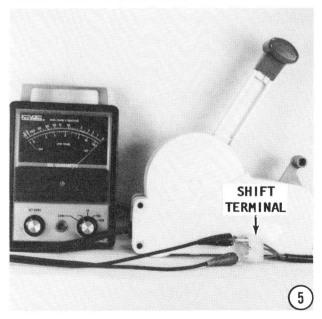

7-14 ACCESSORIES

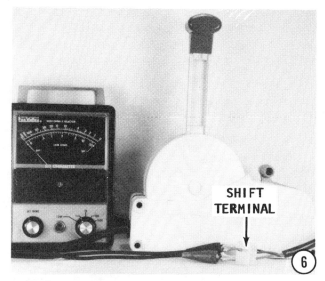

If the switch is defective and requires replacement, proceed to Section 7-6, Single Lever Shift Box Service.

7- Cranking System Inoperative

If the starter fails to crank the engine, check to be sure the throttle lever is in the idle position. If the throttle is advanced more than 1/4 forward, the cutout switch attached to the armature plate will open the circuit to the starter solenoid. If the cranking system fails to operate the starter properly when the throttle lever is in the **IDLE** position, check the 20-ampere fuse between the ingition switch **BAT** terminal and the ammeter **GEN** terminal.

If the starter operates in full throttle (which it should not do), check for a short between the two white leads in the shift box wiring.

Further problems in the cranking system may indicate more serious problems. See Chapter 6, starter motor sections.

DISASSEMBLING

GOOD WORDS

On Johnson units, a friction screw is installed in the bottom side of the shift box. This screw allows friction adjustment of the throttle handle. This arrangement prevents the throttle handle from "creeping" after the operator releases his grip on the handle. The box has a maximum advance screw. The adjustment is made through movement of a screw in the warmup lever. If the engine shuts down when the throttle lever is moved back, then an adjustment must be made at the engine. This is accomplished through an adjustment knob at the engine. Movement of this knob will actually lengthen or shorten the cable slightly for proper operation.

1- Remove the attaching screws; pull the shift box clear; and then disconnect the shift wire under the dash. Remove the screws from the back side of the box.

2- CAREFULLY separate the two halves. Check the shift box for salt water corrosion, worn bushings, and general condition.

Throttle Cable

Notice how the throttle cable enters the throttle half of the shift box through the idle link. On the top side of the shift box,

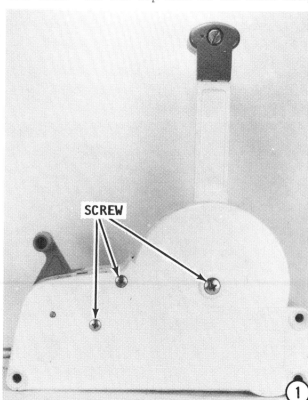

SINGLE LEVER SHIFT BOX 7-15

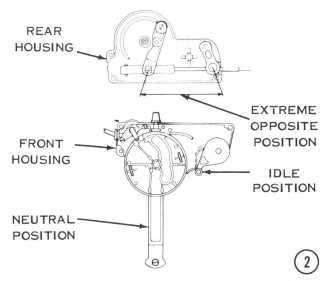

observe the screw and the concave washer. The washer must be installed with the concave side toward the screw to allow the screw to seat properly.

3- Remove the bushing from the shift rod. Observe the two slots in the bushing and how the flat area without a hole faces toward you. The bushing must be installed in this same position.

4- Remove the screw and washer from the top of the shift box half. This is the screw and washer described in the previous paragraph. Lift out the cam lever and the idle link as an assembly.

5- Remove the screws from the end of the sleeve on the end of the cable, and then pull the throttle cable free of the link and sleeve.

6- If the switch fails to check out, as described in the previous tests, the switch and cable assembly **MUST** be replaced. The switch is easily removed by simply removing the attaching screws and lifting the switch free of the shift box.

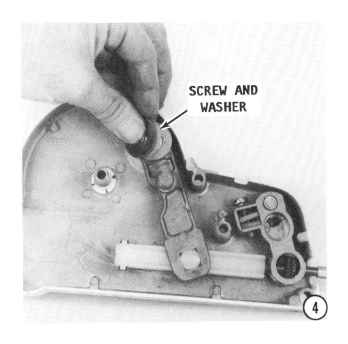

CLEANING AND INSPECTING

Clean the box halves thoroughly inside and out with solvent and blow them dry with compressed air. Apply a thin coat of engine oil on all metal parts. The three-position switch installed in the gear box cannot be repaired. Therefore, if a problem is isolated to the switch, it must be replaced.

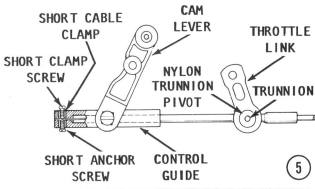

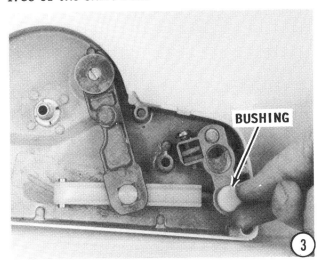

7-16 ACCESSORIES

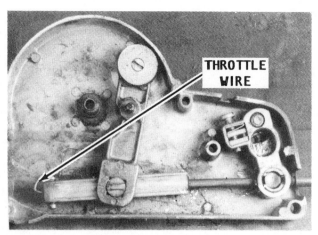

Interior view of a used shift box showing the results of an improper installation. The inner wire was not crimped to hold the adjustment. The wire, therefore, slipped through and was bent as the casting struck the shift box during operation.

Throttle Cable Lubrication

If the throttle cable is not to be replaced, now is an excellent time to lubricate the inner wire.

7- To lubricate the inner wire, remove the casing guide from the cable at both ends. Attach an electric drill to one end of the wire. Momentarily turn the drill on and off to rotate the wire and at the same time allow lubricant to flow into the cable, as shown.

ASSEMBLING

CRITICAL WORDS

Check the end of the cable to determine if the temper has been removed. If the end has a bluish appearance, it has been heated at an earlier date and the temper removed. The temper **MUST** be removed to permit the holding screw to make a crimp in the wire

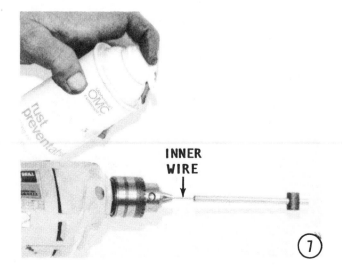

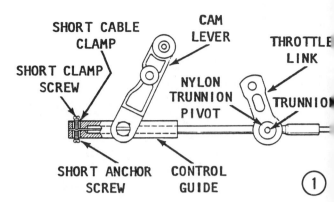

to hold an adjustment. If the wire has not been tempered, heat the end, but not enough to melt the wire.

Tighten the top screw in the sleeve until the screw makes a crimp in the wire. The screw must be tightened to this degree to prevent the wire from slipping during operation. Bring the bottom screw up tight against the wire.

1- Start the two cable retaining screws into the sleeve. These screws are different sizes. On some models, Allen screws are used. Install the short screw on the bottom to prevent the sleeve from rubbing on the shift box. Slide the sleeve onto the cable with the hole aligned with the hole in the plastic sleeve. Feed the wire on through until the end of the wire is flush with the end of the white plastic sleeve.

2- Lower the shift link and throttle link into the box half and secure the throttle link with the screw and washer. Check to be sure the concave side of the washer is facing toward the screw side to permit the screw to seat properly.

3- Slide the throttle cable through the idle link, and then slide the bushing down over the cable.

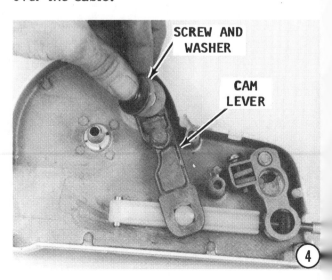

SINGLE LEVER SHIFT BOX 7-17

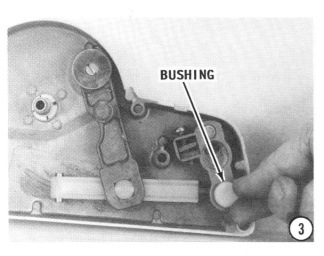

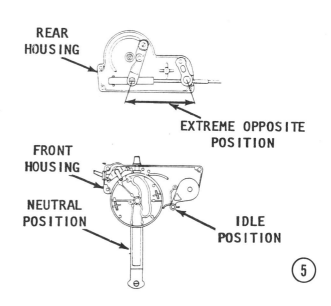

4- ALWAYS TAKE CARE when assembling the shift box, not to damage the remote-control unit. The arm on the switch **MUST** lay in the cut-out portion of the throttle cam.

5- Carefully work the two halves of the box together with the cam lever fitting into the recess of the throttle handle and the throttle link fitting into the warm-up lever.

6- Secure the two halves together with the screws into the side of the box. Secure the shift box to the side of the boat with the attaching hardware. Bolts with self-locking nuts **SHOULD ALWAYS BE USED** because a loose shift box during high speed operation could be extremely dangerous. Connect the shift wire under the dash.

7- The tension of the throttle lever is adjusted by the friction knob under the shift box. Turn the knob clockwise to increase friction and counterclockwise to decrease friction.

8- Remote-Control Cable Installation In the Boat

The remote-control cable must be installed properly for satisfactory operation. The clamp nearest the shift box **MUST** be positioned correctly as follows:

First, move the warm-up lever on the shift box to full advance. Now, measure 36" (actually this measurement could range

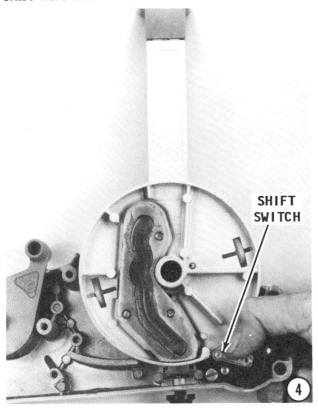

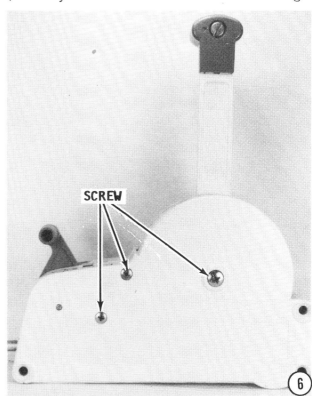

7-18 ACCESSORIES

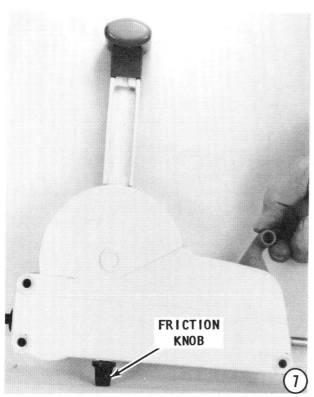

FRICTION KNOB

⑦

from 33" to 42") on the cable from the shift box. **BE SURE** there is no slack in the cable, and then secure the clamp to the boat at the measured position.

Next, place the warm-up lever in the slow position and observe the amount of slack in the cable between the shift box and the first clamp. The slack should not be more than 1/2 inch. **AVOID SHARP TURNS** in the cable. The radius of any bend **MUST** not be less than 5".

ALWAYS use the correct length of cable when replacing the assembly.

9- Adjusting Starter Lockout Switch

Move the shift lever to the **NEUTRAL** position and the warm-up lever to the **START** position. Now, turn the ignition switch to the **START** position in an attempt to crank the engine. If the starter fails to crank the engine, move the warm-up lever to the **IDLE** position, and then rotate the setscrew counterclockwise one-half turn. Move the auxiliary throttle to the **START** position, and then turn the ignition switch to the **START** position again. If the starter still fails to crank the engine with the warm-up lever in the full **ADVANCE** position, back off the warm-up lever to determine the point at which the starter ceases to operate. Make the adjustment of the setscrew clockwise a half-turn at a time until the starter operates only with the warm-up lever in the full **START** position.

7-6 PUSHBUTTON SHIFT BOX SERVICE EVINRUDE UNITS ONLY

GOOD WORDS

A friction screw is installed in the bottom side of the shift box. This screw allows friction adjustment of the throttle handle. This arrangement prevents the throttle handle from "creeping" after the operator releases his grip on the handle. A thumbscrew on the front of the box permits adjustment of the throttle handle to prevent movement past a satisfactory idle position and subsequent shutdown of the engine. If adequate adjustment cannot be made at the shift box, and the the engine continues to

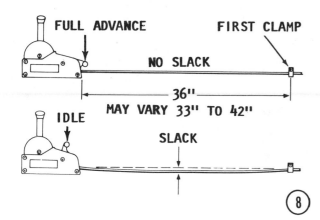

⑧

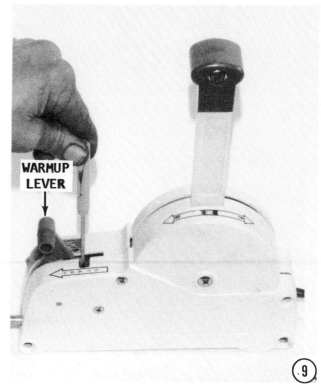

WARMUP LEVER

⑨

shut down when the throttle lever is moved back, then an adjustment must be made at the engine. This is accomplished through an adjustment knob at the engine. Movement of the engine knob will actually lengthen or shorten the cable slightly for proper operation.

TROUBLESHOOTING

The following paragraphs provide a logical sequence of tests, checks, and adjustments, designed to isolate and correct a problem in the Johnson single lever shift box with the warm-up lever to the rear and the Evinrude single lever pushbutton shift box operation.

The procedures and suggestions are keyed by number to matching numbered illustrations as an aid in performing the work.

Many times this type of problem is the result of incorrect cable installation -- the cable is not the proper length or there are too many bends or kinks in the routing. Such an installation will cause the inner cable to travel much further than necessary and therefore, wear on the outer cable. Over a period of time, inner cable wear will result is difficult shifting or throttle operation.

BE SURE to cycle all three shift pushbuttons to the full shift position in both directions, when making any test on the shift box. The shift switch may have a dead spot and will not indicate the switch is defective unless the three buttons are fully cycled for each test.

1- Amp Draw Test

Turn the ignition switch to the ON position and note the ammeter reading. Now, depress the shift button for the FORWARD position, and then for the REVERSE position. Note how much the ammeter reading increased each time a shift button was depressed. If the reading was more than 2.5 amperes for either shift positions, continue with the following checks. If the boat is not equipped with an ampere gauge, then temporarily disconnect the GREEN and BROWN (or RED) wires from the back side of the key switch and temporarily install an amp gauge for the test. Replace the wires after the test is completed.

Disconnect the shift leads at the rear of the engine. Temporarily lay a piece of cloth or other insulating material under the wires to prevent them from shorting out during the following tests.

Again operate the shift buttons and note the current loss. If the current draw is still more than 2.5 amperes, then check for a short in the control box switch or wiring. If the current draw is normal with the leads disconnected from the engine, then check for a short in the gear case coil(s) or wiring. If the coil leads are shorted to each other, both shift coils would be energized, stalling the engine or causing serious damage to the driveshaft.

2- Shift Coil Tests

Testing the shifting coils is accomplished by first disconnecting the wires at the rear of the engine. Next, connect an ohmmeter first to one shift coil lead and ground, and then to the other in the same manner. A reading of more than 4.5 to 6.5 ohms indicates a short in the coil or lead. No reading at all indicates an open circuit. If the results of this test indicate a short in the cirucit, the lower unit must be disassembled and inspected. See Chapter 8.

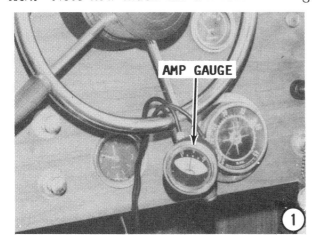

7-20 ACCESSORIES

3- Testing Shift Switch — Forward

To test the switch for the **FORWARD** position, make contact with one probe of a continuity meter (or a test light) to the terminal (purple or red lead) and to the forward (green lead) terminal with the other probe. Now, depress each pushbutton for the **FORWARD**, **NEUTRAL**, and **REVERSE** positions. The meter should indicate continuity (or the test light come on), when the **FORWARD** button is depressed, and indicate an open circuit, for the other two shift positions.

4- Testing Shift Switch — Reverse

To test the shift switch for reverse, make contact with one probe of a continuity meter (or test light), to the terminal (purple or red lead) and to the reverse terminal (blue lead) with the other probe. Again, depress the **FORWARD**, **NEUTRAL**, and **REVERSE** pushbuttons.

The meter should indicate continuity (or the test light come on), for the reverse position and indicate an open circuit, in the other two.

5- Testing Shift Switch — Neutral

After the forward and reverse tests have been completed check for continuity with the **NEUTRAL** pushbutton. Leave the red lead connected and check the blue lead **(REVERSE)** and the green lead **(FORWARD)**. Continuity should not be indicated when the **NEUTRAL** pushbutton is depreessed.

If the switch is defective and requires replacement, proceed to section 7-6, Single Lever Shift Box Service.

6- Cranking System Inoperative

If the starter fails to crank the engine, check to be sure the throttle lever is in the idle position. If the throttle is advanced

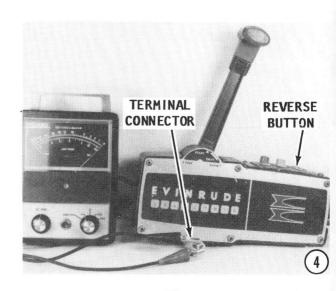

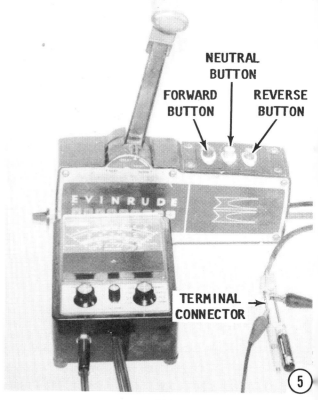

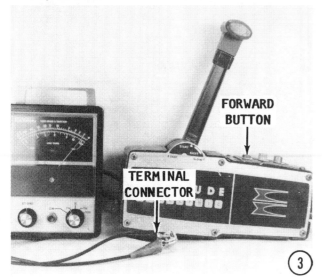

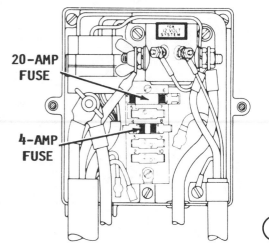

PUSHBUTTON SHIFT BOX 7-21

more than 1/4 forward, the cutout switch attached to the armature plate will open the circuit to the starter solenoid. If the cranking system fails to operate the starter properly when the throttle lever is in the **IDLE** position, check the 20-ampere fuse between the ingition switch **BAT** terminal and the ammeter **GEN** terminal.

If the starter operates at full throttle (which it should not do), check for a short between the two white leads in the shift box wiring.

Further problems in the cranking system may indicate more serious problems. See Chapter 6, starter motor sections.

DISASSEMBLING

The throttle cable and switch box may be replaced without removing the shift box from the boat. If the only service to be performed is replacement of the cable, leave the shift box in place.

1- Remove the Phillips screws on the side plate of the shift box. Remove the front side cover.

2- Notice the screw and retainer at the forward end of the casing guide and just below the throttle lever. Remove the screw and retainer. Pull the throttle cable and casting guide free of the shift box. Take care not loose the trunion caps, one on the top and another on the bottom.

3- Remove the screws from the end of the casting guide and then pull the throttle cable free. **TAKE CARE** not to loose the screws and sleeve from the end of the guide.

Shift Switch Removal

4- Remove the four Phillips screws from the top of the shift box, and then lift off the shift box cover around the push buttons.

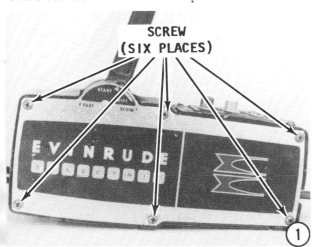

5- Pull upward and remove the three push buttons. New buttons are not supplied with replacement switches. Therefore, **SAVE** the **THREE BUTTONS** for installation with the new switch.

6- Pull the red, green, and blue wires from the bottom of the switch box.

7- Notice the two small Phillips screws on top of the switch box holding the switch to the retainer. Remove these two screws.

8- Work the switch out of the switch box.

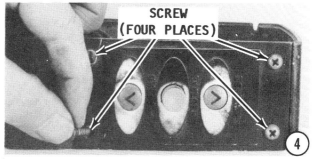

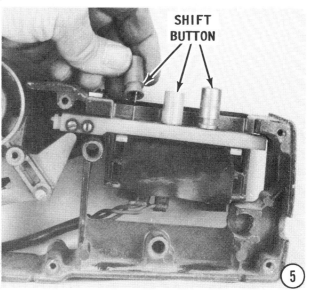

CLEANING AND INSPECTING

Clean the box halves thoroughly inside and out with solvent and blow them dry with compressed air. Apply a thin coat of engine oil on all metal parts. The three-position switch installed in the gear box cannot be repaired. Therefore, if a problem is isolated to the switch it must be replaced.

Throttle Cable Lubrication

If the throttle cable is not to be replaced, now is an excellent time to lubricate the inner wire.

9- To lubricate the inner wire, remove the casing guide from the cable at both ends. Attach an electric drill to one end of the wire. Momentarily turn the drill on and

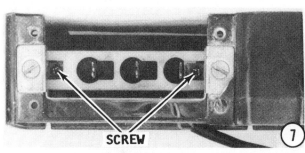

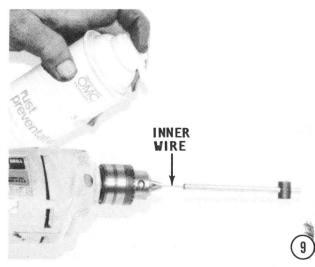

off to rotate the wire and at the same time allow lubricant to flow into the cable, as shown.

ASSEMBLING

Switch Installation

1- Position the switch box inside the shift box underneath the retainer and slider and secure it in place with the two Phillips screws. Check to be sure the two terminals on the bottom side of the switch are towards you. This will place the forward button closest to the throttle handle.

2- Install the two small Phillips screws into the top of the shift box. These two screws secure the switch box to the retainer.

3- Connect the wires to the bottom of the switch. Connect the red wire to the **POSITIVE** terminal; the green to the **FORWARD** terminal; and the blue wire to the **REVERSE** terminal.

4- Slide the buttons down over the protrusions of the switch and seat them in place.

PUSHBUTTON SHIFT BOX 7-23

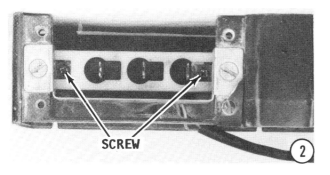

Adjustment

5- Temporarily install the side plate, and then move the throttle hand forward until the boss mark on the bottom of the throttle hand aligns with the mark on the side of the shift box panel.

6- Remove the panel and depress the three buttons one at-a-time. If it is not possible to depress the buttons, loosen the two screws on the selector bracket. Move the bracket forward or aft until the buttons can be depressed. Tighten the screws to secure the bracket in the proper position.

7- Install the shift box cover and secure it in place with the four screws.

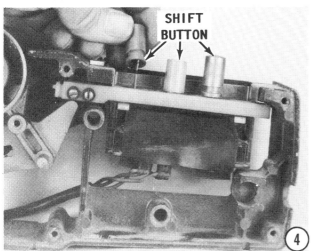

CRITICAL WORDS

If the throttle adjustment is not properly performed, the circuit to the starter solenoid will be opened preventing the starter motor from cranking the engine. Adjustment is made by moving the throttle cable adjustment knob in the trunion on the side of the engine.

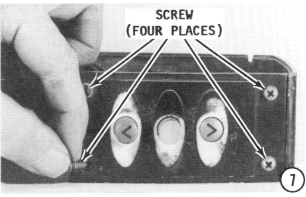

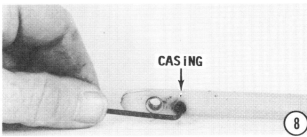

7-24 ACCESSORIES

Check the end of the cable to determine if the temper has been removed. If the end has a bluish appearance, it has been heated at an earlier date and the temper removed. The temper **MUST** be removed to permit the holding screw to make a crimp in the wire to hold an adjustment. If the wire has not been tempered, heat the end, but not enough to melt the wire.

8- Feed the inner cable into the casing guide and align it with the hole in the sleeve. Tighten the two Allen screws in the sleeve until one screw makes a crimp in the wire. The screw must be tightened to this degree to prevent the wire from slipping during operation. Bring the other Allen screw up tight against the wire.

9- Install the cable and cable end into the shift box with the trunion cap on the bottom side. Lower the cable trunion retainer into the recess and at the same time install the end of the shift cable sleeve over the end of the protrusion. Install the other trunion cap over the top of the cable retainer.

Slide the retaining clip over the end of the guide. The guide slips over a pin and the retainer has a hole in the end. The retainer fits over the pin and holds the end of the throttle cable onto the pin. Install the side plate with the attaching Phillips screws. Start the engine and run it at 700 rpm.

CAUTION: Water must circulate through the lower unit to the engine any time the engine is run to prevent damage to the water pump in the lower unit. Just five seconds without water will damage the water pump.

Now, adjust the slide yoke to allow the pushbuttons to be depressed at 700 rpm, but not at 750 rpm. If it is not possible to depress the buttons at 700 rpm, remove the side panel and loosen the two screws on the selector bracket. Move the bracket forward or aft until the buttons can be depressed. To adjust the friction knob under the shift box, turn the knob clockwise to increase friction and counterclockwise to decrease friction.

7-7 CABLE END FITTING INSTALLATION AT THE ENGINE END

FIRST, THESE WORDS

In the early days, the throttle and shift cables were installed using non-adjustable trunions. The trunions were installed on the ends of the cables and formed the connection for the cables to the engine. The inner cable (wire) moved in both directions inside the outer cable and actuated the mechanism at the engine.

The anchor on the engine, to which the trunion is attached, has a **"P"** and an **"S"** stamped on the inside diameter or inside edge of the trunion retainers. These letters identified **PORT** and **STARBOARD**.

As improvements and refinements were incorporated over the years, new cables and trunions became adjustable through the trunion.

The non-adjustable unit is totally obsolete and no longer available. Therefore, if the old-style cable with the non-adjustable trunion requires replacement, the new adjustable type will be installed.

When the new cable and trunion are to be installed, a new trunion retainer **MUST** be purchased and installed port and starboard

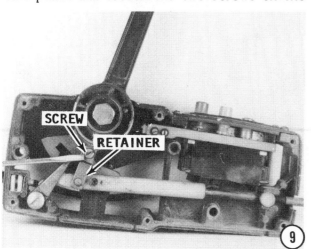

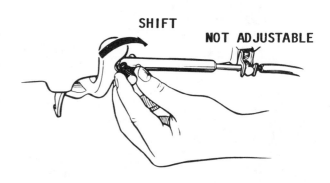

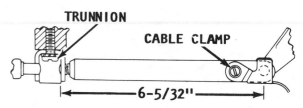

Attaching the non-adjustable shift cable to the shift arm, top, and connecting the throttle cable to the non-adjustable trunnion, bottom.

CABLE END FITTING 7-25

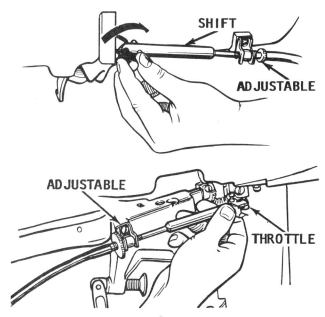

Attaching a shift cable with adjustable trunnion to the shift arm, top, and connecting the throttle cable with adjustable trunnion, bottom.

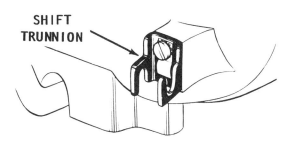

Detailed drawing of the shift trunnion at the engine.

on the engine. The new cable and trunion **CANNOT** be connected to the old-style retainer.

INSTALLATION

Shift Cable End

1- Move the control lever at the shift control box to the **NEUTRAL** position. Slide the gear shift fitting onto the control wire. Check to be sure the inner wire passes completely through the small holes in the cable clamp. Clamp the anchor screws to prevent twisting the cable. The clamp and the anchor screws **MUST** be parallel to the trunion on the gear shift cable.

2- Notice the flat and rounded areas of the casting guide. The flat edge **MUST** face **TOWARD** the engine. In this position, there is a flat area for the lever to ride during the shifting action. After the cable is in place in the casting guide, tighten the top screw until a definite crimp is made in the cable. If the screw is not tightened enough, the inner wire will slip during operation and the adjustment will be lost.

CRITICAL WORDS

Check the end of the cable to determine if the temper has been removed. If the end has a bluish appearance, it has been heated at an earlier date and the temper removed. The temper **MUST** be removed to permit the holding screw to make a crimp in the wire to hold an adjustment. If the wire has not been tempered, heat the end, but not enough to melt the wire. Bring the second screw up tight against the wire.

3- Insert the shift cable control vertically into the trunion bracket and turn the cable to a horizontal position, as indicated by the arrows in the accompanying illustration.

4- Attach the shift cable end to the shift lever on the engine by inserting the fitting into the shift control lever, and then

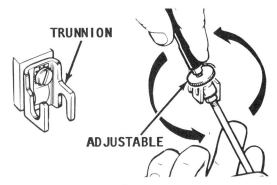

Proper installation of the throttle cable into the trunnion at the engine end.

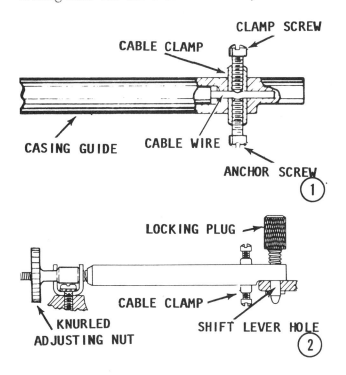

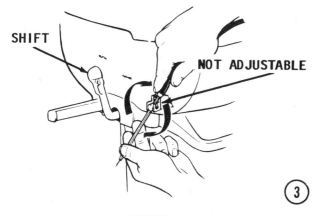

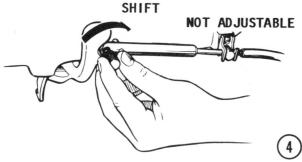

pushing inward, and at the same time rotating the fitting 1/2-turn. This action will lock the fitting in the shift lever.

Throttle Cable End Installation

5- Install the throttle lock pin spring over the casting guide. Start the screws into the small cylinder, and then slide the cylinder down through the pin spring and into the casting guide. Notice how the cylinder has a hole. This hole should be positioned vertically with the casting to align with the hole in the guide. Slide the casing guide down over the throttle cable and insert the end of the wire through the sleeve. Tighten the top screw until a definite crimp is made in the wire.

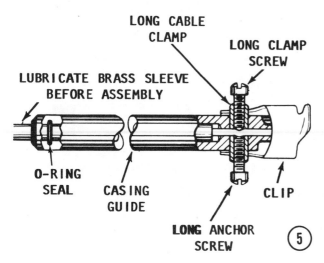

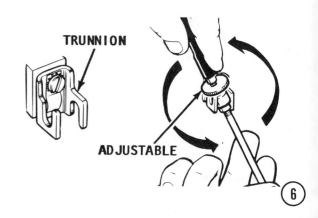

CRITICAL WORDS

Check the end of the cable to determine if the temper has been removed. If the end has a bluish appearance, it has been heated at an earlier date and the temper removed. The temper **MUST** be removed to permit the holding screw to make a crimp in the wire to hold an adjustment. If the screw is not tightened to this degree, the wire will slip during operation and the adjustment will be lost. If the wire has not been tempered, heat the end, but not enough to melt the wire. Bring the bottom screw up tight against the wire.

6- Install the trunion retainers to the engine, if necessary. Check to be sure the retainer with **"P"** stamped on the inside is installed on the **PORT** side of the engine and the retainer with the **"S"** installed on the **STARBOARD** side. Connect the trunion cap to the trunion retainer. This is accomplished by holding the trunion in a vertical position; inserting it into the retainer; and then turning it to the horizontal position, as shown.

7- Slide the guide over the pin onto the engine, and then snap the retainer clip over the end of the guide to lock it in place.

Cable Adjustments

See Chapter 8, Lower Unit, to properly adjust the shift cable and to adjust the throttle cable.

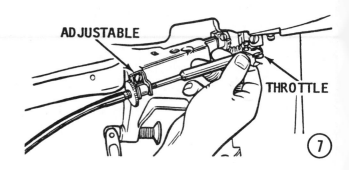

8
LOWER UNIT

8-1 DESCRIPTION

The lower unit is considered as that part of the outboard below the exhaust housing. The unit contains the propeller shaft, the driven and pinion gears, the driveshaft from the powerhead and the water pump. On models equipped with shifting capabilities, the forward and reverse gears, together with the clutch, shift assembly, and related linkage, are all housed within the lower unit.

The lower unit is removed by one of four methods depending on the model year and the engine horsepower.

1- Removal of the powerhead is necessary before the lower unit can be serviced.

2- The window on the exhaust housing may be opened and the shift coupler disconnected.

3- The lower unit is lowered a couple inches and the shift coupler removed.

4- The lower unit does not have shifting capabilities, therefore, removal of the unit is not an involved procedure.

To determine the proper procedures to follow, simply check the Appendix for the engine being serviced.

CHAPTER COVERAGE

Four different lower units are covered in this chapter with separate sections for each.

The first type unit, presented in Section 8-5, covers the 1.5 hp to 4 hp engines with no shift capabilities.

The second type unit, presented in Section 8-6, covers the 5.0 hp to the 25 hp engines with mechanical shift.

The third unit, presented in Section 8-7, covers the 28 hp to 40 hp engines with mechanical shift. These units are very similar to those covered in Section 8-6

Non-shift lower unit with the drive gear, cap, and propeller shaft shown.

Mechanical shift lower unit for a 5 hp to 25 hp engine with the cap removed, exposing the internal parts.

except for the type of bearings installed and therefore slightly different assembling procedures.

The fourth type units, presented in Section 8-8, are the lower units with electric shift.

Water pump service work is by far the most common reason for removal of the lower unit. Each lower unit service section contains complete detailed procedures to rebuild the water pump. The instructions given to prepare for the water pump work must be performed as listed. However, once the pump is ready for installation, if no other work is to be performed on the lower unit, the reader may jump to the pump assembling procedures and proceed with installation of the water pump.

Each section is presented with complete detailed instructions for removal, disassembly, cleaning and inspecting, assembling, adjusting, and installation of only one type unit.

ILLUSTRATIONS

Because this chapter covers such a wide range of models over an extended period of time, the illustrations included with the procedural steps are those of the most popular lower units. In some cases, the unit being serviced may not appear to be identical with the unit illustrated. However, the step-by-step work sequence will be valid in all cases. If there is a special procedure for a unique lower unit, the differences will be clearly indicated in the step.

SPECIAL WORDS

All threaded parts are right-hand unless otherwise indicated.

If there is any water in the lower unit or metal particles are discovered in the gear lubricant, the lower unit should be completely disassembled, cleaned and inspected.

Gear arrangement used on the 40 hp manual shift lower unit.

Gear arrangement used on the 40 hp electromatic shift lower unit.

Actually, problems in the lower unit can be classified into three broad areas.

1- Lack of proper lubrication in the lower unit. Most often, this is caused by failure of the operator to check the gear oil level frequently and to add lubricant when required.

2- Water entering the lower unit through a faulty seal. Water allowed to remain in the lower unit over a period of non-use time will separate from the oil and can be destructive.

3- Excessive clutch dog and clutch ear wear on the forward and reverse gears. This condition is caused by excessive wear in the bellcrank under the powerhead. A worn bellcrank will result in sloppy shifting of the lower unit and cause the clutch components to wear and develop shifting problems. Improper shifting techniques at the shift box will also result in excessive wear to the clutch dog and clutch ears of the forward and reverse gears.

8-2 TROUBLESHOOTING MANUAL SHIFT

Troubleshooting **MUST** be done **BEFORE** the unit is removed from the powerhead to permit isolating the problem to one area. Always attempt to proceed with troubleshooting in an orderly manner. The shotgun approach will only result in wasted time, incorrect diagnosis, replacement of unnecessary parts, and frustration.

The following procedures are presented in a logical sequence with the most prevalent, easiest, and less costly items to be checked listed first.

Arrangement of propeller shaft parts used on most 1.5 hp to 4 hp engines.

Unable to Shift into Forward or Reverse

Remove the propeller and check to determine if the shear pin has been broken. If the unit being serviced has the shear pin at the rear of the propeller, the propeller should be removed and the shear pin checked at the rear of the propeller shaft.

Check the bellcrank under the powerhead. This is accomplished by first checking in the Appendix to determine the type of shift mechanism installed on the unit being serviced. If the Appendix check reveals the powerhead must be removed in order to check the shift mechanism, then the powerhead must be removed. If the unit has a window in the exhaust housing, then the window must be removed. Hold the shift rod with a pair of pliers and at the same time attempt to move the shift lever on the starboard side of the engine. If it is possible to move the shift lever, the bellcrank is worn.

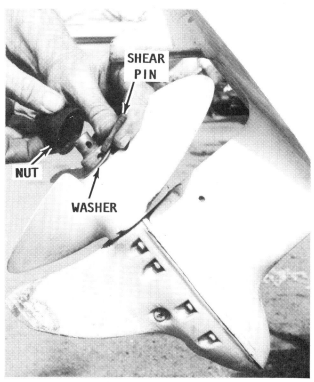

Arrangement of the shear pin, washer, and propeller nut on a typical lower unit. In this case, the shear pin is installed just behind the washer.

If the engine is the type requiring the lower unit to be lowered slightly to gain access to the shift rod: Lower the lower unit slightly, and then hold the shift rod with a pair of pliers and attempt to move the shift lever on the starboard side of the engine. If the lever can move, the bellcrank is worn and must be repaired.

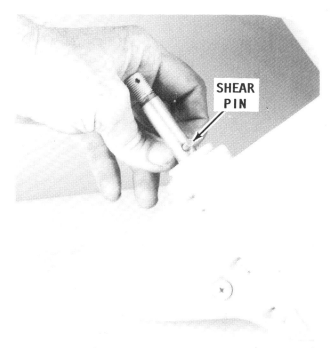

Location of the shear pin through the propeller shaft just ahead of the propeller.

View into the exhaust housing after the powerhead has been removed. The bellcrank part of the upper mechanical shift mechanism is clearly visible.

8-4 LOWER UNIT

View of a badly corroded lower unit. Water entered and was allowed to remain over an extended period of time causing extensive damage.

Water in the Lower Unit

Water in the lower unit is usually caused by fish line becoming entangled around the propeller shaft behind the propeller and damaging the propeller seal. If the line is not removed, it will cut the propeller shaft seal and allow water to enter the lower unit. Fish line has also been known to cut a groove in the propeller shaft.

The propeller should be removed each time the boat is hauled from the water at the end of an outing and any material entangled behind the propeller removed before it can cause expensive damage. The small amount of time and effort involved in pulling the propeller is repaid many times by reduced maintenance and service work, including the replacement of expensive parts.

Slippage in the Lower Unit

If the shift seems to be slipping as the boat moves through the water: Check the

Entangled fish line finally removed from ahead of the propeller. The line cut the seal. The damaged seal allowed water to enter the lower unit causing the disaster shown in the illustration at the top of this column. The propeller should be removed frequently and debris removed from the propeller shaft to prevent such costly repairs.

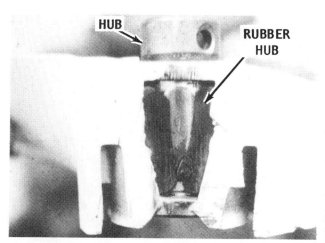

Cut-a-way view of a propeller showing the propeller, rubber hub, and metal hub.

propeller and the rubber hub. If the propeller has been subjected to many strikes against underwater objects, it could slip on its hub. If the hub is damaged or excessively worn on the small propellers, it is not economical to have the hub or propeller rebuilt. A new propeller may be purchased for considerably less than meeting the expense of rebuilding an old worn propeller.

Difficult Shifting

Verify that the ignition switch is **OFF**, or better still, disconnect the spark plug wires from the plugs, to prevent possible personal injury, should the engine start. Shift the unit into **REVERSE** gear at the shift control box, and at the same time have an assistant turn the propeller shaft to ensure the clutch is fully engaged. If the shift handle is hard to move, the trouble may be in the lower unit, with the shift cable, or in the shift box, if used.

Isolate the problem: Disconnect the shift cable, if used, at the engine. Operate the shift lever. If shifting is still hard, the problem is in the shift cable or control box, see Chapter 7. If the shifting feels normal

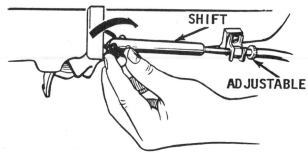

Diagram to illustrate the shift cable hookup quick disconnect at the engine.

with the shift cable disconnected, the problem must be in the lower unit. To verify the problem is in the lower unit, have an assistant turn the propeller and at the same time move the shift cable back-and-forth. Determine if the clutch engages properly.

Jumping out of Gear

If a loud thumping sound is heard at the transom while the boat is underway, the unit is jumping out of gear, the propeller does not have a load, therefore the rushing water under the hull forces the lower unit in a backward direction. The unit jumps back into gear; the propeller catches hold; the lower unit is forced forward again, and the result is the thumping sound as the action is repeated. Normally this type of action occurs perhaps once a day, then more frequently each time the clutch is operated, until finally the unit will not stay in gear for even a short time.

The following areas must be checked to locate the cause:

1- Check the bellcrank under the powerhead. This is accomplished by first checking in the Appendix to determine the type of shift mechanism installed on the unit being

View into the exhaust housing after the powerhead has been removed. The bellcrank part of the upper mechanical shift mechanism is clearly visible.

serviced. If the Appendix check reveals the powerhead must be removed in order to check the shift mechanism, then the powerhead must be removed. If the unit has a window in the exhaust housing, then the window must be removed. Hold the shift rod with a pair of pliers and at the same time attempt to move the shift lever on the starboard side of the engine. If it is possible to move the shift lever, the bellcrank is damaged.

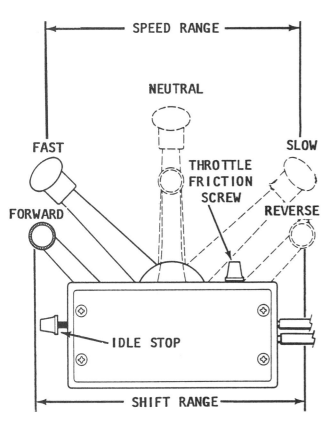
Diagram to illustrate the shift positions and throttle range of a dual lever shift control box.

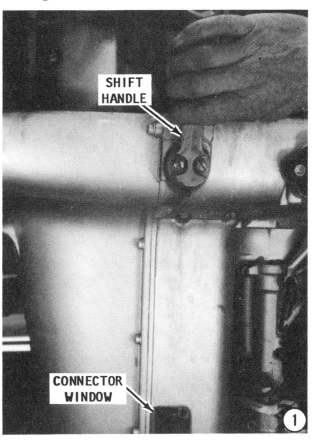

8-6 LOWER UNIT

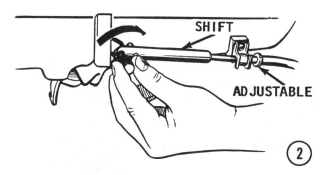

Powerhead of a low horsepower engine removed exposing the interior of the exhaust housing. This bellcrank is quite different than the one shown on the previous page.

If the engine is the type requiring the lower unit to be lowered slightly to gain access to the shift rod: Lower the lower unit slightly, and then hold the shift rod with a pair of pliers and attempt to move the shift lever on the starboard side of the engine. If the lever can move, the bellcrank is damaged and must be repaired.

2- Disconnect the shift cable at the engine. Attempt to shift the unit into forward gear with the shift lever on the starboard side of the engine and at the same time rotate the propeller in an effort to shift into gear. Shift the control lever at the control box into forward gear. Move the shift cable at the engine up to the shift handle and determine if the cable is properly aligned. The control lever may have jumped a tooth on the slider or on the shift lever arc. If a tooth has been jumped, the cable would lose its adjustment and the unit would fail to shift properly. If the inner cable should slip on the end cable guide, the adjustment would be lost.

3- Move the shift lever at the engine into the neutral position and the shift lever at the control box to the neutral position. Now, move the shift cable up to the shift lever and see if it is aligned. Shift the unit into reverse at the engine and shift the control lever at the control box into reverse. Move the cable up and see if it is aligned. If the cable is properly aligned, but the unit still jumps out of gear when the cable is connected, one of three conditions may exist.

a- The bellcrank is worn excessively or damaged.

b- The coupler at the connector at the shift rod is misaligned. This coupler is used

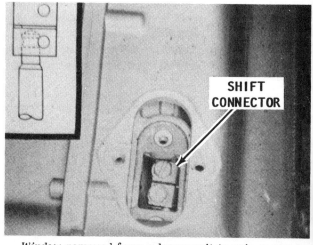

Window removed from a lower unit to gain access to the shift connector, as explained in the text. The detail drawing, upper left, illustrates the relationship of the bolt to the shift rod.

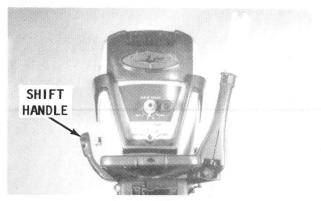

Frontal view of a low horsepower engine showing the mechanical shift lever on the starboard side.

to connect the upper shift rod with the lower rod. If the coupler has not been installed properly, any shifting will be difficult.

c- Parts in the lower unit are worn from extended use.

Frozen Powerhead

This condition is suggested when the operator unsuccessfully attempts to crank the engine, either with a hand starter or with a starter motor. The flywheel will not rotate. Do not assume the engine is "frozen" until the lower unit has been removed and thoroughly checked. If the lower unit is "locked" (the drivehsaft or propeller shaft will not rotate), the powerhead will have the indication of being "frozen" (failure to rotate the flywheel).

The first step to perform under these conditions is to "pull" the lower unit, and then again attempt to crank the engine. If the attempt is successful with the lower unit disconnected, the problem is in the lower unit. If the attempt to crank the engine is still unsuccessful, the problem is in the powerhead.

8-3 PROPELLER REMOVAL

The shear pin on most outboard units is installed through the propeller shaft between the propeller washer and the propeller nut.

To remove the propeller, first pull the cotter key, and then remove the propeller nut, shear pin, and washer. Because the

Checking for a "frozen" powerhead or "locked" lower unit. If the crankshaft can be rotated back-and-forth slightly with the hand starter, the problem is in the lower unit, not the powerhead.

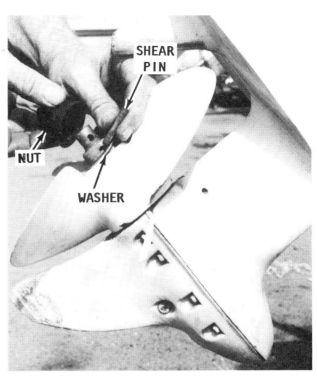

Arrangement of the shear pin, washer, and propeller nut on a typical lower unit. In this case, the shear pin is installed after the propeller and washer are in place.

shear pin is not a tight fit, the propeller is able to move on the pin and cause burrs on the hole. The propeller may be difficult to remove because of these burrs. To overcome this problem, the propeller hub has

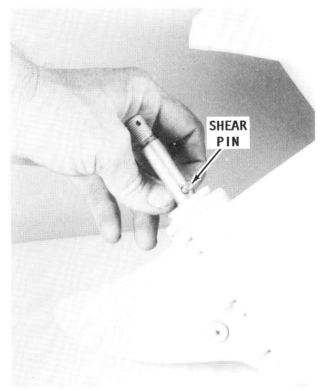

Installation of the shear pin before the propeller, and washer are installed.

8-8 LOWER UNIT

Propeller showing the grooves in the hub to ease removal from the propeller shaft.

two grooves running the full length of the hub. Hold the shaft from turning, and then rotate the propeller 1/4 turn to position the grooves over the shear pin holes. The propeller can then be pulled straight off the shaft. After the propeller has been removed, file the shear pin holes on both sides of the shaft to remove the burrs.

If the propeller is the type with the shear pin installed ahead of the propeller next to the bearing carrier, first remove the cotter key, then the propeller nut. Next, slide the propeller free of the shaft, and remove the shear pin.

Propeller installation procedures are outlined at the end of each lower unit section.

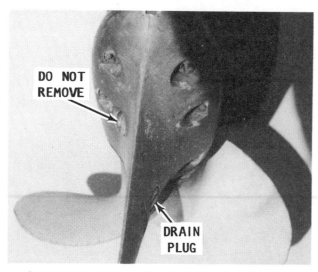

*Lower unit showing the drain plug and the Phillips screw which **MUST NOT** be removed by mistake, as explained in the text.*

8-4 DRAINING LOWER UNIT

Position a suitable container under the lower unit, and then remove the **FILL** screw and the **VENT** screw.

CRITICAL WORD

The Phillips screw securing the shift fork in place is located very close to the vent screw. If the wrong screw is removed, **BAD NEWS, VERY BAD NEWS**. The lower unit will have to be disassembled in order to return the shift fork to its proper location.

Allow the gear lubricant to drain into the container. As the lubricant drains, catch some with your fingers, from time-to-time, and rub it between your thumb and finger to determine if any metal particles are present. If metal is detected in the lubricant, the unit must be completely disassembled, inspected, and the damaged parts replaced.

Check the color of the lubricant as it drains. A whitish or creamy color indicates the presence of water in the lubricant. Check the drain pan for signs of water separation from the lubricant. The presence of any water in the gear lubricant is **BAD NEWS**. The unit must be completely disassembled, inspected, the cause of the problem determined, and then corrected.

Filling instructions are outlined at the end of each lower unit section.

8-5 LOWER UNIT SERVICE
1.5 HP TO 4.0 HP ENGINES — NO SHIFT

Description

This is a very simple direct drive unit without any shift capabilities. Reverse is obtained by rotating the engine 180° and holding that position while the boat is moved sternward.

Propeller Removal

Remove the propeller according to the procedures outlined in Section 8-3.

Draining the Lower Unit

Drain the lower unit according to the procedures outlined in Section 8-4.

GOOD WORDS

If water is discovered in the lower unit and the propeller shaft seal is damaged and requires replacement, the lower unit does **NOT** have to be removed in order to accomplish the work.

SERVICE NO SHIFT 8-9

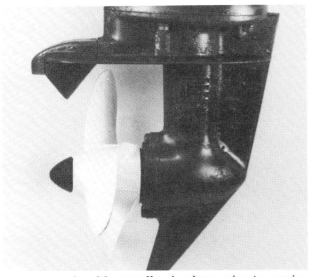

Lower unit with propeller in place prior to service work.

The seal may be replaced by first removing the two screws securing the cap in place and then tapping on the cap with a soft-headed mallet to jar it loose. The cap is then removed, the seal removed and replaced, and the cap installed and secured.

LOWER UNIT REMOVAL

ADVICE

If the only work to be performed is service of the water pump, be extremely **CAREFUL** to prevent the driveshaft from being pulled up and free of the pinion gear in the lower unit. **NEVER** carry the lower unit by the driveshaft. If the shaft should be released from the pinion, the lower unit **MUST** be disassembled to align the pinion gear and driveshaft, then the driveshaft installed.

1- Disconnect the spark plug wire from the plug. Remove the retaining bolts securing the lower unit to the exhaust housing. **CAREFULLY** pull directly downward, to prevent damage to the water tube, and remove the lower unit.

WATER PUMP REMOVAL

2- Remove the screws securing the water pump to the lower unit housing. It is very possible corrosion will cause the screw heads to break-off when an attempt to remove them is made. If this should happen, use a chisel and break-a-way the water pump housing from the lower unit. **EXERCISE CARE** not to damage the lower unit housing.

3- After the screws have been removed, slide the water pump, impeller, the impeller key, and the lower water pump plate upward and free of the driveshaft.

If the only work to be performed is service of the water pump, proceed directly to Page 8-14, Water Pump Installation.

LOWER UNIT DISASSEMBLING

4- Remove the gearcase head and the two screws. Pull on the propeller shaft or tap on the gearcase head to separate the gearcase head from the lower unit housing. The driven gear is pressed onto the propeller shaft. Therefore, the propeller shaft and gear are considered as a complete assembly. If either is damaged and requires replacement, the two are purchased as an assembly.

5- Pull upward on the driveshaft, and at the same time, reach inside the lower unit and remove the pinion gear.

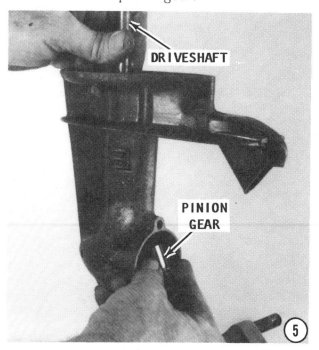

SPECIAL WORDS

On the Weedless type lower unit, a thrust bearing is installed under the pinion gear. This thrust bearing can only be removed by tapping it out in the following manner: Turn the lower unit so the propeller shaft opening is facing downward. Now, gently rap the unit on a work bench or block of wood. The thrust bearing and pinion gear will be dislodged and fall free.

6- If the seal at the top of the lower unit housing under the water pump is to be replaced, remove the seal using any type seal remover. The OMC tool number is 377565. To remove the seal in the gearcase head, work the seal free by using a punch and mallet from the back side. Remove the O-ring.

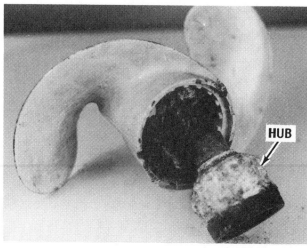

Example of a damaged propeller. This unit should have been replaced long before this amount of damage was sustained.

SERVICE NO SHIFT 8-11

CLEANING AND INSPECTING

Clean all water pump parts with solvent, and then dry them with compressed air. Inspect the water pump cover and base for cracks and distortion, possibly caused from overheating. Inspect the face plate and water pump insert for grooves and/or rough surfaces. If possible, **ALWAYS** install a complete new water pump while the lower unit is disassembled. A new impeller will ensure extended satisfactory service and give "peace of mind" to the owner. If the old impeller must be returned to service, **NEVER** install it in reverse to the original direction of rotation. Installation in reverse will cause premature impeller failure.

Inspect the impeller side seal surfaces and the ends of the impeller blades for cracks, tears, and wear. Check for a glazed or melted appearance, caused from operating without sufficient water. If any question exists, and as previously stated, install a new impeller if at all possible.

Clean all parts with solvent and dry them with compressed air. **DISCARD** all O-rings and gaskets. Inspect and replace the driveshaft if the splines are worn. Inspect

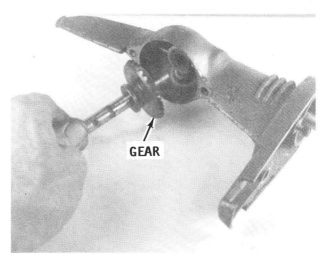

Propeller shaft and gear removal from a low horsepower lower unit. All parts should be thoroughly cleaned and inspected.

the gearcase and exhaust housing for damage to the machined surfaces. Remove any nicks and refurbish the surfaces on a surface plate. Start with a No. 120 Emery paper and finish with No. 180.

Check the water intake screen and passages. Inspect the drive gear, pinion gear, and thrust washers. Replace these items if they appear worn. If any of the surfaces are

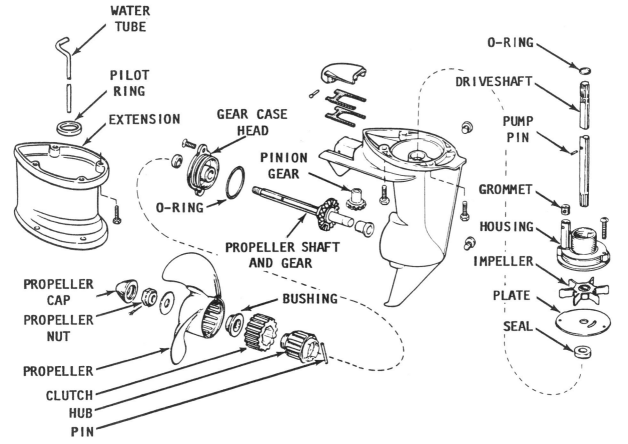

Exploded drawing of the non-shifting lower unit, with major parts identified.

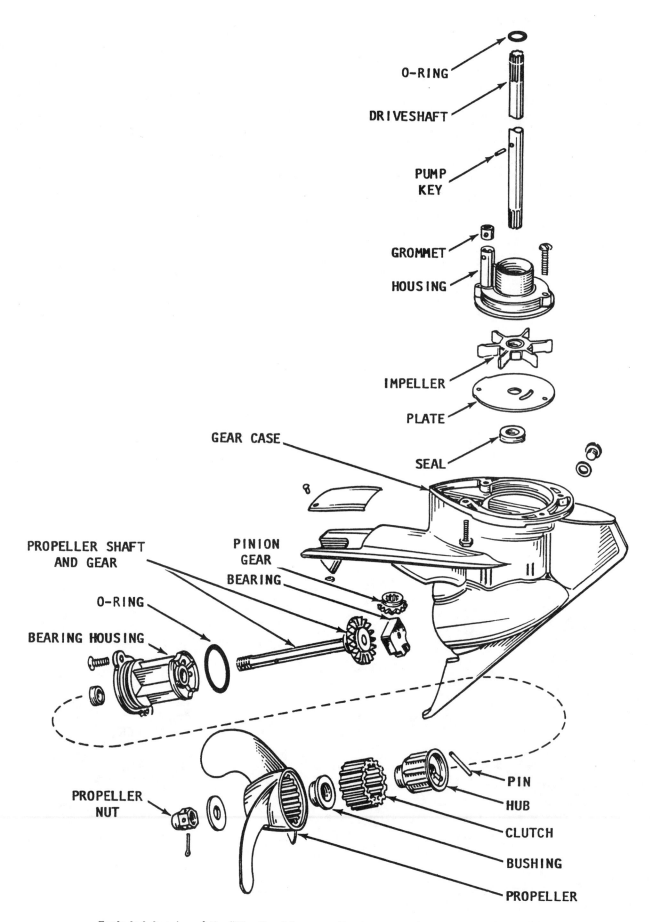

Exploded drawing of the "Weedless" lower unit gear case, with major parts identified.

nicked, chipped, or the edges rounded, the operator may be performing the shift operation improperly. These items **MUST** be replaced if they are damaged.

LOWER UNIT ASSEMBLING

1- Tap a **NEW** seal into place on top of the lower unit housing.

2- Tap a **NEW** seal into place in the gearcase head. Install a **NEW** O-ring into the groove in the gearcase.

3- Install the pinion gear into the recess in the lower unit housing. If the unit being serviced is the "Weedless" type gearcase, install the thrust bearing with the bosses on the bearing indexed between the two bosses in the gearcase. Hold the pinion gear in place with one hand, and with the other hand install the driveshaft down into the lower unit. Continue to hold the pinion gear, and at the same time, rotate the driveshaft slightly after it makes contact with the pinion gear to allow the splines on the shaft to index with the splines in the gear.

SPECIAL WORDS

After the driveshaft is installed, care must be exercised **NOT** to allow the driveshaft to slip out of position in the pinion gear. This is especially important during water pump installation work. If the driveshaft should come free, the lower unit must be disassembled in order to install the driveshaft.

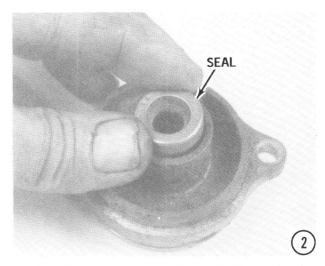

4- Coat the propeller shaft and the gearcase O-ring with oil as an aid to installation. Install the gearcase head over the propeller shaft. Slide the propeller shaft through the lower unit with the driven gear teeth indexed with the teeth of the pinion gear. It may be necessary to rotate the propeller shaft slightly in order to index the driven and pinion gear teeth. The teeth **MUST** engage fully and properly or the gearcase head will be damaged when the attaching screws are installed. Coat the screws securing the head to the lower unit with sealer, and then install the screws. **CAREFULLY** tap on the gearcase head with a

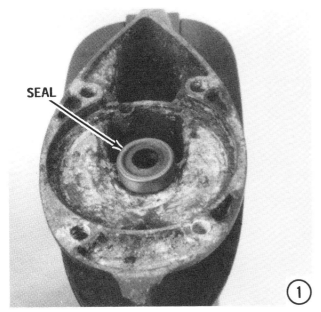

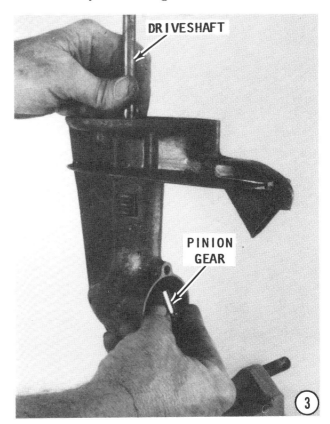

soft-headed mallet and tighten the screws **EVENLY** and **ALTERNATELY**.

CRITICAL WORDS

If the screws are not tightened evenly, or the driven gear and pinion gear teeth are not fully and properly engaged, the gearcase head will be thrown out of line just a whisker, and the ears through which the bolts pass may snap off. **BAD NEWS!** A new gearcase head would have to be purchased.

If the unit being serviced is the "Weedless" type, two sets of matching marks on the gearcase head and the lower unit **MUST** be aligned when the head is installed.

If the unit being serviced is the "High-Thrust" type, the gearcase has a hole which **MUST** face upward when the head is installed.

WATER PUMP INSTALLATION

5- Lay down a bead of sealer No. 1000 onto the lower unit surface. Slide the water pump plate down the driveshaft and onto the lower unit surface.

6- Insert the water pump impeller pin into the driveshaft hole.

7- Slide the water pump impeller down the driveshaft and into place on top of the water pump base plate with the pump pin indexed in the impeller. Lubricate the inside surface of the water pump with lightweight oil.

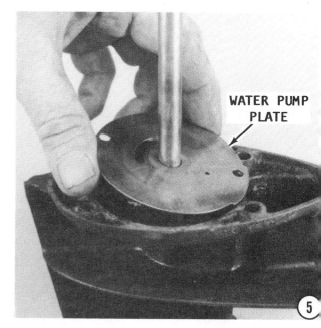

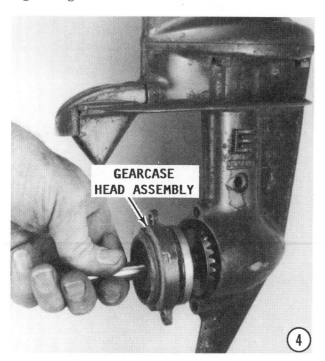

8- Lower the water pump housing down the driveshaft and over the impeller. Rotate the driveshaft **CLOCKWISE** as the water pump housing is lowered to allow the impeller blades to assume their natural and proper position inside the housing. Continue to rotate the driveshaft and work the water pump housing downward until it is seated on the lower unit upper housing surface.

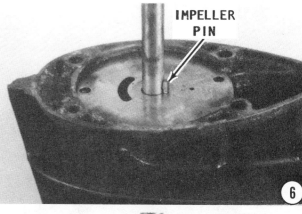

SERVICE NO SHIFT 8-15

9- **ALWAYS** rotate the driveshaft **CLOCKWISE** while the screws are tightened to prevent damaging the impeller vanes. If the impeller is not rotated, the housing could damage or cut the end of the vanes as the screws are brought up tight. The rotation allows them to spring back in a natural position. Place a **NEW** grommet into the water pump housing for the water pickup tube. If a new water pump was installed, this seal will already be in place. Install a **NEW** O-ring on the top of the driveshaft.

LOWER UNIT INSTALLATION

10- Clean and shine the water pump tube with lightweight sandpaper, and then coat it with oil as an aid to installation. Apply oil to the grommet in the water pump housing as a further aid to installation. This tube is very small in size and will bend easily during installation if it has even a little difficulty passing through the rubber grommet in the water pump housing.

Bring the lower unit together to mate with the exhaust housing. Guide the water tube into the water pump housing grommet, and at the same time rotate the propeller shaft **CLOCKWISE**. Rotating the propeller shaft will also rotate the driveshaft and allow the splines on the driveshaft to index with the splines of the engine crankshaft.

Continue to work the lower unit closer to the exhaust housing until the mating surfaces make contact. Coat the retaining screws with sealer to prevent corrosion, and then start them in place. Tighten the retaining screws **EVENLY** and **ALTERNATELY**.

8-16 Lower Unit

GEAR LUBRICANT

FILLING THE LOWER UNIT

11- Fill the lower unit with lubricant. Insert the lubricant tube into the bottom opening, and then fill the unit until lubricant is visible at the vent hole. Install the vent plug. Remove the gear lubricant tube, and install the drain/fill plug.

12- After the lower plug has been installed, remove the vent plug again and using a squirt-type oil can, add lubricant through this vent hole. A squirt-type oil can must be used to allow the trapped air in the lower unit to escape at the same time the final lubricant is added. Once the unit is completely full, install and tighten the vent plug.

PROPELLER INSTALLATION

FIRST, THESE GOOD WORDS

The propellers used on the outboards covered in this section have a removable clutch ring and a clutch hub, and bushing. Under normal conditions, these items are **NOT** removed from the propeller. However, if they have been removed for any number of reasons, they should be coated with OMC Type "A" lubricant prior to installation. The bushing is installed first, then the clutch hub, and finally the clutch ring.

13- Install the shear pin. Apply a light coating of anti-corrosive lubricant onto the propeller shaft. Slide the propeller onto the shaft, then the washer, and finally the propeller nut, with the flange on the nut **TOWARDS** the propeller. Tighten the nut securely. Install a cotter pin to prevent the nut from backing out. Slip the rubber cap over the propeller nut.

14- Perform a functional check of the completed work by mounting the engine in a test tank, in a body of water, or with a flush attachment connected to the lower unit. If the flush attachment is used, **NEVER** operate the engine above an idle speed, because the no-load condition on the propeller would allow the engine to **RUNAWAY** resulting in serious damage or destruction of the engine.

Installation of the propeller on a low horsepower engine lower unit.

MANUAL SHIFT 5 HP TO 25 HP 8-17

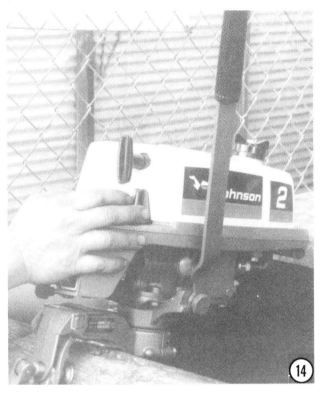

CAUTION: Water must circulate through the lower unit to the engine any time the engine is run to prevent damage to the water pump in the lower unit. Just five seconds without water will damage the water pump.

Start the engine and observe the tattle-tale flow of water from idle relief in the exhaust housing. The water pump installation work is verified. If a "Flushette" is connected to the lower unit, **VERY LITTLE** water will be visible from the idle relief port.

8-6 LOWER UNIT SERVICE MANUAL SHIFT — 5 HP TO 25 HP ENGINES

Propeller Removal
Remove the propeller according to the detailed procedures outlined in Section 8-3.

Draining Lower Unit
Drain the lower unit according to the detailed procedures outlined in Section 8-4.

GOOD WORDS
Before the lower unit can be removed, the type of shift disconnect must be determined. The different shift type disconnects are lettered A thru E for simplicity. The type used on the engine being serviced may quickly be identified from the Appendix. The model year and horsepower must be known in order to use the Appendix.

Type A — No Shift Unit
This type is used on the direct drive engines, without a reverse gear. The engine is rotated 180° with the steering lever to move the boat sternward. Because a shift rod is not used, naturally there is no requirement to disconnect this item. The necessary procedure to remove the lower unit is to simply remove the bolts attaching the unit to the exhaust housing, and then "drop" the lower unit.

Type B — Pin In Upper Driveshaft
This type is used in conjunction with the Type A, C, and E. "Pin in the upper driveshaft", means that the pin holds and pushes seal and spring assembly against the powerhead and thus provides a bottom seal for the powerhead. After the lower unit attaching bolts have been removed, the flywheel must be rotated (to rotate the driveshaft) until the pin is aligned with two slots in the upper portion of the exhaust housing. The lower housing can then be separated from the exhaust housing. If an attempt is made to force the lower unit from the exhaust housing without aligning the driveshaft pin, as just described, the pin may be broken and other items damaged.

Type C — Shift Disconnect Under Powerhead
Remove the powerhead. The shift rod is attached to the shift lever underneath the powerhead. Disconnect the shift rod from

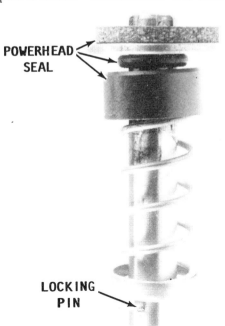

Location of the locking pin in the shift shaft. The parts above the pin are contained in the exhaust housing.

8-18 LOWER UNIT

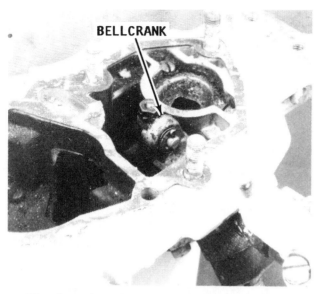

View into the exhaust housing after the powerhead has been removed. The bellcrank part of the upper mechanical shift mechanism is clearly visible.

the shift lever. Remove the attaching bolts, and then separate the lower unit from the exhaust housing.

Type D — Window Removal to Gain Access

1- Remove the metal plate from the port side of the engine. Access to the shift coupler is gained through the opening.

2- Disconnect the shift rod from the exhaust housing by removing the bottom bolt from the shift coupler.

Type E — Shift Disconnect Coupler

Loosen the attaching screws securing the lower unit to the powerhead. Allow the lower unit to drop approximately one inch, and then remove the bottom bolt in the shift

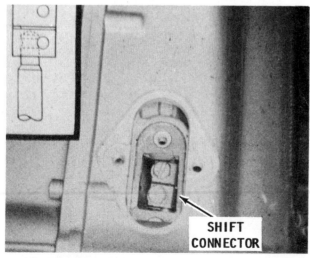

Window removed from a lower unit to gain access to the shift connector, as explained in the text. The detail drawing, upper left, illustrates the relationship of the bolt to the shift rod.

Removing the bolt through the shift connector after the lower unit has been separated from the exhaust housing enough to allow access for the socket wrench.

coupler. The lower unit may then be completely separated from the exhaust housing.

GOOD WORDS

In **MOST** cases, if any unit being serviced has the 6-inch extension, it is **NOT** necessary to remove the extension in order to "drop" the lower unit. However, as in most things in life, there are rare exceptions and here is one. If the lower unit is separated from the extension and the driveshaft connection is not accessible, then the extension will have to be removed to gain access to the coupler.

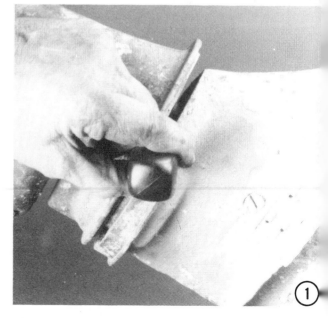

LOWER UNIT REMOVAL

1- After the shift rod has been disconnected, as described in the previous paragraphs, remove the bolts securing the lower unit to the housing. Some units may have an additional bolt on each side and one at the rear of the engine. Work the lower unit loose from the exhaust housing. It is not uncommon for the water tube to be stuck in the water pump making separation of the lower unit from the exhaust housing difficult. However, with patience and persistence, the tube will come free of the pump and the lower unit separated from the exhaust housing.

MORE GOOD WORDS

Position the lower unit in a vertical position on the edge of the work bench resting on the cavitation plate. Secure the lower unit in this position with a C-clamp. The lower unit will then be held firmly in a favorable position for further service work. An alternate method is to cut a groove in a short piece of 2" x 6" wood to accommodate the lower unit with the cavitation plate resting on top of the wood. Clamp the wood in a vise and service work may then be performed with the lower unit erect (in its normal position), or inverted (upside down). In both positions, the cavitation plate is the supporting surface.

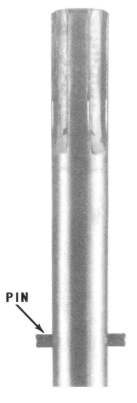

2- Remove the O-ring from the top of the driveshaft. Some units may have a pin installed in this location, instead of an O-ring. In this case, remove the pin from the driveshaft. The washer, springs, and other parts will have remained in the exhaust housing.

WATER PUMP REMOVAL

3- Remove the screws securing the water pump to the lower unit housing. It is very possible corrosion will cause the screw heads to break-off when an attemept to remove them is made. If this should happen, use a chisel and breakaway the water pump housing from the lower unit. **EXERCISE CARE** not to damage the lower unit housing.

4- After the screws have been removed, slide the water pump, impeller, the impeller key, and the lower water pump plate upward and free of the driveshaft.

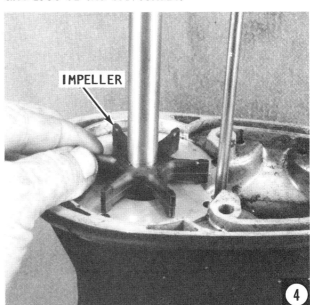

8-20 LOWER UNIT

ADVICE

If the only work to be performed is service of the water pump, be extremely **CAREFUL** to prevent the driveshaft from being pulled up and free of the pinion gear in the lower unit. **NEVER** carry the lower unit by the driveshaft. If the shaft should be released from the pinion, the lower unit **MUST** be disassembled to align the pinion gear and the driveshaft, then the driveshaft installed. To install the water pump, proceed directly to Page 8-31, Water Pump Installation

LOWER UNIT DISASSEMBLING

GOOD WORDS

One of two type of driveshafts may be installed in the lower units covered in this section. One has a spline on the lower end of the shaft to index with the splines in the pinion gear. The other type driveshaft has a key and keyway in the lower end of the driveshaft. The key indexes with a matching keyway in the pinion gear.

5- **CAREFULLY** pull upward on the driveshaft. If the driveshaft comes free easily, the unit is the type with the splines on the end of the shaft. If the shaft will not come free, it is the type with the key and keyway. Therefore, the driveshaft will be removed later when the lower unit is disassembled. If the driveshaft comes free, remove it at this time.

6- Turn the lower unit upside down and again clamp it in the vise or slide it into the wooden block, if one is used. Carefully examine the lower portion of the unit. The cap is considered that part below the split

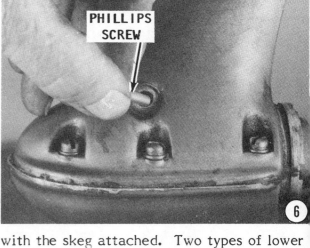

with the skeg attached. Two types of lower unit caps are installed on engines covered in this section. One has a Phillips screw installed, and the other does not. If the Phillips screw is installed, remove the screw.

7- Remove the attaching screws around the cap. These screws may be slotted-type or Phillips screws. **CAREFULLY** tap the cap to jar it loose, and then separate it from the lower unit housing. If the cap did not have a Phillips screw on the outside, observe the two slots inside the cap.

TAKE TIME

Before proceeding with the disassembly work, take time to study the arrangement of parts in the lower unit. You may elect to follow the practice of many professional mechanics and take a polaroid picture of the unit as an aid during the assembly work. Several engineering and production changes

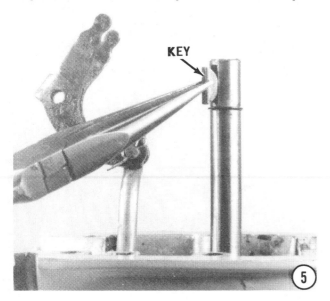

MANUAL SHIFT 5 HP TO 25 HP 8-21

Lower unit with the cap removed exposing internal parts before disassembling.

have been made to the lower unit over the years. Therefore, the positioning of the gears, shims, bearings, and other parts may vary slightly from one unit to the next.

To show each and every arrangement with a picture in this manual would not be practical. Even if it were done, the ability to associate the unit being serviced with the illustration would be almost impossible. Therefore, take time to make notes, scribble out a sketch, or take a couple photographs.

8- Lift the shift lever out of the cradle, and then remove the cradle from the shift dog. Raise the propeller shaft and at the same time tap with a soft-headed mallet on the bottom side to jar it loose, then remove the shaft assembly from the lower unit. The forward and reverse gear including the bearings will all come out with the propeller shaft. The forward gear is the gear at the opposite end of the shaft from the propeller.

GOOD WORDS

Notice the bearing split on the back side of the forward gear. Also observe the pin in the housing and a matching slot in the bearing. The pin must index in the slot during installation. With the reverse gear, some bearing heads have a hole and a matching pin installed in the housing indexes in this hole. Some other reverse gears have a tab protruding from the bearing head.

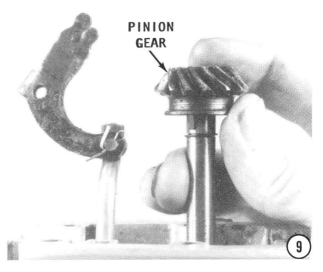

This tab indexes in a slot in the housing during installation. By taking note at this time of the particular type of installation for the unit being serviced, the task of installation will progress more smoothly. If the installation work is not performed properly, the lower unit housing will quickly be damaged requiring the purchase of a new unit.

9- If the unit being serviced is the type with a keyway in the driveshaft, remove the pinion gear from the shaft, then the key, and snap ring, before attempting to remove the driveshaft shaft.

10- Slide the forward gear, babbitt bearing, and washer free of the propeller shaft. Remove the clutch dog. Remove the reverse gearcase head, reverse gear, and washer from the shaft.

11- Turn the lower unit housing right side up and again clamp it in the vise. Remove the bearing carrier and bearing assembly. This is accomplished by using a bearing carrier puller, as shown. An alternate method is to use two screwdrivers to remove the carrier from the lower unit. Sometimes the bearing carrier is difficult to remove. One effective method to release a

8-22 LOWER UNIT

Bearing carrier prior to removal. The two screws shown are used to align the carrier with the gasket during installation.

stubborn bearing carrier is to heat the lower unit housing while attempting to remove the carrier. If this method is employed, **TAKE CARE** not to overheat the lower unit. Excessive heat may damage internal parts. Remove the gasket from underneath the bearing carrier housing.

12- Clean the upper part of the shift rod as an aid to pulling it through the bushing and O-ring. Pull the shift rod from the lower unit housing. The shift rod passes through an O-ring and bushing in the lower unit housing. These two items prevent water from entering the lower unit. A special tapered punch is required to remove the bushing from the lower unit housing. Obtain the special punch, and then remove the bushing, and the O-ring.

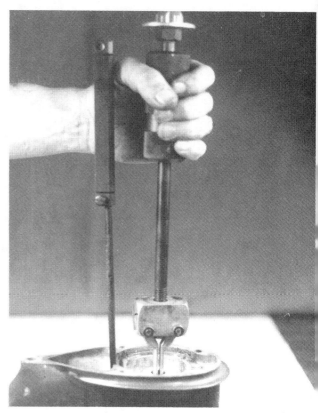

Using a slide hammer to "pull" the bearing carrier from the upper portion of the lower unit.

CLEANING AND INSPECTING

Clean all water pump parts with solvent, and then dry them with compressed air. Inspect the water pump cover and base for cracks and distortion, possibly caused from overheating. Inspect the face plate and water pump insert for grooves and/or rough surfaces. If possible, **ALWAYS** install a complete new water pump while the lower unit is disassembled. A new impeller will ensure extended satisfactory service and give "peace of mind" to the owner. If the old impeller must be returned to service, **NEVER** install it in reverse to the original direction of rotation. Installation in reverse will cause premature impeller failure.

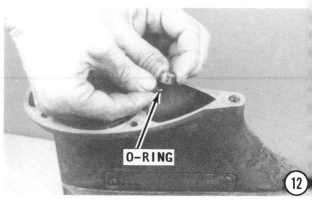

Worn clutch dog ears. This clutch dog must be replaced.

A new shift cradle (right) compared with one badly worn and unfit for service (left).

Inspect the impeller side seal surfaces and the ends of the impeller blades for cracks, tears, and wear. Check for a glazed or melted appearance, caused from operating without sufficient water. If any question exists, and as previously stated, install a new impeller if at all possible.

Clean all parts with solvent and dry them with compressed air. **DISCARD** all O-rings and gaskets. Inspect and replace the driveshaft if the splines are worn. Inspect the gearcase and exhaust housing for damage to the machined surfaces. Remove any nicks and refurbish the surfaces on a surface plate. Start with a No. 120 Emery paper and finish with No. 180.

Check the water intake screen and passages by removing the bypass cover, if one is used. Inspect the clutch dog, drive gears, pinion gear, and thrust washers. Replace these items if they appear worn. If the

Bearing head with a new seal and O-ring installed.

Badly worn propeller shaft. This damage may have started from an entangled fish line that was not removed, then aggravated when water entered the lower unit.

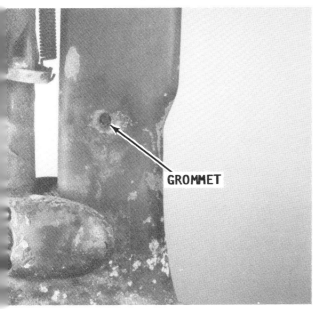

Grommet installed about halfway up the exhaust housing where the upper and lower water tubes meet. Deposits of foreign material have a habit of forming in this area and restricting the flow of water. Therefore, this grommet should be removed and the area cleaned.

Distorted tapered bearing unfit for further service.

8-24 LOWER UNIT

Badly worn water pump impeller. The impeller should be inspected thoroughly. The best practice is to replace the impeller each time the lower unit is disassembled for service.

clutch dog and drive gear arrangement surfaces are nicked, chipped, or the edges rounded, the operator may be performing the shift operation improperly or the controls may not be adjusted correctly. These items **MUST** be replaced if they are damaged.

Inspect the dog ears on the inside of the forward and reverse gears. The gears must be replaced if they are damaged.

Check the cradle that rides on the inside diameter of the clutch dog. The sides of the cradle must be in good condition, free of any damage or signs of wear. If damage or wear has occurred, the cradle must be replaced.

Check the shift lever and the two prongs that fit inside the cradle. Check to be sure the prongs are not worn or rounded. Damage or wear to the prongs indicates the lever must be replaced.

Badly worn pinion gear from a lower unit. The teeth of all gears must be carefully inspected for wear and damage.

READ AND BELIEVE

The three numbered illustrations in the right column on this page clearly show a lower unit that has been assembled **INCORRECTLY**.

Illustration No. 1 Insepct very closely, the bushing bearing on the forward gear for any kind of indication the pin was missed when the housing was installed during the last repair work. Check the reverse gearcase head and if there is any indication of a pin mark, then check the housing for evidence the pin has been driven into the housing. Some forward bearing carries have a lip that indexes with a slot in the lower unit housing. Check for evidence the lip did not index properly.

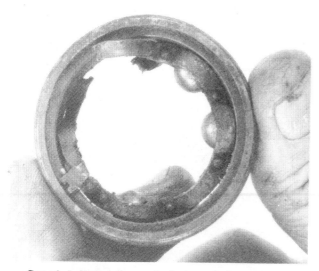

Caged ball bearing set destroyed due to lack of lubrication, vibration, corrosion, metal particles, or all of the above.

Babbitt bearing unfit for further service. This type bearing must not show any signs of wear.

MANUAL SHIFT 5 HP TO 25 HP 8-25

Standard water pump (left) and a new type (right). The newer pump is only used with an updated lower unit.

Illustration No. 2 If the pin has been driven down into the lower unit housing, it must be drilled out as described in the next paragrph. The accompanying illustration compares a proper and an improper installation.

Illustration No. 3 If the pin must be drilled, **EXERCISE CARE** not to drill to deeply. If a hole is drilled deeper than necessary, then insert a couple drops of melted solder into the hole, and set the new pin in place with the **LARGE** portion of the pin flush with the housing. If the pin is not flush, remove it, drop more soldered into the hole and make another test. Continue to drop solder into the hole and test until the pin is flush with the housing when it is installed. If more soldered is inserted into the hole than necessary, the pin may be tapped into the solder while it is still warm and the pin made flush with the housing.

Indent mark on a forward gear bearing indicating the unit was not assembled properly. The pin in the housing caused this damage.

Pin pressed into the housing because of improper installation. This pin must be drilled out before the unit can be returned to service.

A rusted and corroded gear. Water was allowed to enter the lower unit through a bad seal and cause this damage to the gear and other expensive parts.

Drilling the pin out of the housing. A new pin must be installed and the unit assembled properly.

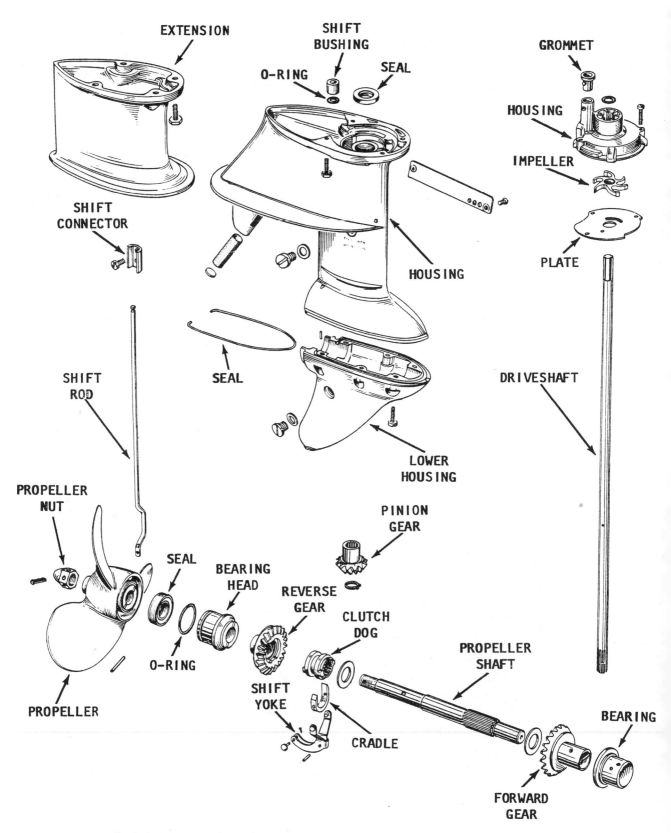

Exploded drawing of a lower unit with a window for access to the shift connector.

MANUAL SHIFT 5 HP TO 25 HP   8-27

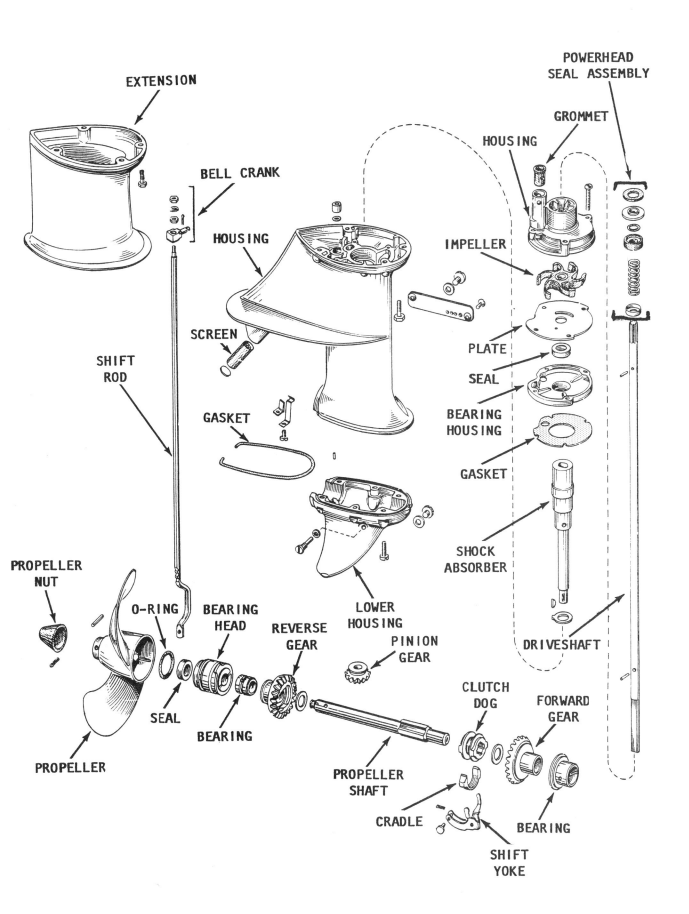

Exploded drawing of a lower unit requiring the powerhead to be removed to gain access to the shift disconnect.

ASSEMBLING

1- Place the lower unit on the work bench with the water pump recess facing upward. Install a **NEW** O-ring into the shift cavity. Work the bushing into place on top of the O-ring with a punch and mallet. Inject just a couple drops of oil into the bushing and O-ring as an assist during installation of the shift rod.

2- Turn the exhaust housing upside down. If the unit being serviced uses a Woodruff key, to secure the pinion gear to the driveshaft, install the driveshaft through the housing. Install the snap ring into the groove near the end of the driveshaft, and then the Woodruff key, and finally the the pinion gear onto the driveshaft. Lower the assembled driveshaft into place in the lower unit housing. If the driveshaft is the splined-type, lower the pinion gear into place at this time. The driveshaft will be installed later.

Assembling the Propeller Shaft

3- Slide the clutch dog onto the propeller shaft splines. Apply a light coating of lubricant to the washer and then insert it into the center of the forward gear. Slide the forward gear onto the end of the propeller shaft. Slide the forward gear bearing onto the shaft and into the forward gear.

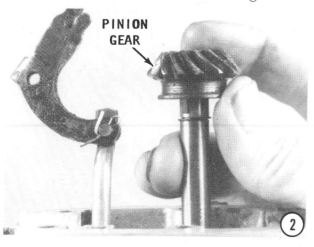

4- Apply a light coating of lubricant to the washer, and then insert it into the center of the reverse gear. Slide the reverse gear onto the propeller shaft from the propeller end. Check to be sure a new O-ring and bearing seal has been installed into the gear case head, and then install the gearcase head assembly onto the propeller shaft.

CRITICAL WORDS

Look into the front part of the lower unit housing. Notice the pin protruding up from the housing. Now, observe the slot in the forward gear bearing. When the propeller shaft assembly is installed into the lower unit, this pin **MUST** index into the hole

Locating pins in the lower unit housing. A hole in the forward gear bearing must index with the right pin and the hole in the gearcase head bearing with the left pin when the bearings are installed.

MANUAL SHIFT 5 HP TP 25 HP 8-29

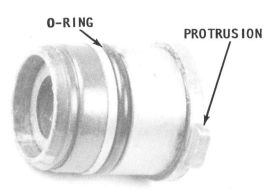

An assembled bearing head without the index hole. This bearing has a protrusion which indexes with a cutout in the lower unit housing.

in the forward gear bearing. Also notice the protrusion on the end of gearcase head. This protrusion **MUST** index with the slot in the housing when the propeller shaft assembly is installed. Some gearcase units may have a hole. A pin imbedded in the lower unit housing indexes with the hole during installation.

5- Check to be sure the pinion gear is properly located. Check to be sure the shift rod is clean and smooth (free of any burrs or corrosion). Coat the shift rod and the O-ring with oil as an aid to installation. Slide the shift rod down through the O-ring and bushing into the gear case.

6- Slide the propeller shaft assembly into the lower unit housing. Check to be sure the slot in the forward gear bearing indexes with the pin in the lower unit, and the protrusion on the end of the gear case head indexes in the slot in the housing. On some models, a pin in the lower unit housing **MUST** index with a hole in the gearcase head.

7- Lubricate the cradle, and then slip it into the clutch dog groove.

8- Bring the shift lever down over the cradle and snap the fingers of the lever into the cradle. Check to be sure the clutch dog is in the **NEUTRAL** position. Push or pull on the shift rod to move it up or down until the clutch dog is in the center between the forward and reverse gears. If the lever uses the pin-type holding mechanism, install the pin into the lever.

9- Lay down a bead of No. 1000 Sealer into the groove of the cap in preparation to installing the seal.

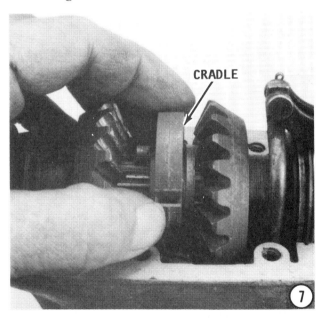

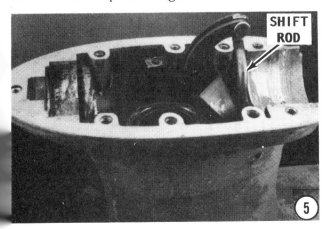

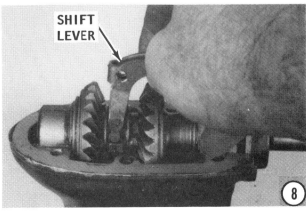

8-30 LOWER UNIT

Two shift lever screws. The right screw has been ground to assist installation, as explained in the text.

10- Place a **NEW** seal in the lower cap and hold the seal in the groove with sealer. Apply a small amount of silicone sealer on each side of the bearing gear case head. This sealer will form a complete seal when the lower unit cap is installed.

GOOD WORD

If time is taken to grind the end of the screw to a **SHORT** point, it will make the task of installation much easier. If the cap and shift lever are not aligned exactly, the screw will "seek" and make the alignment as it passes through. However, do not make a long point or the screw will not have enough support and would bend during operation of the shift lever.

11- Position the lower unit cap over the gear assembly onto the lower unit housing. If the unit being serviced uses a shift lever pin, work the cap until the pin indexes into the recess of the cap.

12- Apply a drop of sealer into the opening for each cap retaining screw to ensure a complete seal between the cap and the lower unit housing. Install the screws securing the cap to the lower unit housing. Tighten the screws **ALTERNATELY** and **EVENLY**.

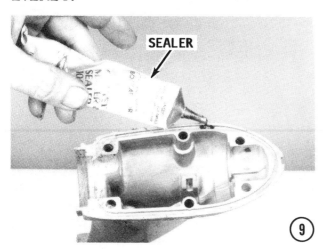

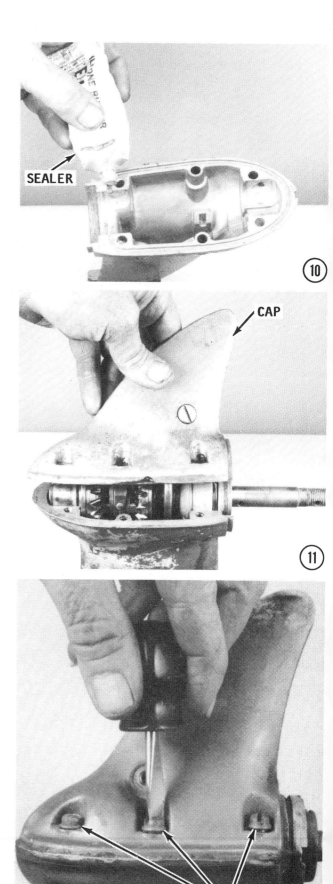

MANUAL SHIFT 5 HP TO 25 HP   8-31

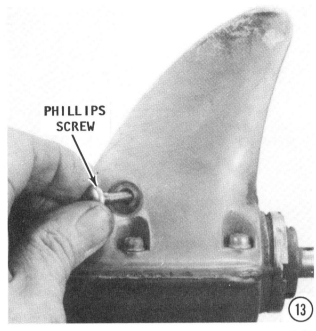

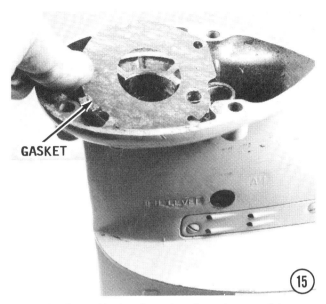

13- If the screw-type cap is used, with the Phillips screw in the side of the cap, use a flashlight and align the hole in the cap with the hole in the shift lever. Install the tapered Phillips screw into the housing and through the lever.

14- Apply a drop of sealer to the threads, and then tighten the screw securely.

15- Turn the lower unit rightside up. Install a **NEW** gasket onto the upper surface of the lower unit.

16- Install the bearing housing and bearing assembly by sliding a couple bolts through the housing to align the base gasket.

WATER PUMP INSTALLATION

17- Apply a coating of sealer to the upper surface of the lower unit. Install the water pump base plate.

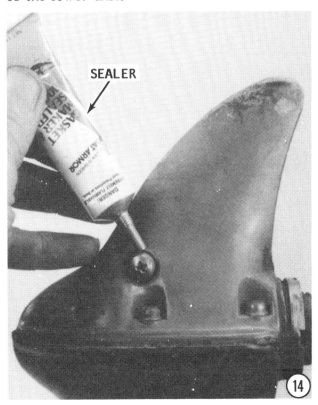

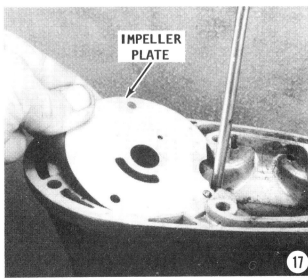

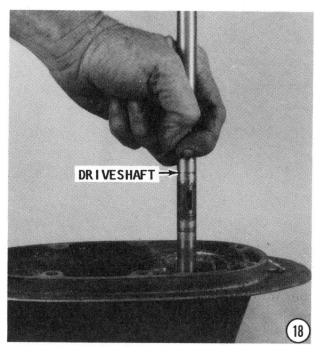

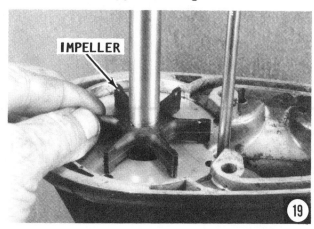

18- If the unit being serviced, uses the splined-type driveshaft, slide the driveshaft into the lower unit, and then rotate the shaft very slowly. When the splines of the driveshaft index with the pinion gear, the shaft will drop slightly. Install the water pump pin or key.

19- Slide the water pump down the driveshaft and into place on top of the water pump base plate with the pump pin or key indexed in the impeller. Lubricate the inside surface of the water pump with lightweight oil.

20- Lower the water pump housing down the driveshaft and over the impeller. Rotate the driveshaft **CLOCKWISE** as the water pump housing is lowered to allow the impeller blades to assume their natural and proper position inside the housing. Continue to rotate the driveshaft and work the water pump housing downward until it is seated on the lower unit upper housing.

21- **ALWAYS** rotate the driveshaft **CLOCKWISE** while the screws are tightened to prevent damaging the impeller vanes. If the impeller is not rotated, the housing could damage or cut the end of the vanes as the screws are brought up tight. The rotation allows them to spring back in a natural position.

22- Place a **NEW** grommet into the water pump housing for the water pickup. If a new water pump was installed, this seal will already be in place. Install a **NEW** O-ring on the top of the driveshaft.

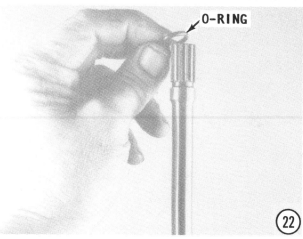

MANUAL SHIFT 5 HP TO 25 HP

23- If the lower unit being serviced uses a pin on the top of the driveshaft, install the pin at this time. Shift the lower unit into **FORWARD** gear and at the same time rotate the propeller shaft **CLOCKWISE**. The lower unit assembling is now complete and ready to mate with the exhaust housing.

LOWER UNIT INSTALLATION

GOOD WORDS

If the unit being serviced uses the shift rod coupler arrangement either through the window or before the lower unit and exhaust housings are fully mated, these words are extremely critical. Connecting the shift rod with the coupler is not an easy task but can be accomplished as follows: First, notice the cutout area on the end of the shift rod. This area permits the bolt to pass through the coupler, past the shift rod, and into the other side of the coupler. It is this bolt that holds the shift rod in the coupler. Now, in order for the bolt to be properly installed, the cutout area on the shift rod **MUST** be aligned in such a manner to allow the bolt to be properly installed. Therefore, as the lower unit is mated with the exhaust housing, exercise patience as the two units come together, to enable the bolt to be installed at the proper time. If the rod is allowed to

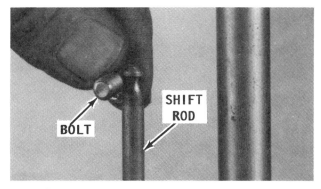

Shift rod and the screw through the connector. This illustration clearly shows a proper installation.

move too far into the coupler before the bolt is installed, it may be possible to force the bolt into place, past the shift rod. The threads on the bolt will be stripped, and the shift rod will eventually come out of the coupler.

24- Install the coupler onto the lower unit shift rod, with the **NO THREAD** section facing towards the window. With the coupler in this position, the bolt may be inserted through the coupler and "catch" the threads on the far side. Install the coupler bolt in the manner described in the previous paragraph.

Shift rod connector with the clear hole through the connector and shift rod, indicating the rod is properly located.

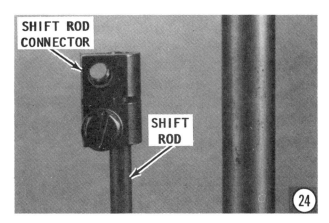

8-34 Lower Unit

SPECIAL WORDS

If the unit being serviced uses the shift rod coupler arrangement, then the coupler must be connected to the shift rod **BEFORE** the lower unit is fully mated with the exhaust housing, as described in the previous step and the "Good Words".

If the unit being serviced uses the driveshaft with the pin, extreme care must be exercised as the shaft is guided into the exhaust housing to allow the pin to index with the groove in the housing.

If the unit being serviced uses the bolt through the window arrangement, insert the bolt into the connector. **TAKE TIME** to read and understand the "Good Words" just before Step 24, before making this connection. After the bolt is in place, install and secure the window with the attaching hardware.

If the unit being serviced has the shift connector requiring the powerhead to be removed, install the shift rod into the shift linkage of the shift handle. Install the powerhead.

25- Check to be sure the water pickup tube is clean, smooth, and free of any corrosion. Coat the water pickup tube and grommet with lubricant as an aid to installation. Guide the lower unit up into the exhaust housing with the water tube sliding into the rubber grommet of the water pump. Continue to work the lower unit towards the exhaust housing, and at the same time rotate the propeller shaft as an aid to indexing the driveshaft splines with the crankshaft. Start the bolts securing the lower unit to the exhaust housings together. Tighten the bolts **EVENLY** and **ALTERNATELY**.

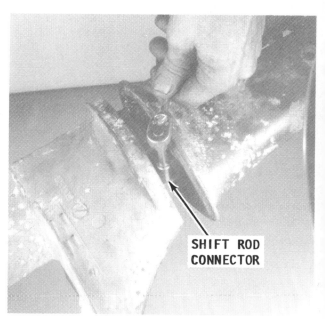

Connecting the shift rod before the lower unit is mated fully to the exhaust housing.

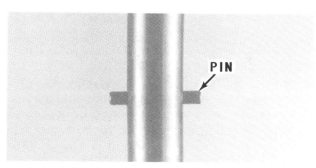

Driveshaft with the pin installed and ready for installation into the exhaust housing. As the shaft is raised, it must be rotated slowly to allow the pin to pass a slot in the housing. After the pin is past the slot, it must be rotated again to allow the splines to index with the splines in the crankshaft.

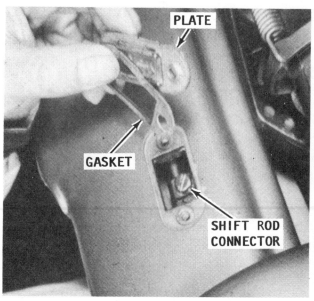

Window, ready to be installed after the shift rod has been properly connected.

MANUAL SHIFT 5 HP TO 25 HP   8-35

Shift rod connected on a lower unit requiring the powerhead to be removed to gain access to the connection.

26- Fill the lower unit with lubricant. Insert the lubricant tube into the bottom opening, and then fill the unit until lubricant is visible at the vent hole. Install the vent plug. Remove the gear lubricant tube, and install the drain/fill plug.

27- After the lower plug has been installed, remove the vent plug again and using a squirt-type oil can, add lubricant through this vent hole. A squirt-type oil can must be used to allow the trapped air in the lower unit to escape at the same time the final lubricant is added. Once the unit is completely full, install and tighten the vent plug.

A FEW GOOD WORDS

The propeller washer, if used, and drive pin play an extremely important role. When shifting gears during normal operation, or if the propeller should hit an underwater obstacle, the propeller is subjected to considerable shock. A washer is installed between the propeller and drive pin. This washer **MUST** always be in place for proper operation. If the hub should slip, the propeller will move back towards the propeller nut and lock against the drive pin. The washer is designed to stop propeller movement so the drive pin can be easily removed for service. Now, on with the installation.

28- Install the propeller. Coat the propeller shaft with an anti-corrosion grease. Install the propeller with the drive pin holes aligned. Install the washer and drive pin. Slide the propeller cap into place and secure it with the cotter pin.

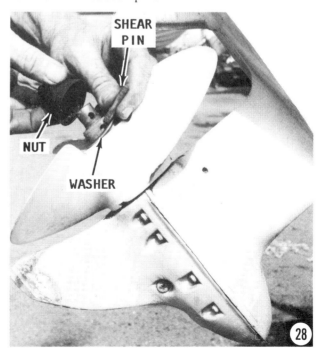

If the unit being serviced has a propeller using the rear-type shear pin arrangement: Install the shear pin and then coat the propeller shaft with anti-corrosion grease. Install the propeller, propeller nut, and then the cotter pin.

29- Final adjustment for remote control units: Shift the lower unit into **NEUTRAL** gear. At the shift box, move the shift lever to the **NEUTRAL** position. If the pin on the end of the shift cable, does not align with the shift handle, move the adjusting knob until the pin aligns and will move into the shift handle. With the shift cable removed, move the lower unit into **FORWARD** gear and at the same time rotate the propeller **CLOCKWISE** to ensure the gears are fully indexed. At the control box, move the shift lever into the **FORWARD** position. Again check to be sure the pin on the end of the shift cable aligns with the hole in the shift lever. Adjust the knob on the shift cable until the pin does align with the hole in the shift lever.

Perform a functional check of the completed work by mounting the engine in a test tank, in a body of water, or with a flush attachment connected to the lower unit. If the flush attachment is used, **NEVER** operate the engine above an idle speed, because the no-load condition on the propeller would allow the engine to **RUNAWAY** resulting in

in serious damage or destruction of the engine.

CAUTION: Water must circulate through the lower unit to the engine any time the engine is run to prevent damage to the water pump in the lower unit. Just five seconds without water will damage the water pump.

Start the engine and observe the tattle-tale flow of water from idle relief in the exhaust housing. The water pump installation work is verified. If a "Flushette" is connected to the lower unit, **VERY LITTLE** water will be visible from the idle relief port. Shift the engine into the three gears and check for smoothness of operation and satisfactory performance.

8-7 LOWER UNIT SERVICE MANUAL SHIFT -- 28 HP TO 40 HP

Propeller Removal
Remove the propeller according to the detailed procedures outlined in Section 8-3.

Draining Lower Unit
Drain the lower unit according to the detailed procedures outlined in Section 8-4.

GOOD WORDS
Only one type of shift mechanism and removal procedures are used on the engines covered in this section. The mechanism installed is referred to as Type D in the Appendix. Access to the shift disconnect is through a window on the starboard side of the engine.

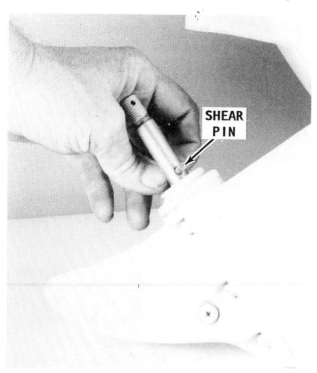

Location of the shear pin through the propeller shaft just ahead of the propeller.

MANUAL SHIFT 28 HP TO 40 HP

LOWER UNIT REMOVAL

Type D — Window Removal to Gain Access

1- Remove the metal plate from the port side of the engine. Access to the shift coupler is gained through the opening. Disconnect the shift rod from the exhaust housing by removing the bottom bolt from the shift coupler.

GOOD WORDS

In **MOST** cases, if any unit being serviced has the 6-inch extension, it is **NOT** necessary to remove the extension in order to "drop" the lower unit.

2- Remove the rear exhaust housing cover installed on some 35 hp models and on all 40 hp models. This cover **MUST** be removed to gain access to one of the bolts securing the lower unit to the exhaust housing. Remove the bolts securing the lower unit to the housing. Some units may have an additional bolt on each side and one at the rear of the engine. Work the lower unit loose from the exhaust housing. It is not uncommon for the water tube/s to be stuck in the water pump making separation of the lower unit from the exhaust housing difficult. However, with patience and persistence, the tube/s will come free of the pump and the lower unit separated from the exhaust housing.

MORE GOOD WORDS

Position the lower unit in a vertical position on the edge of the work bench

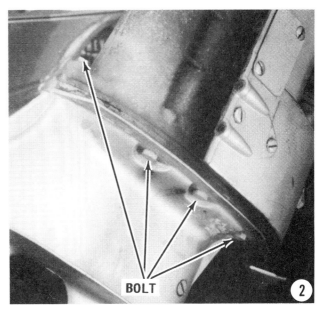

resting on the cavitation plate. Secure the lower unit in this position with a C-clamp. The lower unit will then be held firmly in a favorable position for further service work. An alternate method is to cut a groove in a short piece of 2" x 6" wood to accommodate the lower unit with the cavitation plate resting on top of the wood. Clamp the wood in a vise and service work may then be

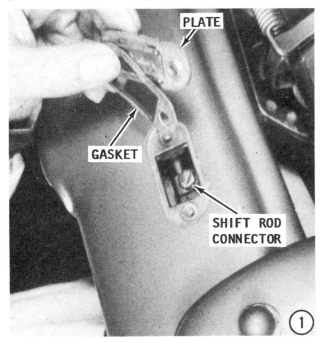

A piece of 2" x 6" wood with a cutout to accept the lower unit housing. The piece of wood is clamped in a vise and the lower unit placed in position with the cavitation plate resting on the wood. The lower unit is then ready for service work.

8-38 LOWER UNIT

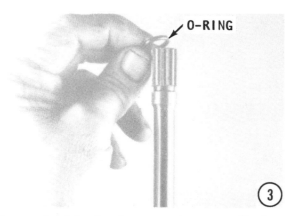

performed with the lower unit erect (in its normal position), or inverted (upside down). In both positions, the cavitation plate is the supporting surface.

3- Remove and **DISCARD** the O-ring from the top of the driveshaft.

WATER PUMP REMOVAL

AUTHOR'S APOLOGY

The photographs taken for this section involved a water pump from a electric shift lower unit. However, the pump and the service procedures are identical for the manual shift. Therefore, disregard any wiring shown in the photographs.

4- Remove the O-ring from the top of the water pump. Remove the screws securing the water pump to the lower unit housing. It is very possible corrosion will cause the screw heads to break-off when an attempt to remove them is made. If this should happen, use a chisel and breakaway the water pump housing from the lower unit. **EXERCISE CARE** not to damage the lower unit housing.

5- After the screws have been removed, slide the water pump, impeller, the impeller key, and the lower water pump plate upward and free of the driveshaft.

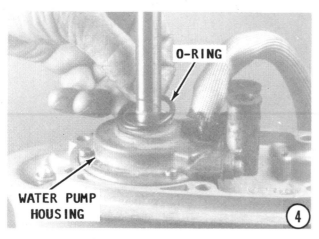

GOOD NEWS

If the only work to be performed is service of the water pump, disregard the following steps, and proceed directly to the Water Pump Assembling portion of this section on Page 8-48.

LOWER UNIT DISASSEMBLING

6- **CAREFULLY** pull upward on the driveshaft and remove it from the lower unit.

7- Turn the lower unit upside down and again clamp it in the vise or slide it into the wooden block, if one is used. Carefully examine the lower portion of the unit. The cap is considered that part below the split

MANUAL SHIFT 28 HP TO 40 HP

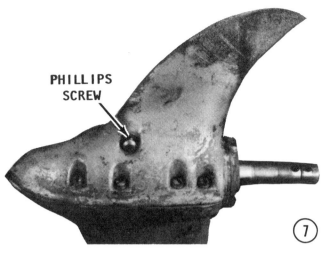

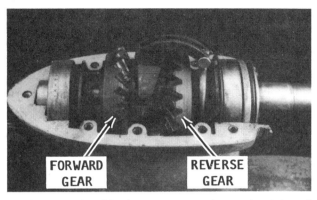

Lower unit with the cap removed exposing internal parts. Take time to make a sketch or take a photograph showing the arrangement of parts as an assist during the installation work.

with the skeg attached. Remove the Phillips screw from the starboard side of the lower housing. This screw passes through the shift yoke and threads into the other side of the housing.

8- Remove the attaching screws around the cap. These screws may be slotted-type or Phillips screws. **CAREFULLY** tap the cap to jar it loose, and then separate it from the lower unit housing. If the cap did not have a Phillips screw on the outside, observe the two slots inside the cap.

TAKE TIME

Before proceeding with the disassembly work, take time to study the arrangement of parts in the lower unit. You may elect to follow the practice of many professional mechanics and take a polaroid picture of the unit as an aid during the assembly work. Several engineering and production changes have been made to the lower unit over the years. Therefore, the positioning of the gears, shims, bearings, and other parts may vary slightly from one unit to the next.

To show each and every arrangement with a picture in this manual would not be practical. Even if it were done, the ability to associate the unit being serviced with the illustration would be almost impossible. Therefore, take time to make notes, scribble out a sketch, or take a couple photographs.

9- Lift the shift lever out of the cradle, and then remove the cradle from the shift dog. Raise the propeller shaft and at the same time tap with a soft-headed mallet on the bottom side to jar it loose, then remove the shaft assembly from the lower unit. The forward and reverse gear including the bearings will all come out with the propeller shaft. The forward gear is the gear at the opposite end of the shaft from the propeller.

GOOD WORDS

Notice that the forward gear bearing is a tapered bearing with a race and that the taper faces outward, **AWAY** from the gear.

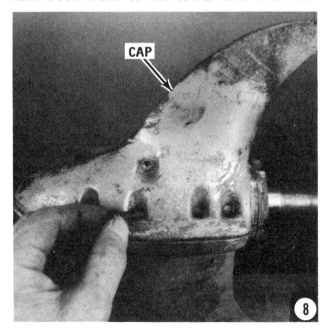

8-40 LOWER UNIT

Arrangement of parts after the propeller shaft has been removed.

Also observe the seal retainer on the propeller end of the propeller shaft. Now, notice the matching pin in the lower unit housing. During the installation work, the retainer **MUST** be installed with the pin indexed in the hole. Take note of the snap ring installed between the thrust washer and the reverse gear bearing. One more item of particular interest. Notice the two sides of the thrust washer. One side is as a normal washer, but the other side is a babbitt. The babbitt side **MUST** face toward the reverse gear during installation. The washer also has two dog ears, one facing upward and the other downward.

By taking note at this time of these items and exactly how they are installed, the task of assembling and installation will progress more smoothly.

10- Remove the attaching screws, and then the U-shaped bracket from the top of the pinion gear

11- Reach into the lower housing and remove the pinion gear.

12- Slide the tapered bearing, forward gear, washer, and clutch dog off the propeller shaft.

13- Remove the seal retainer, reverse bearing, snap ring, washer, reverse gear and bearing, washer from the propeller end of the shaft.

14- Turn the lower unit housing right side up and again clamp it in the vise.

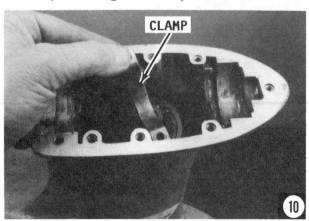

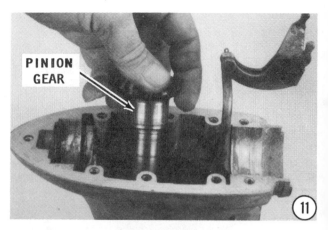

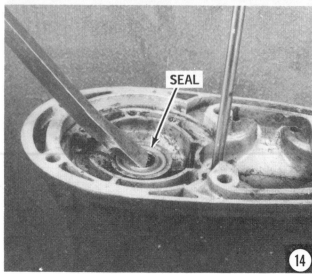

MANUAL SHIFT 28 HP TO 40 HP 8-41

Using a slide hammer to remove the seals or bearings from underneath the water pump housing.

Remove the upper seal using a seal puller. An alternate method, if the puller is not available, is to use two screwdrivers and work the seal out of the housing. **TAKE CARE** not to damage the seal recess as the seal is being removed. A babbitt bearing is installed under the seal. Late model units may have caged neddle bearings installed. Normally, it is **NOT** necessary to remove this bearing. However, check the bearing surface with a finger and if any roughness is felt, the bearing **MUST** be replaced.

15- Clean the upper portion of the shift rod as an aid to pulling it through the bushing and O-ring. Pull the shift rod from the lower unit housing.

16- The shift rod passes through an O-ring and bushing in the lower unit housing. These two items prevent water from entering the lower unit. A special tapered

punch is required to remove the bushing from the lower unit housing. Obtain the special punch, and then remove the bushing, and the O-ring.

CLEANING AND INSPECTING

Clean all water pump parts with solvent, and then dry them with compressed air. Inspect the water pump cover and base for cracks and distortion, possibly caused from overheating. Inspect the face plate and water pump insert for grooves and/or rough surfaces. If possible, **ALWAYS** install a complete new water pump while the lower unit is disassembled. A new impeller will ensure extended satisfactory service and give "peace of mind" to the owner. If the old impeller must be returned to service, **NEVER** install it in reverse to the original direction of rotation. Installation in reverse will cause premature impeller failure.

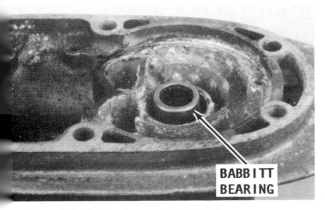

Caged ball bearing set installed on some newer model lower units instead of the babbitt bearing.

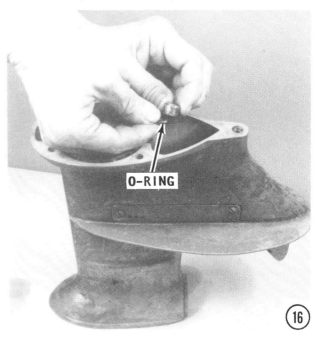

Babbitt bearing unfit for further service. This type bearing must not show any signs of wear.

View of a badly corroded lower unit. Water entered and was allowed to remain over an extended period of time causing extensive damage.

Inspect the impeller side seal surfaces and the ends of the impeller blades for cracks, tears, and wear. Check for a glazed or melted appearance, caused from operating without sufficient water. If any question exists, and as previously stated, install a new impeller if at all possible.

Clean all parts with solvent and dry them with compressed air. **DISCARD** all O-rings and gaskets. Inspect and replace the driveshaft if the splines are worn. Inspect the gearcase and exhaust housing for damage to the machined surfaces. Remove any nicks and refurbish the surfaces on a surface plate. Start with a No. 120 Emery paper and finish with No. 180.

Badly worn pinion gear. The teeth of all gears must be carefully inspected for wear and damage.

Caged ball bearing set destroyed due to lack of lubrication, vibration, corrosion, metal particles, or all of the above.

A rusted and corroded gear. Water was allowed to enter the lower unit through a bad seal and cause this damage to the gear and other expensive parts.

MANUAL SHIFT 28 HP TO 40 HP 8-43

A new shift cradle (right) compared with one badly worn and unfit for service (left).

Check the water intake screen and passages by removing the bypass cover, if one is used. Inspect the clutch dog, drive gears, pinion gear, and thrust washers. Replace these items if they appear worn. If the clutch dog and drive gear arrangement surfaces are nicked, chipped, or the edges rounded, the operator may be performing the shift operation improperly or the controls may not be adjusted correctly. These items **MUST** be replaced if they are damaged.

Inspect the dog ears on the inside of the forward and reverse gears. The gears must be replaced if they are damaged.

Check the cradle that rides on the inside diameter of the clutch dog. The sides of the

Distorted tapered bearing unfit for further service.

cradle must be in good condition, free of any damage or signs of wear. If damage or wear has occurred, the cradle must be replaced.

Check the shift lever and the two prongs that fit inside the cradle. Check to be sure the prongs are not worn or rounded. Damage or wear to the prongs indicates the lever must be replaced.

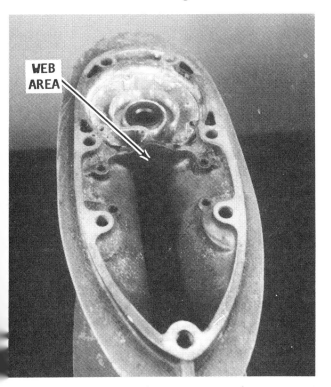

The web area indicated should be inspected carefully for the slightest sign of a hairline crack. The smallest evidence of a crack is cause to replace the housing because exhaust gases will find their way into the water pump area when the engine is operating and the boat planing. These exhaust gases will cause the engine to run hot.

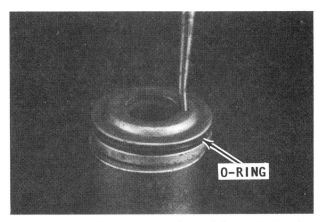

Using a punch to remove the seal from the bearing retainer.

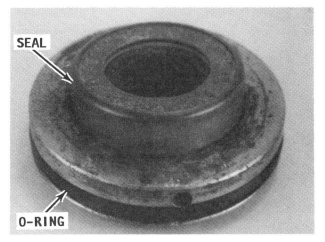

Installing a seal into the bearing retainer. Notice the O-ring has been installed.

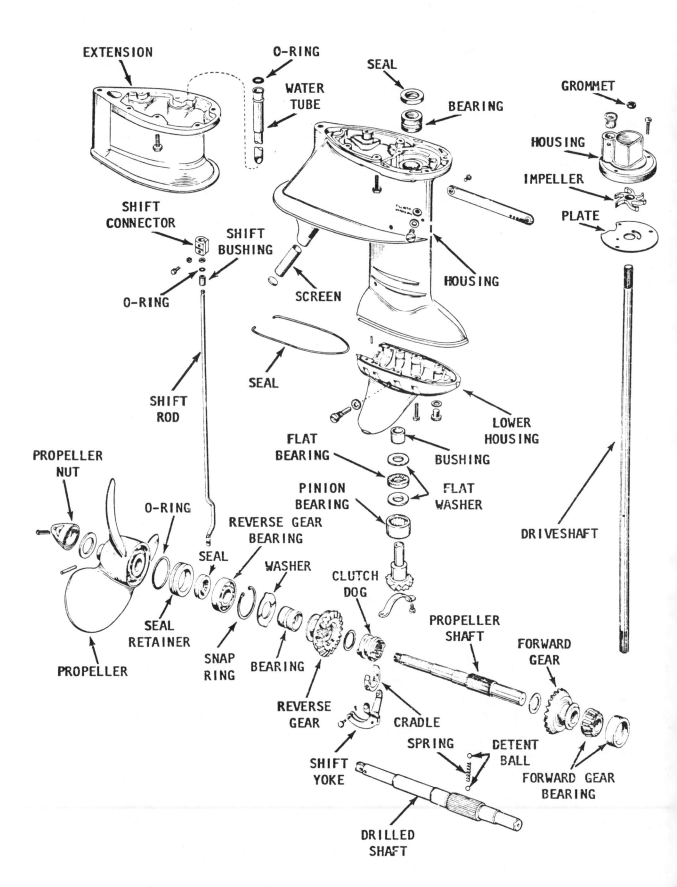

Exploded drawing of a typical manual shift lower unit installed on 35 hp and 40 hp engines. Two propeller shafts are shown. The second shaft with the two balls and spring may be installed as a conversion. This new type shaft arrangement holds the shift dog more securely in all gear positions. Conversion parts are available as separate items.

MANUAL SHIFT 28 HP TO 40 HP   8-45

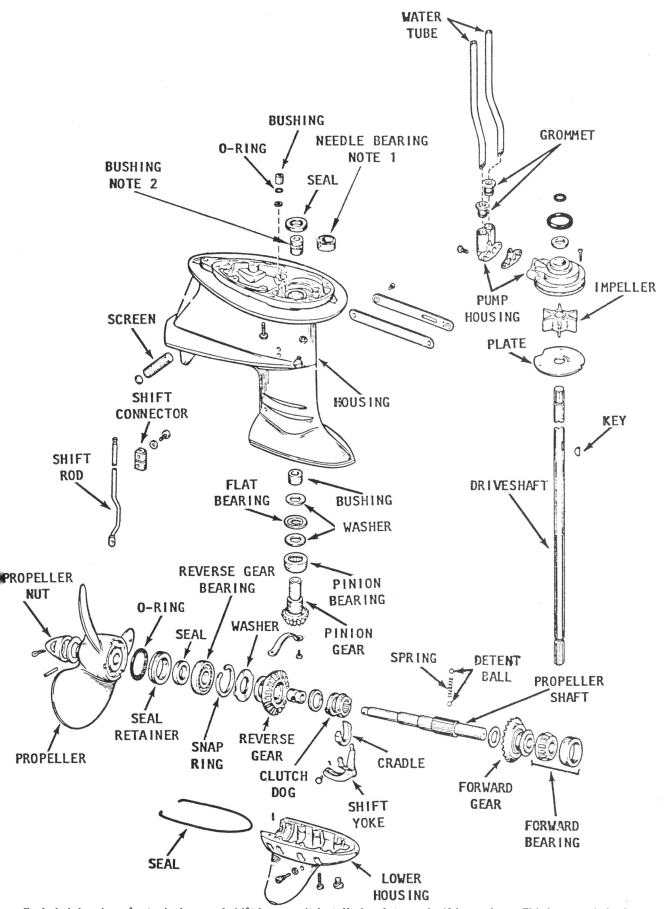

Exploded drawing of a typical manual shift lower unit installed on late mode 40 hp engines. This lower unit is almost identical to the one shown on Page 8-44, except the water pump has a two-line recirculating system. The propeller conversion parts shown are available as separate items.

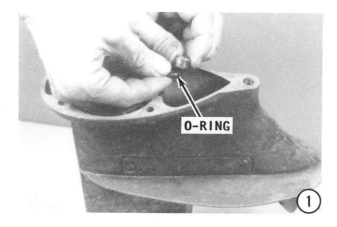

ASSEMBLING

1- Place the lower unit on the work bench with the water pump recess facing upward. Install a **NEW** O-ring into the shift cavity. Work the bushing into place on top of the O-ring with a punch and mallet. Inject just a couple drops of oil into the bushing and O-ring as an assist during installation of the shift rod.

2- Lower the pinion gear into the housing. Check to be sure it seats properly.

3- Lower the U-shaped pinion gear retaining bracket into position and secure it in place with the attaching screws.

4- Check to be sure the pinion gear is properly located. Check to be sure the shift rod is clean and smooth (free of any burrs or corrosion). Coat the shift rod and the O-ring with oil as an aid to installation. Slide the shift rod down through the O-ring and bushing into the gear case.

Assembling the Propeller Shaft

5- Apply a light coating of lubricant to the washer, and then insert it into the center of the reverse gear. Slide the reverse gear onto the propeller shaft from the propeller end. Install the thrust washer with

the babbitt side **TOWARDS** the reverse gear. Install the snap ring, the bearing. Check to be sure a **NEW** seal and O-ring has been installed into the seal retainer, and then install the retainer.

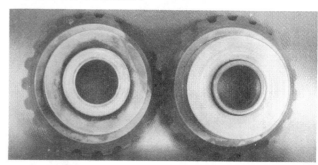

Forward bearing (left) with a non-replaceable babbitt bearing. Reverse bearing (right) with a sliding babbitt bearing that is replaceable.

6- Slide the clutch dog onto the propeller shaft splines. Apply a light coating of lubricant to the washer and then insert it into the center of the forward gear. Slide the forward gear onto the end of the propeller shaft. Slide the forward gear bearing onto shaft with the large end of the taper **TOWARDS** the forward gear. Move the bearing into place on the forward gear.

Check to be sure a new O-ring and bearing seal has been installed into the gear case head, and then install the gearcase head assembly onto the propeller shaft.

CRITICAL WORDS

The seal retainer has a hole and the lower housing of the lower unit has a pin. This pin **MUST** index into the hole in the retainer when the propeller shaft is installed. If the pin is not seated properly in the hole, the seal retainer will work part way out of the housing and the lubricant in the lower unit will be lost.

7- Slide the propeller shaft assembly into the lower unit housing. Check to be sure the forward and reverse gear index with the pinion gear and the hole in the seal retainer indexes with the pin in the lower unit housing. Lubricate the cradle, and then slip it into the clutch dog groove.

8- Bring the shift lever down over the cradle and snap the fingers of the lever into the cradle. Check to be sure the clutch dog is in the **NEUTRAL** position. Push or pull on

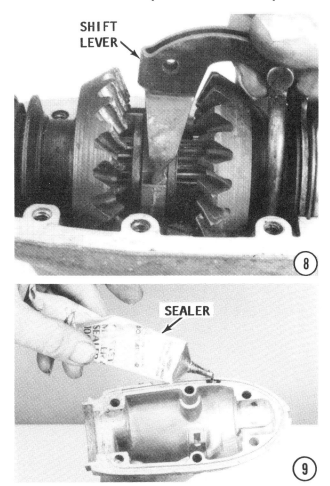

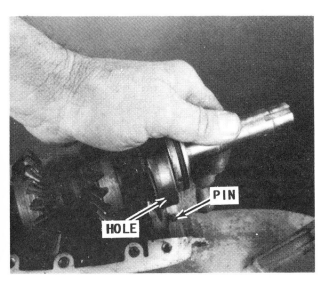

The seal retainer ready to be installed. The hole in the retainer must index over the pin in the housing.

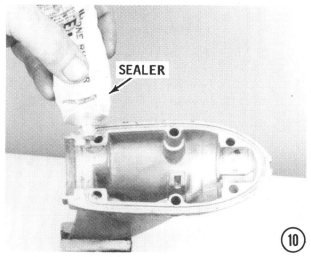

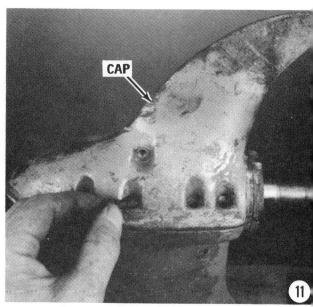

the shift rod to move it up or down until the clutch dog is in the center between the forward and reverse gears.

9- Lay down a bead of No. 1000 Sealer into the groove of the cap in preparation to installing the seal.

10- Place a **NEW** seal in the lower cap and hold the seal in the groove with sealer. Apply a small amount of silicone sealer on each side of the bearing gear case head. This sealer will form a complete seal when the lower unit cap is installed. Position the lower unit cap over the gear assembly onto the lower unit housing.

11- Apply a drop of sealer into the opening for each cap retaining screw to ensure a complete seal between the cap and the lower unit housing. Install the screws securing the cap to the lower unit housing. Tighten the screws **ALTERNATELY** and **EVENLY**.

GOOD WORD

If time is taken to grind the end of the screw to a **SHORT** point, it will make the task of installation much easier. If the cap and shift lever are not aligned exactly, the screw will "seek" and make the alignment as it passes through. However, do not make a long point or the screw will not have enough support and would bend during operation of the shift lever.

12- Use a flashlight and align the hole in the cap with the hole in the shift lever. Install the tapered Phillips screw into the housing and through the lever. Tighten the screw securely.

13- Install the babbitt or neddle bearing, if it was removed. The babbitt bearing may be installed using the proper size socket and hammer. If the caged needle bearing is installed tap on the numbered side of the bearing.

14- Coat the outside edge of a **NEW** seal with No. 1000 sealer, and then tap the seal into place in the top of the upper lower unit housing.

Two shift lever screws. The right screw has been ground to assist installation, as explained in the text.

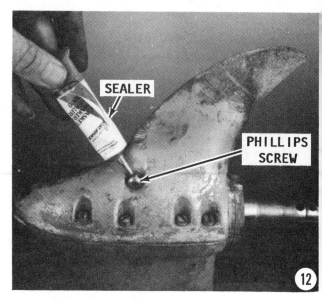

MANUAL SHIFT 28 HP TO 40 HP

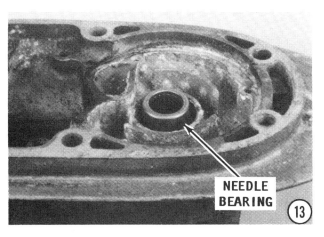

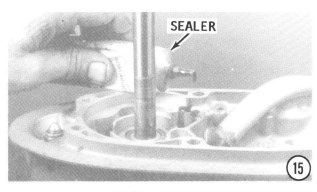

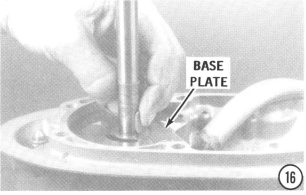

WATER PUMP INSTALLATION

15- Apply a coating of sealer to the upper surface of the lower unit.

16- Install the water pump base plate. Slide the driveshaft into the lower unit, and then rotate the shaft very slowly. When the splines of the driveshaft index with the pinion gear, the shaft will drop slightly. Install the water pump pin or key.

17- Slide the water pump down the driveshaft and into place on top of the water pump base plate with the pump pin or key indexed in the impeller. Lubricate the inside surface of the water pump with lightweight oil.

18- Lower the water pump housing down the driveshaft and over the impeller. Rotate the driveshaft **CLOCKWISE** as the water pump housing is lowered to allow the impeller blades to assume their natural and proper position inside the housing. Continue to rotate the driveshaft and work the water pump housing downward until it is seated on the lower unit upper housing.

ALWAYS rotate the driveshaft **CLOCKWISE** while the screws are tightened to prevent damaging the impeller vanes. If the impeller is not rotated, the housing could

Standard water pump (left) and a new type (right). The newer pump is only used with an updated lower unit.

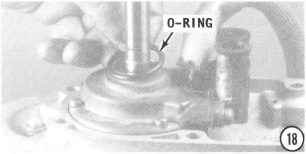

8-50 LOWER UNIT

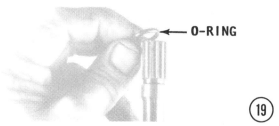

damage or cut the end of the vanes as the screws are brought up tight. The rotation allows them to spring back in a natural position.

Place **NEW** grommets into the water pump housing for the water pickup. If a new water pump was installed, this seal will already be in place.

19- Install a **NEW** O-ring on the top of the driveshaft.

LOWER UNIT INSTALLATION

GOOD WORDS

Connecting the shift rod with the coupler is not an easy task but can be accomplished as follows: First, notice the cutout area on the end of the shift rod. This area permits the bolt to pass through the coupler, past the shift rod, and into the other side of the coupler. It is this bolt that holds the shift rod in the coupler. Now, in order for the bolt to be properly installed, the cutout area on the shift rod **MUST** be aligned in such a manner to allow the bolt to be properly installed. Therefore, as the lower unit is mated with the exhaust housing, exercise patience as the two units come together, to enable the bolt to be installed at the proper time. If the rod is allowed to move too far into the coupler before the bolt is installed, it may be possible to force

Shift rod connector with the clear hole through the connector and shift rod, indicating the rod is properly located.

the bolt into place, past the shift rod. The threads on the bolt will be stripped, and the shift rod will eventually come out of the coupler.

20- Install the coupler onto the lower unit shift rod, with the **NO THREAD** section facing towards the window. With the coupler in this position, the bolt may be inserted through the coupler and "catch" the threads on the far side. Install the coupler bolt in the manner described in the previous paragraph.

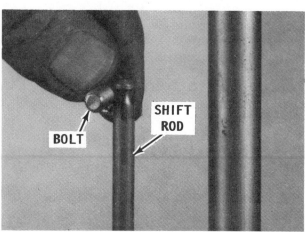

Shift rod with the bolt alongside to illustrate how the bolt must fit into the shift rod groove when the bolt is installed through the connector.

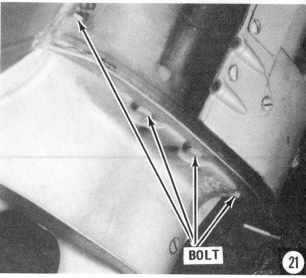

MANUAL SHIFT 28 HP TO 40 HP   8-51

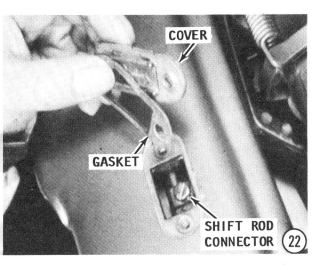

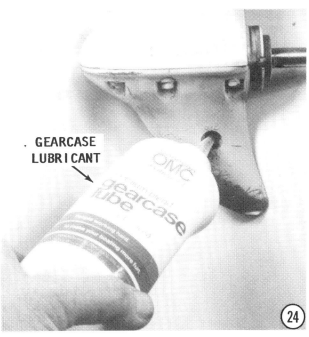

21- Check to be sure the water pickup tubes are clean, smooth, and free of any corrosion. Coat the water pickup tubes and grommets with lubricant as an aid to installation. Guide the lower unit up into the exhaust housing with the water tube sliding into the rubber grommet of the water pump. Continue to work the lower unit towards the exhaust housing, and at the same time rotate the propeller shaft as an aid to indexing the driveshaft splines with the crankshaft.

22- Insert the bolt into the connector. **TAKE TIME** to read and understand the "Good Words" just before Step 20, before

making this connection. After the bolt is in place, install and secure the window with the attaching hardware.

Start the bolts securing the lower unit to the exhaust housing. Tighten the bolts **EVENLY** and **ALTERNATELY**.

23- Install the rear cover over the exhaust housing on some 35 hp and all 40 hp engines. When the cover is installed, check to be sure the idle relief rubber tube on the upper side underneath the powerhead fits into the recess of the cover. Secure the cover in place with the attaching hardware.

24- Fill the lower unit with lubricant. Insert the lubricant tube into the bottom opening, and then fill the unit until lubricant is visible at the vent hole. Install the vent

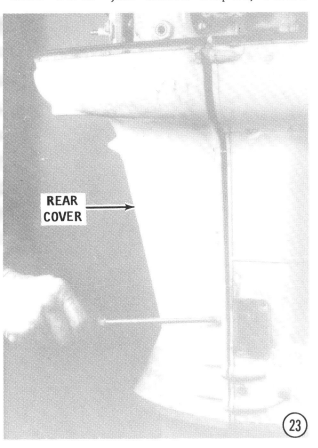

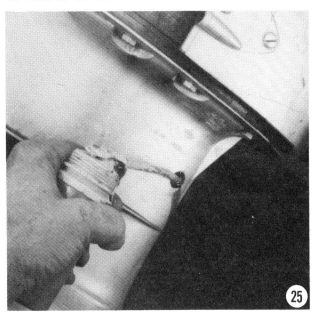

8-52 LOWER UNIT

plug. Remove the gear lubricant tube, and install the drain/fill plug.

25- After the lower plug has been installed, remove the vent plug again and using a squirt-type oil can, add lubricant through this vent hole. A squirt-type oil can must be used to allow the trapped air in the lower unit to escape at the same time the final lubricant is added. Once the unit is completely full, install and tighten the vent plug.

A FEW GOOD WORDS

The propeller washer, if used, and drive pin play an extremely important role. When shifting gears during normal operation, or if the propeller should hit an underwater obstacle, the propeller is subjected to considerable shock. A washer is installed between the propeller and drive pin. This washer **MUST** always be in place for proper operation. If the hub should slip, the propeller will move back towards the propeller nut and lock against the drive pin. The washer is designed to stop propeller movement so the drive pin can be easily removed for service. Now, on with the installation.

26- Install the propeller. Coat the propeller shaft with an anti-corrosion grease. Install the propeller with the drive pin holes aligned. Install the washer and drive pin. Slide the propeller cap into place and secure it with the cotter pin.

27- Final adjustment for remote control units: Shift the lower unit into **NEUTRAL** gear. At the shift box, move the shift lever to the **NEUTRAL** position. If the pin on the end of the shift cable, does not align with the shift handle, move the adjusting knob until the pin aligns and will move into the shift handle. With the shift cable removed, move the lower unit into **FORWARD** gear and at the same time rotate the propeller **CLOCKWISE** to ensure the gears are fully indexed. At the control box, move the shift lever into the **FORWARD** position. Again check to be sure the pin on the end of the shift cable aligns with the hole in the shift lever. Adjust the knob on the shift cable until the pin does align with the hole in the shift lever.

Perform a functional check of the completed work by mounting the engine in a test tank, in a body of water, or with a flush attachment connected to the lower unit. If the flush attachment is used, **NEVER** operate the engine above an idle speed, because the no-load condition on the propeller would allow the engine to **RUNAWAY** resulting in serious damage or destruction of the engine.

CAUTION: Water must circulate through the lower unit to the engine any time the engine is run to prevent damage to the water pump in the lower unit. Just five seconds without water will damage the water pump.

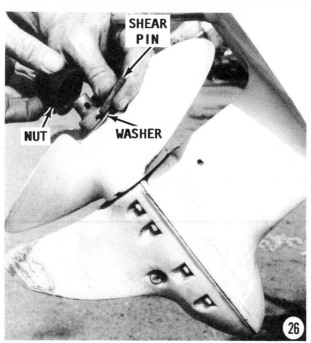

Start the engine and observe the tattle-tale flow of water from idle relief in the exhaust housing. The water pump installation work is verified. If a "Flushette" is connected to the lower unit, **VERY LITTLE** water will be visible from the idle relief port. Shift the engine into the three gears and check for smoothness of operation and satisfactory performance.

8-8 ELECTROMATIC LOWER UNIT

DESCRIPTION

Electromatic gearcases were used on all Johnson 40 hp units from 1962 thru 1970. Prior to 1962, the 40 hp unit was equipped with a manual shift, as covered in the previous section.

When the unit is shifted to the forward position, an electric switch in the shift box closes the circuit to the forward electromagnetic coil in the gearcase. After the coil is energized, magnetism attracts and anchors the free end of the clutch spring to the flange of the clutch hub. The revolving gear causes the spring to wrap around the hub, creating a direct coupling with the propeller shaft.

Power is transmitted through the pinion gear, forward gear, and propeller shaft to the propeller.

When the lower unit is shifted to the reverse position, the reverse coil is energized, and the same sequence of events takes place. The reverse gear assembly is **AL-WAYS** the one nearest the propeller.

The boat battery provides 12-volt power for operation. Therefore, all engines covered in this manual are equipped with a generator to maintain battery amperage and voltage for efficient operation of the shift mechanism. When the key is in the **ON** position, power moves through the ignition switch to the switch in the shift box, and on to the lower unit.

The necessary wiring is routed from the dash to the shift box; then to the rear of the engine to a knife-disconnect fitting; and then down to the lower unit. The forward shift wire is green and the reverse wire is blue. An easy way to remember the color code is green for go, forward that is.

TROUBLESHOOTING

In order to prevent unnecessary service work, specific troubleshooting should be performed. The following steps present a logical sequence of tests and checks to pinpoint problems in the lower unit.

1- Check the quantity of lubricant in the lower unit and top it off, if necessary. The unit will not operate properly if a lubricant other than OMC Type C is used. If any doubt exists as to the type of lubricant in the lower unit, drain the unit, refill with the Type C material, and then check operation of the shift mechanism. At the same time the quantity of lubricant is being checked, observe the material carefully for any sign of water. Position a suitable container under the lower unit, and then remove the **FILL** plug and the **VENT** plug. Allow the lubricant to drain into the container. As the lubricant drains, catch some with your fingers, from time-to-time, and rub it between your thumb and finger to determine if any metal particles are present. If metal is detected in the lubricant, the unit must be

Gear arrangement used on the 40-hp electromatic shift lower unit.

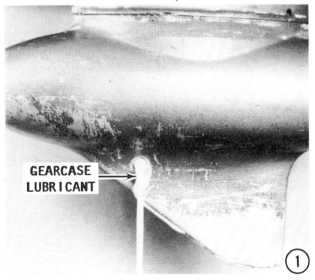

8-54 LOWER UNIT

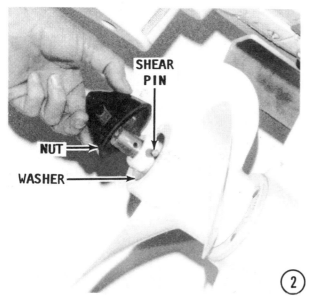

completely disassembled, inspected, the cause of the problem determined, and then corrected.

2- Check for a broken shear pin by removing the propeller. First pull the cotter key, and then remove the propeller nut, drive pin, and washer. Because the drive pin is not a tight fit, the propeller is able to move on the pin and cause burrs on the hole. Propeller removal may be difficult because of these burrs. To overcome this problem, the propeller hub has two grooves running the full length of the hub. Hold the shaft from turning, and then rotate the propeller 1/4 turn to position the grooves over the drive pin holes. The propeller can then be pulled straight off the shaft. After the

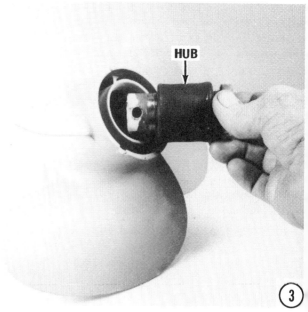

propeller has been removed, file the drive pin holes on both sides of the shaft to remove the burrs.

3- Check the propeller and the rubber hub. See if the hub is shredded. If the propeller has been subjected to many strikes against underwater objects, it could slip on its hub. For this size engine, in most cases, it is less expensive to purchase a new propeller instead of having it rebuilt.

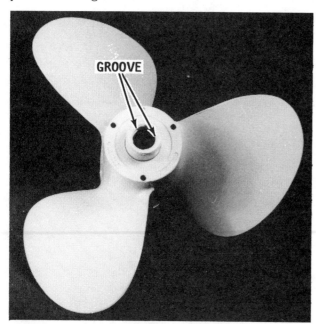

Propeller with the grooves in the hub as an assist to removal from the shaft.

ELECTROMATIC SHIFT

4- Battery Check: Begin with a thorough check of the battery. Measure the gravity of the electrolyte in each cell by withdrawing only enough to lift the float. Take the reading at eye level. A fully charged battery cell should read 1.280; at half-charge, 1.210; and a dead battery will read only 1.150. If the electrolyte level is low, bring it up to full level with clean clear water. **NEVER ADD ACID** to a battery cell. If water is added, it is not possible to take an accurate reading until the battery has been charged for a few hours.

5- Battery Voltage: Check the total battery voltage for a full 12 volts. Clean any corrosion from on or around the cables and terminals. Remove the cables, clean the posts until bright metal is visible. Scrape out the inside of the battery terminals, then connect and tighten them securely.

6- Amperage Draw Check: Turn the ignition switch to the **ON** position and observe the amperage reading on the dash ammeter. If an ammeter is not installed on the dash, one must be temporarily connected to the system for this test by first removing the wire marked **BAT** from the key switch, and then connecting the ammeter in

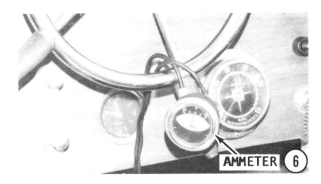

series with this wire and the key switch terminal marked **BAT**.

Check the current draw. If the draw exceeds 2.5 amps in either gear, disconnect the shift wires at the shift wire disconnects on the port side of the engine. Again, check the current draw. A higher reading than 2.5 amps indicates a short in the wiring; in the shift switch; or in the shift box. If the readings are within acceptable limits, reconnect the shift wires at the engine, shift the unit into forward and then reverse gear. Check the current draw in each gear. A high-amp draw, indicates a shorted wire to the lower unit, or a short in one of the coils.

A broken driveshaft from the powerhead to the lower unit indicates both forward and reverse gears were energized at the same time. Check the shift box, shift switch, and the wiring to the lower unit.

7- Defective Wiring Check: Leave the wire marked **BAT** disconnected from the key switch for this test. Check the wiring

8-56 Lower Unit

leading to the ignition switch and from the switch for 12 volts. If the reading is less than 12 volts, the key switch is defective and should be replaced.

8– Shift Box and Coil Tests: Disconnect the blue and green shift wires at the rear of the engine. Connect one lead of a voltmeter to the green wire of the shift box, and the other lead to a good ground. Turn the ignition switch to **ON**, and move the shift lever into forward gear. The voltmeter must indicate 12 volts. Next, connect the voltmeter to the blue wire, shift into reverse gear, and the voltmeter should indicate 12 volts. If the voltmeter fails to indicate 12 volts during either one of these tests, the shift box requires service, see Chapter 7, Accessories.

Leave the shift wires disconnected; turn the ignition switch **OFF**; move the shift lever to the **NEUTRAL** position; and connect one lead of an ohmmeter to the green (forward) wire leading from the rear of the engine to the lower unit, and the other lead to a good ground. The ohmmeter should indicate from 4.5 to 6.5 ohms. Make the same test for reverse gear, the blue wire leading from the rear of the engine to the lower unit, and check for the same reading. If the ohmmeter fails to indicate the required resistance, a wire is broken, or the coil in the lower unit is shorted.

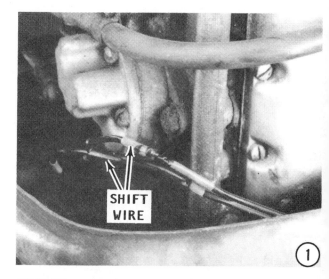

ELECTROMATIC REMOVAL

1– Disconnect the spark plug wires to prevent engine start while performing the work. At the rear of the engine: Slide the insulating sleeve back on the shift cable wires. Disconnect the shift and engine shift terminals.

2– Remove the Phillips screws along each side of the exhaust cover.

3– Remove the attaching hardware from the exhaust plate on the port side of the engine. Remove the two screws from the inner plate and the clamp on the shift wire.

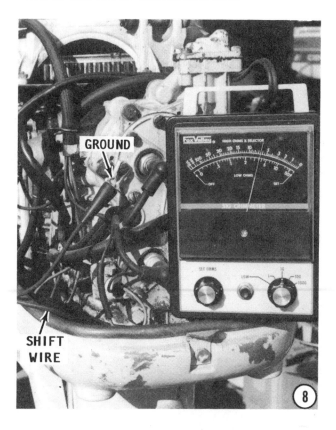

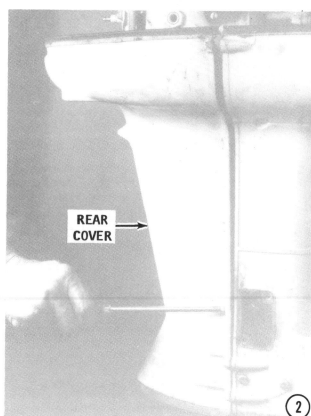

ELECTROMATIC SHIFT 8-57

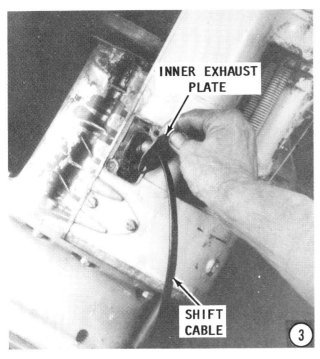

Pull the shift wire down through the exhaust cover. Apply oil or soap onto the cable and remove the inner plate from the cable.

4- If the lower unit being serviced has the 6-inch extension above the lower unit, remove the screws from the bottom side of the extension. It is not necessary to remove the extension in order to remove the lower unit. Remove screws securing the lower unit to the exhaust housing or to the extension. Work the lower unit away from the exhaust housing or extension. The water tubes may be stuck in the water pump. Therefore, if difficulty is encountered in freeing the lower unit, force a wide blade chisel, stiff scraper, or other suitable tool, between the two surfaces and work each side of the lower unit away from the exhaust housing.

DISASSEMBLING

5- Set the cavitation plate on the edge of the work bench or other suitable surface, and secure it firmly with a C-clamp. An alternate method is to cut a deep "V" in a piece of 2" x 6" of wood, and then slide the lower unit into the "V" resting it on the cavitation plate.

6- Remove the O-ring from the top of the driveshaft.

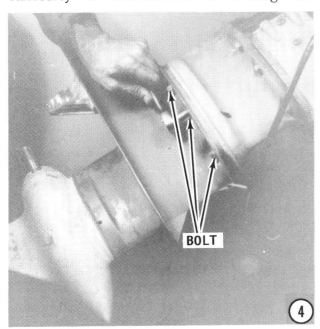

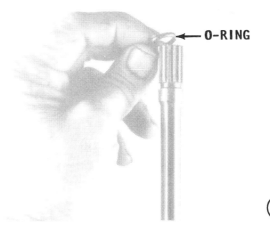

8-58 LOWER UNIT

7- Remove the screws attaching the water pump to the lower housing. If the screws are stubborn and refuse to release, or if they break off due to corrosion, it may be necessary to actually chisel the water pump free of the housing. Lift the water pump and impeller up off the driveshaft.

If the only work to be performed is service of the water pump, proceed directly to Page 8-72, Water Pump Installation.

8- Remove the pin or Woodruff key from the driveshaft, and then remove the base plate from the lower unit housing.

9- Lift the driveshaft straight up and out of the lower unit.

10- Lay the lower unit flat on the bench because when the gearcase nuts are removed, the lower section could drop to the floor and be severely damaged. Use a 9/16" deep-well socket and remove the lower unit gearcase stud nuts, then the washers, and shift cable retainer. **DISCARD** the nuts because they are the self-locking type and **MUST NOT** be used a second time.

11- Tap the front cone with a soft-headed mallet to separate the lower housing from the upper housing a distance of about 3".

After the upper and lower housings have been separated about 3", slide the terminal sleeves on the forward and reverse wires back and disconnect the quick-disconnects of the shift cable to the coil.

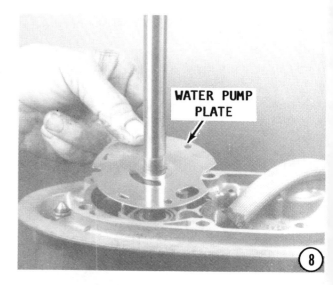

VERY GOOD WORDS

If the lower housing cannot be dislodged from the upper housing, because the long bolts extending through the upper housing into the lower housing are badly corroded, a decision must be made. Something will be destroyed in order to proceed with the work. In almost all cases the sacrificed piece is the less expensive upper housing. Therefore, cut through the upper housing on both

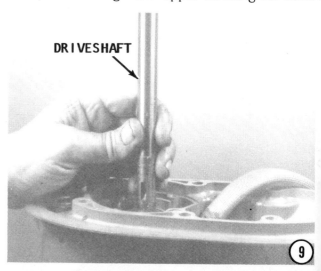

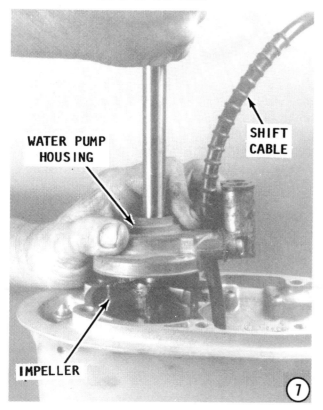

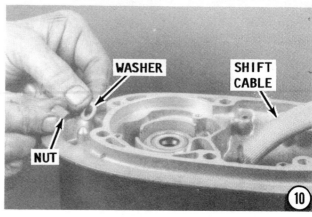

ELECTROMATIC SHIFT 8-59

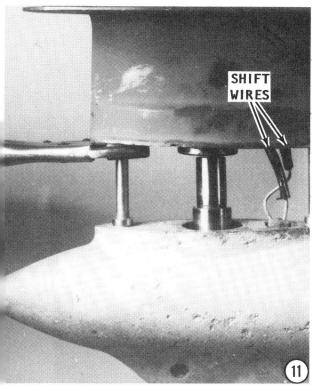

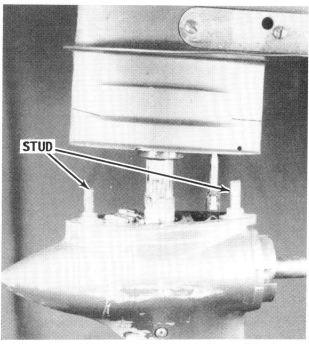

Separating the upper and lower portions of the lower unit after the retaining bolts have been cut.

sides, as shown. The lower housing can then be separated from the upper housing. The studs can then be pressed out of the upper housing and the hacksaw cut welded shut and the housing returned to service. If the studs cannot be pressed free, the upper housing must be replaced.

SPECIAL NOTE

The upper housing has a bearing to accommodate the pinion gear. This bearing is comprised of 20 individual needles. **TAKE CARE** not to lose any of the needles to the bearing set.

12- Clamp the lower unit by the skeg in a vise equipped with soft jaws, as shown. Remove the retainer screw and washer, then lift the nylon coil lead retainer from the lower housing and at the same time work the forward and reverse wires free of the retainer.

13- Remove the four screws in the bearing head. This is the cap the propeller shaft passes through. Notice how each screw has an O-ring behind the head. These O-rings **MUST** be in place during installation to maintain a water-tight unit.

14- With a small chisel and mallet, work the cap free of the propeller shaft. The chisel is to be worked on the cap, **NOT** in the groove between the cap and the lower housing. If the cap is damaged it may be replaced without great expense, but damage to the lower unit is **BAD NEWS**.

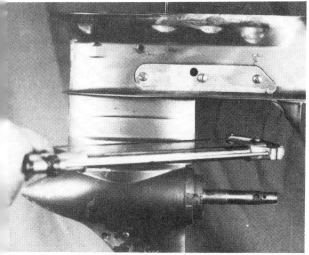

Using a hacksaw to cut through the retaining bolts in order to separate the upper and lower portions of the lower unit.

8-60 Lower Unit

WARNING

The next step involves a dangerous procedure and should be executed with care while wearing **SAFETY GLASSES**. The retaining ring is under tremendous tension in the groove and while it is being removed. If it should slip off the Truarc pliers, it will travel with incredible speed causing personal injury if it should strike a person. Therefore, continue to hold the ring and pliers firm after the ring is out of the groove and clear of the lower unit. Place the ring on the floor and hold it securely with one foot before releasing the grip on the pliers. An alternate method is to hold the ring inside a trash barrel, or other suitable container, before releasing the pliers.

15- Obtain a pair of Truarc pliers. Insert the tips of the pliers into the holes of the retaining ring. Now, **CAREFULLY** remove the retaining ring from the groove and gear case without allowing the pliers to slip. Release the grip on the pliers in the manner described in the above **WARNING**.

16- Install the two screws into the coil. These are the two screws that were removed from the gear case earlier. Rock the coil out and at the same time feed the reverse coil blue wire down into the recess in the lower unit. Continue working the coil out and down the propeller shaft until it is clear. The thrust washer will come off with the coil.

17- Hold the propeller shaft firmly and pull it free from the lower unit housing. The reverse gear, spring, and hub will come out with the shaft as an assembly.

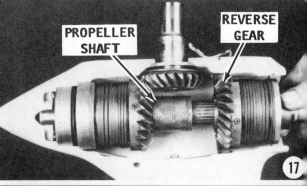

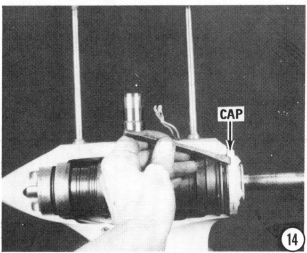

ELECTROMATIC SHIFT 8-61

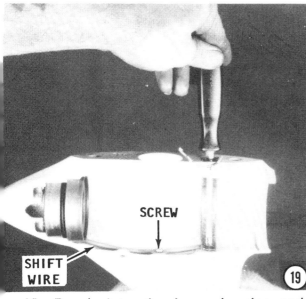

Using a slide hammer to remove the forward bearing race from the lower unit.

professional mechanics use heat and drive the seal out. This method is very effective and adjacent parts are not harmed. Remove the O-ring from the bearing case head.

WORD OF ADVICE

Professional mechanics have discovered that when the lower unit is being rebuilt, a

18- Reach into the lower housing and remove the pinion gear. Remove the forward gear, spring, and hub assembly.

OBSERVE

Notice the metal guard extending down the inside of the lower unit. This guard protects the forward coil wire. Identify the Phillips screw in the bottom of the lower housing.

19- Remove the screw and metal tab in the bottom of the lower housing.

20- Using a special coil remover, tool No. 379784 and Kit No. 379843, remove the forward coil. If these tools are not available, you may heat the outside edge of the lower housing with a torch. A considerable amount of heat will be required. After the housing is heated, grasp the studs with a gloved hand and at the same time tap the housing on a board. The forward coil and bearing race will be released from the housing. If the tools are available, the coil and race can be removed with no sweat.

21- Remove the seal from the gear case head, by driving it out from the backside with a screwdriver or punch and mallet. A special seal remover, is available, but most

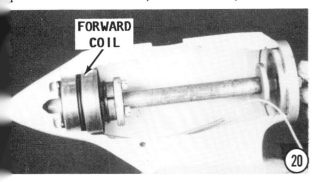

8-62 Lower Unit

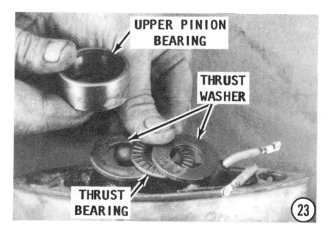

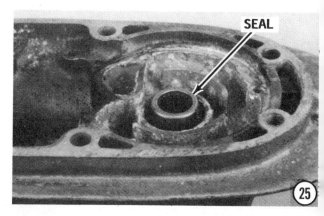

new gear case head should be purchased and installed. The new head will have a new bearing, O-ring, and seal installed as an assembly.

22- On the upper housing of the lower unit, use a slide hammer with the adaptor fingers fitting behind the pinion gear bearing, and remove the bearing from the housing.

23- After the upper bearing has been removed, reach in and remove the washer, thrust bearing, and another washer. Notice the different size holes in the two washers.

24- Remove the upper driveshaft seal using a special puller No. 377565 or a standard seal remover.

25- Remove the upper bearing using special tool No. 309916.

GOOD WORDS

In most cases, it is not necessary to remove the upper bearing. A couple of quick checks can be made to determine the condition of the bearing. One is to feel the driveshaft in the area where it passes through the bearing. If the shaft is smooth with no indication of roughness, the bearing is usually considered fit for further service. Another method is to insert a finger into the center of the bearing and determine if the needles roll freely and smoothly. If there is no evidence of binding or roughness, the bearing does not have to be removed and replaced. In other words, let a sleeping dog lie.

MORE GOOD WORDS

The forward and reverse gear assemblies look almost identical. However, there is a difference. The forward gear uses a babbitt

Reverse (left) and forward (right) gear, bearing, and hub assemblies. The reverse gear has a caged needle bearing arrangement; the forward gear has a babbitt bearing.

Reverse (left) and forward (right) gear hubs. The reverse hub has a smooth exterior surface and the forward gear is knurled.

bearing and the hub is knurled to provide more positive engagement. The reverse gear assembly uses needle bearings and the hub is smooth. The greatest percentage of motor operation is in forward gear with the reverse gear turning in the opposite direction, the needle bearings are used in the reverse gear hub for more satisfactory operation. Proper gear installation is extremely critical for satisfactory performance. The hubs are different and are easily identified. After the gears are assembled it is very easy, in a moment of haste, to pick a gear assembly from the bench and install it in the wrong location. Therefore, after the service work on the gears has been completed and they are ready for installation, identify one with a felt pen marking or with a tag to ensure proper installation.

STILL MORE GOOD WORDS

If the clutch has been slipping, replace the spring and the hub. It is very difficult to accurately determine what is considered "excessive" wear on these two items. Therefore, if the clutch has been slipping, the modest cost of the spring and hub is justified in eliminating this area as a possible source of shifting problems.

26- Obtain and wear a pair of **SAFETY GLASSES** while working with the Truarc pliers in this step. Practice the same **SAFETY** precautions given in Step 15 and the **WARNING** just before that step. Disassemble the forward and reverse gear

assembly using a pair of Truarc pliers and carefully remove the snap ring from the hub on the front of the gear.

27- Lift the gear and spring assembly from the hub. As mentioned earlier in the "More Good Words", the forward gear has a babbitt bearing and the reverse gear a needle bearing arrangement. Therefore, **EXERCISE CARE** when lifting the reverse gear from the hub, not to lose any of the needles.

28- To remove the springs from the forward and reverse gears, first remove the Allen screws around the outside diameter of each gear. Next, pull the spring from the gear. Notice the nylon tapered spacer installed under the spring. This washer **MUST** be installed properly to permit the spring to seat level in the gear.

Perform Steps 26 thru 28 for the other gear.

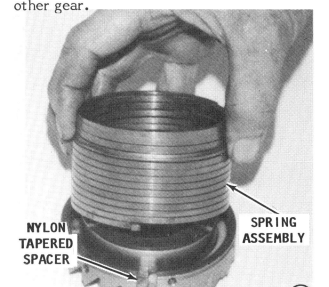

8-64 LOWER UNIT

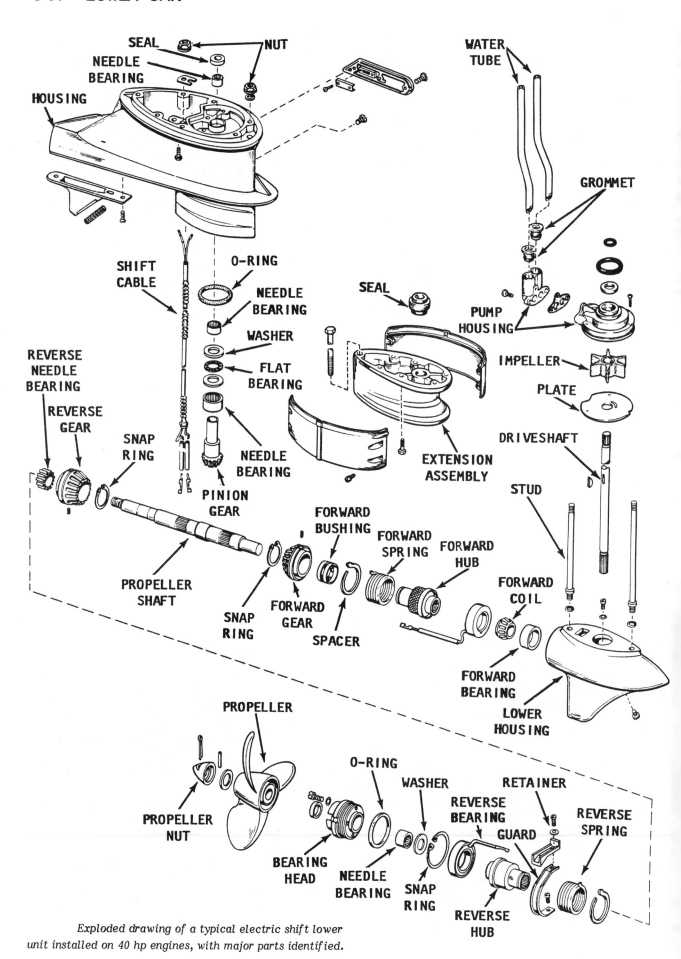

Exploded drawing of a typical electric shift lower unit installed on 40 hp engines, with major parts identified.

CLEANING AND INSPECTING

Clean the parts with solvent and blow them dry with compressed air. Remove all seal and gasket material from mating surfaces. Blow all water and oil passages, and screw holes clean with air.

After the parts are clean and dry, apply a coating of light engine oil to the bearings and bright mating surfaces of the shafts and gears as a prevention against corrosion.

Inspect the shaft bearing surfaces, splines, and keyways for wear and burrs. Check for evidence of an inner bearing race turning on the shaft. Check for damaged threads. Measure the runout on all shafts to reveal any bent condition. If necessary, turn the shaft in a lathe as a check for out-of-round.

Carefully check the inside and outside surfaces of the gearcases, housing, and covers for cracks. Pay special attention to the areas around screw and shaft holes. Verify all traces of old gasket material has been removed from mating surfaces. Check O-ring grooves for sharp edges which could cut a new seal. Inspect gear teeth and shaft holes for wear and burrs. Hold the center race of each bearing and turn the outer race to be sure it turns freely without any evidence of rough spots or binding. Inspect the rollers and balls for any sign of pits or flat spots.

Comparison of a new thrust bearing (left) with a worn and broken bearing cage (right).

Inspect the outside diameter of the outer races and the inside diameter of the inner races for evidence of turning in the housing or on the shaft. Any sign of discoloration or scores is evidence of overheating.

Check the thrust washers for wear and distortion. If they do not have uniform thickness and lay flat, they **MUST** be replaced.

Inspect all springs for tension, distortion, corrosion or discoloration.

Inspect the shift cables for broken leads or damaged insulation. Use an ohmmeter and test for continuity. Use the ohmmeter to test the coil resistance which should indicate 4.5 to 6.5 ohms. Check the coil leads for breaks and damaged insulation.

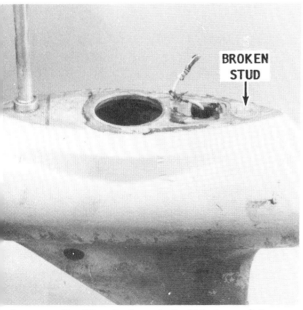

Lower unit with a broken stud. This type of damage is usually the result of not using a NEW nut on the stud the last time the unit was assembled. The nut worked loose, allowed the two sections to vibrate, and the stud was broken. The stud must now be drilled out and replaced.

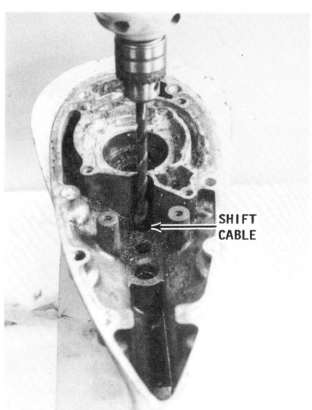

Replacing the shift cable by drilling out the remaining cable with an undersized drill after the old cable has been broken off as close to the housing as possible.

8-66 LOWER UNIT

Check the water pickup screen on the upper housing. Blow air through the screen to dislodge any debris. Clean the area behind the screen.

Inspect the propeller for cracks, gouged, bent, or broken blades. Replace all bent, worn, corroded, or damaged parts. Burrs can be removed with a file.

ALWAYS install **NEW** O-rings, gaskets, and seals during assembling and installation to prevent leaks.

ASSEMBLING ELECTROMATIC

Forward and Reverse Gear Assemblies

1- Insert the spacer, with its key, into the slot in the cupped end of the **FORWARD** gear in such a way that it encircles the gear in the opposite direction to the normal winding of the spring coils. Place the spring in the gear with the spring key beside the spacer key. Now, shift both keys to the side of the slot against which they will pull.

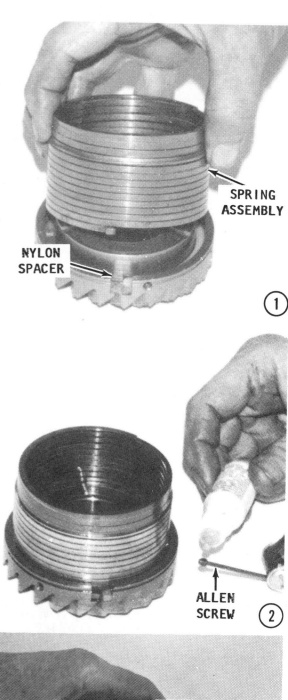

2- Coat three **NEW** setscrews with Loctite TL-242, and install them in the forward gear. Tighten the setscrews in rotation, to a torque value of 30-35 in.-lbs., beginning with the one nearest the spring. Bake the assembly in a 300° oven for 1/2 hour. If an oven is not available, apply Locquic Primer "T" to the screws before tightening them, and then allow them to cure for 4 hours.

3- The tolerance between the clutch hub and the bushing is very close. **CAREFULLY** slide the gear-and-spring assembly onto the hub.

New shift cable ready for installation. Coat the cable with lubricant as an aid to installation.

ELECTROMATIC SHIFT 8-67

WARNING
This next step can be dangerous. The snap ring is placed under tremendous tension with the Truarc pliers while it is being placed into the hub groove. Therefore, wear **SAFETY GLASSES** and exercise care to prevent the snap ring from slipping out of the pliers. If the snap ring should slip out it would travel with incredible speed and cause personal injury if it struck a person.

4- Install the Truarc snap ring into the groove of the forward gear hub.

5- Assemble the **REVERSE** gear by first installing the nylon spacer into the cupped end of the reverse bevel gear, with its key in the slot of the bevel gear. The spacer **MUST** be positioned to encircle the cupped area in the opposite direction to the normal winding of the spring coil. Install the spring with the key indexed in the slot beside the spacer key. Slide both keys against the side of the slot they will pull against when reverse gear is selected.

6- Coat three Allen-head cup-point setscrews with Loctite, and then install them to secure the spring to the bevel gear. Tighten the setscrews in rotation, to a torque value of 30-35 in.-lbs., beginning with the one nearest the spring. Bake the assembly in a $300°$ oven for 1/2 hour. If an oven is not available, apply Locquic Primer "T" to the screws before tightening them, and then allow them to cure for 4 hours.

8-68 LOWER UNIT

7- TAKE NOTE of the ring at the top of the reverse hub used to retain the needle bearings. The needle bearings are held in a cage. The correct number of needles will fill the cage. This ring is the only visible difference between the forward and reverse hubs. Coat the needle bearings with grease or vaseline to hold them in place. **ALWAYS** count and take care to be sure the total number of needle bearings are replaced during installation. **NEVER** use a grease to hold the needles in place which will not dissolve quickly, or the parts will be ruined due to lack of initial lubrication. After the needle bearings are all in place, **CAREFULLY** slide the gear-and-spring assembly down over the hub.

8- A Truarc snap ring secures the gear to the hub. Use a pair of Truarc snap-ring pliers to install this snap ring with the chamfered edge against the bevel gear.

Assembling the Bearing Head

9- If the bearing head seal was removed, press a new seal into the head. The seal can be installed using a block of wood and a mallet. If the bearing was removed, install a new bearing from the back side of the head. A special bearing installer tool, No. -308104 is required to install the bearing. Press against the **LETTERED** side of the bearing, until the bearing is flush with the head surface. If the special bearing tool is not available, a new head must be purchased with the bearing installed. Install a **NEW** O-ring around the head seal.

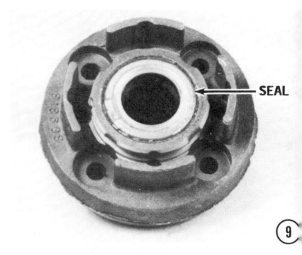

Assembling the Lower Unit

10- Use special bearing installer Tool No. 378737 and No. 308099 and press the upper bearing into the upper housing, if the original was removed. Press against the lettered side of the bearing.

11- Position the upper housing rightside up, and install the seal above the bearing. This seal may be carefully tapped into place with a mallet.

12- To install the lower pinion gear bearing: Position the washer with the large hole in the gear case, then the thrust washer, and then the washer with the small hole. Install the lower pinion gear bearing outer race using special tool No. 378737 and

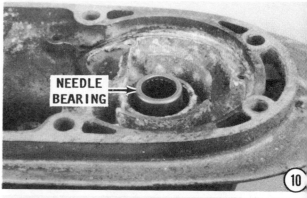

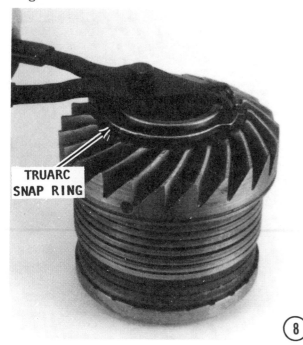

ELECTROMATIC SHIFT 8-69

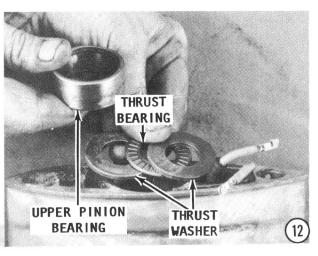

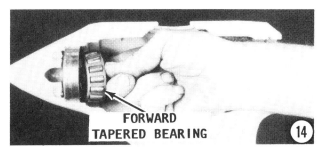

No. 308098. Press against the lettered side of the bearing. If the special tools are not available, the race can be carefully worked into place with a soft-headed mallet.

13- Install the front propeller shaft bearing race, using driver shaft guide plate and bearing race installer, Tool No. 309033, No. 309932, and No. 379247. Place the tapered front portion of the lower unit on a block of wood and drive the race into the housing. If the special tools are not available, the outside diameter of the lower unit may be heated and the race tapped in with a wooden block and mallet.

14- Install the cone-shaped roller bearing into the bearing race. Naturally, the tapered end of the bearing enters the race first.

15- Install the forward coil in the housing using special tool No. 379230. Feed the coil wire into the groove in the bottom of the housing. Use a driver shaft and guide plate to install the coil properly.

16- The wire ends have a plastic covering. This covering must be located behind the metal guard. The guard clamp must fit alongside the housing. Install the Phillips screw in the bottom of the housing with part of the metal tab protruding out of the square hole on top of the lower unit and also with the wire coming out the hole.

17- Lower the forward gear, hub, and spring assembly into the lower unit and over the forward coil.

18- Position the pinion gear into its recess and resting against the forward gear.

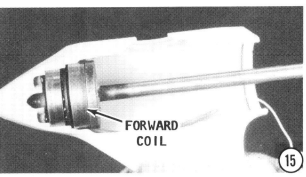

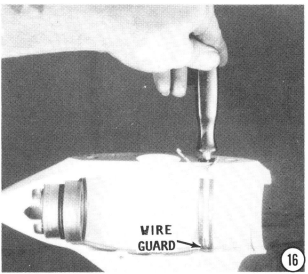

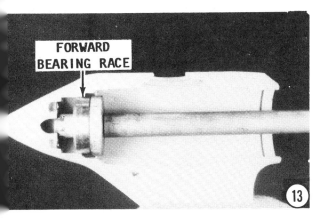

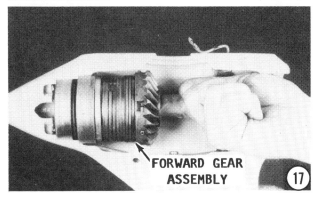

8-70 Lower Unit

GOOD WORDS

The accompanying photographs were taken of a V4 Electromatic lower unit. The 40 hp has two flat washers and a thrust washer installed on top of the pinion bearing in the upper portion of the lower unit. These washer were installed in Step 12. Therefore, disregard the washers shown just above the pinion gear.

19- Hold the pinion gear up and at the same time install the propeller shaft. As the propeller shaft is moved into the lower unit, turn the shaft slowly **CLOCKWISE** to allow the splines on the propeller shaft to engage in the forward gear.

20- Install the reverse gear with the splines of the reverse gear hub engaged with the splines of the propeller shaft.

21- Install the reverse coil and at the same time feed the blue lead through the opening in the lower housing. The lead on the back side of the coil **MUST** be on **TOP**.

WARNING

This next step can be dangerous. The snap ring is placed under tremendous tension with the Truarc pliers while it is being

placed into the groove. Therefore, wear **SAFETY GLASSES** and exercise care to prevent the snap ring from slipping out of the pliers. If the snap ring should slip out, it would travel with incredible speed and cause personal injury if it struck a person.

22- Use a pair of Truarc pliers and install the Truarc snap ring into the groove just forward of the coil. Check to be sure the coil leads are correctly positioned and will not be damaged by any moving part in the lower unit. Double check to be sure the green lead is well protected by the metal guard. Secure the leads in the nylon retainer, with the washer and screw. **DO NOT** overtighten the screw as a precaution against damaging the wire.

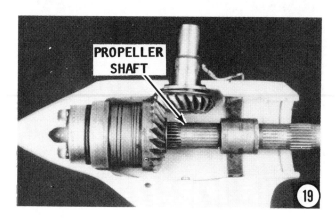

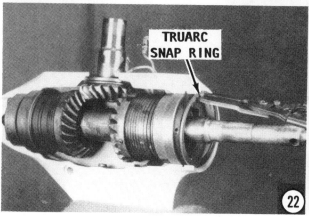

ELECTROMATIC SHIFT 8-71

23- Slide the thrust washer onto the propeller shaft and into the recess of the coil with the babbitt side of the washer FACING the coil.

GOOD WORDS

Alignment of the gear case head holes with the holes in the reverse coil is very difficult because once the head is in place, the O-ring prevents the head from turning. Therefore, before installing the gear case head, insert a guide pin into opposite corner holes in the reverse coil. Check to be sure the large O-ring is properly positioned in the groove in the gear case head.

24- Install the gear case head with the holes in the head indexing over the pins protruding from the reverse coil holes as described in the previous paragraph, "Good Words". Dip the gear case head screws in Perfect Seal No. 4, or equivalent, and then slide an O-ring onto the screw. Now, start the screws through the holes that do not contain pins. **DO NOT** tighten the screws at this time. Remove the two pins and start the other two screws. Tighten the four screws alternately and evenly. Rotate the propeller shaft and check to be sure it turns without excessive drag.

25- Set the wires into the retainer and secure them in place with the screw and washer. The washer will hold the screw.

*Bearing head screw with O-ring. A **NEW** O-ring should be installed each time the screw is removed during service work. The recess in the screw head must be cleaned to allow the O-ring to seat properly. If the ring is left exposed, when the screw is tightened, the head will cut and destroy the sealing ability of the ring.*

26- Check to be sure the coils have not grounded to the lower unit. This can be accomplished by using an ohmmeter to check the forward and reverse coil for resistance. The meter should indicate 4.5 to 6.5 ohms resistance.

27- Turn the upper section of the lower unit upside-down. Apply a heavy coating of

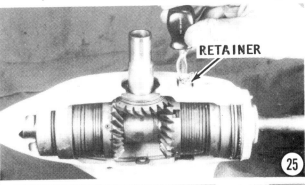

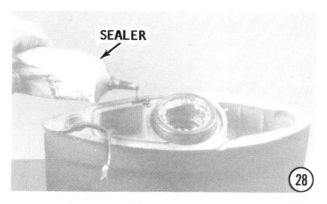

OMC needle bearing lubricant into the pinion gear race. Install the needle bearings in the outer race.

28- Coat the O-ring groove around the pinion gear bearing with sealer. Install the O-ring into the groove.

29- Clamp the skeg of the lower section in a vise. Attach a pair of vise grip pliers about 3-inches up on one of the lower section studs. Lower the upper section down over the studs until the section rests on the vise grip pliers. Connect the wire cable, green-to-green and blue-to-blue. Pull the sleeves down over the connectors. Apply sealer around the surface of the lower section O-ring. Remove the vise grip pliers from the stud and slowly lower the upper housing down onto the lower section.

30- Check to be sure the shift cable retaining fork is over the hole as the stud passes through the retainer washer. Continue to lower the upper housing and at the same time work the shift wires up into the cavity of the upper housing to prevent the wires from being pinched when the upper housing makes contact with the lower section. Replace one washer over each stud with a self-locking nut. Tighten the nuts alternately and evenly.

Obtain an ohmmeter. Check the resistance of the forward and reverse wires for 4.5 to 6.5 ohms resistance. If the ohmmeter does not indicate the proper resistance on each lead, the lower unit MUST be separated and the wiring check for a short or broken wire.

WATER PUMP INSTALLATION

31- Lower the driveshaft down through the water pump seal and into the lower unit. As the driveshaft is lowered, rotate the shaft slowly to permit the splines on the shaft to index with the pinion gear.

32- Apply sealer to the upper housing surface, and then place the water pump plate in position on the upper housing.

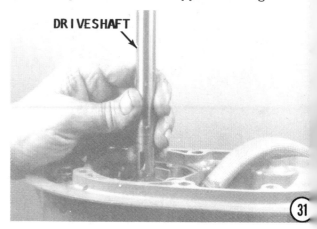

ELECTROMATIC SHIFT 8-73

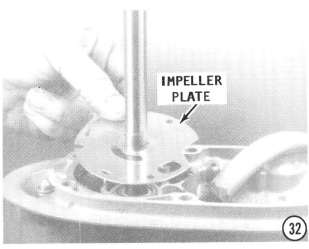

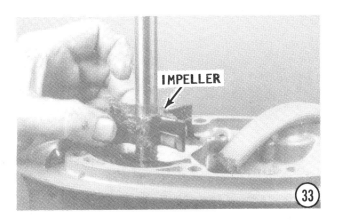

33- Install the Woodruff key into the recess in the driveshaft. Some models may use a pin. If the pin is used, insert it into the hole in the driveshaft. Slide the **NEW** impeller down the driveshaft with the slot in the impeller aligned with the Woodruff key or the pin, if a pin is used. Continue to work the impeller down the driveshaft until it is resting on the surface of the water pump plate.

34- Check to be sure **NEW** seals and O-rings have been installed in the water pump. Lubricate the inside surface of the water pump with light-weight oil. Lower the water pump housing down the driveshaft and over the impeller. Rotate the driveshaft **CLOCKWISE** as the water pump housing is lowered to allow the impeller blades to assume their natural and proper position inside the housing. Continue to rotate the driveshaft and work the water pump housing downward until it is seated on the lower unit upper housing.

Coat the threads of the water pump attaching screws with sealer, and then secure the pump in place with the screws. Tighten the screws alternately and evenly.

35- Fill the lower unit with lubricant. Use only OMC Type C or Premium Lubricant. The unit will not operate properly with any other type or brand of lubricant. Insert the lubricant tube into the bottom opening, and then fill the unit until lubricant is visible at the vent hole. Install

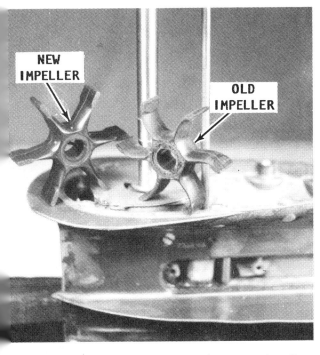

Comparison of a new (left) water pump impeller, with one unfit for service (right).

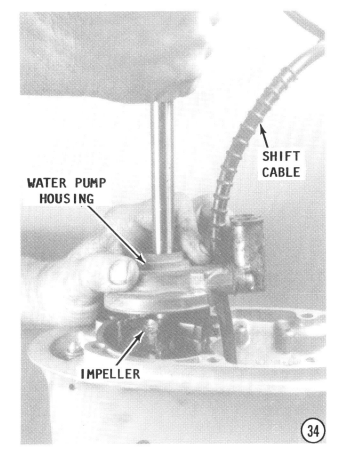

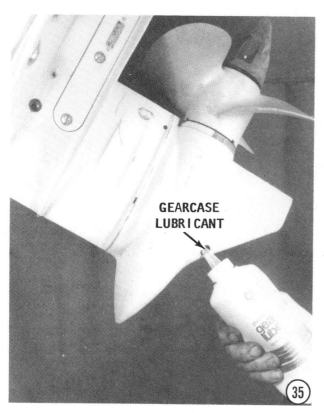

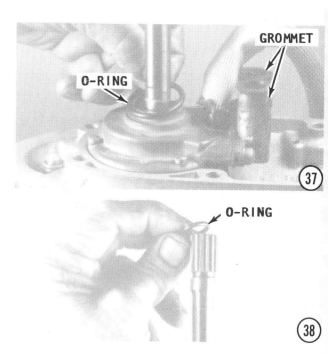

the vent plug. Remove the gear lubricant tube, and install the drain/fill plug.

36- After the lower plug has been installed, remove the vent plug again and use a squirt-type oil can to add lubricant through this vent hole. A squirt-type oil can must be used to allow the trapped air in the lower unit to escape at the same time the final lubricant is added. Once the unit is completely full, install and tighten the vent plug.

37- Check to be sure the O-ring on top of the water pump is in place.

38- Install the O-ring onto the top of the driveshaft just below the splines.

GOOD WORDS

Clean both lower unit water tubes with sandpaper. These tubes should be clean and shiny as an aid to mating the lower unit to the exhaust housing. If these tubes are not thoroughly clean, great difficulty may be encountered in mating the lower unit with the exhaust housing. After the water tubes have been cleaned, apply a light coating of oil or lubricant to the outside surface of the tubes. Apply a light coating of oil to the electric shift cable.

LOWER UNIT INSTALLATION

39- Position the assembled lower unit under the exhaust housing. Work the electric shift cable up through the exhaust housing and out the hole on the port side. Slowly

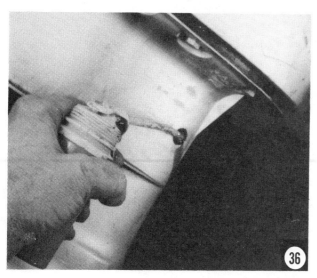

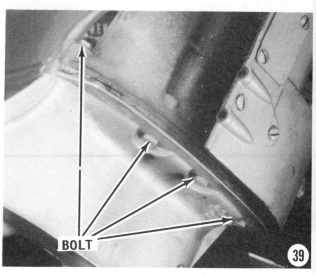

ELECTROMATIC SHIFT 8-75

lift the lower unit into place with the driveshaft indexing with the crankshaft splines and the water tubes entering the exhaust housing grommets. It may be necessary to have an assistant rotate the flywheel ever so slowly **CLOCKWISE** and to pull the electric shift cable through while the lower unit is being mated with the exhaust housing. Rotating the flywheel will permit the driveshaft to index with the splines of the crankshaft. Coat the threads of the attaching screws with sealer. After the mating surfaces of the exhaust housing and the lower unit have made contact, start the screws securing the two units together. Tighten the screws alternately and evenly.

40- Feed the electric shift cable through the inner exhaust cover until the cover is in place on the surface of the exhaust housing. Secure the cover in place with the two attaching screws. **A REMINDER:** The upper screw also holds the clamp used to secure the shift cable in place.

41- Install the rear housing cover with the exhaust relief boot fitting into the recess of the cover. Secure the cover in place with the attaching hardware.

42- Position the outer exhaust cover in place on the port side. Secure the cover in place with the attaching hardware.

43- Connect the shift wires to the harness at the back of the engine, **BLUE**-to-**BLUE** and **GREEN**-to-**GREEN**. Slide the rubber protective sleeves in place over the connectors.

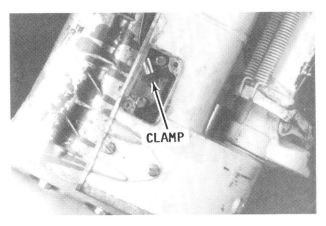

Exhaust housing with the inner plate installed and the shift wire clamp properly installed.

Propeller Installation

A FEW GOOD WORDS

The propeller washer and drive pin play an extremely important role. When shifting gears during normal operation, or if the propeller should hit an underwater obstacle, the propeller is subjected to considerable shock. A washer is installed between the propeller and drive pin. This washer **MUST** always be in place for proper operation. If the hub should slip, the propeller will move back towards the propeller nut and lock against the drive pin. The washer is designed to stop propeller movement so the drive pin can be easily removed for service. Now, on with the installation.

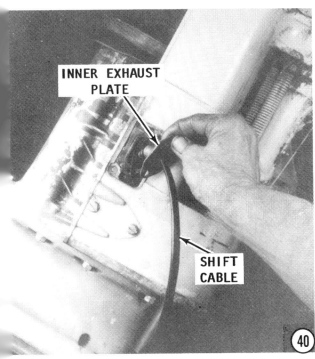

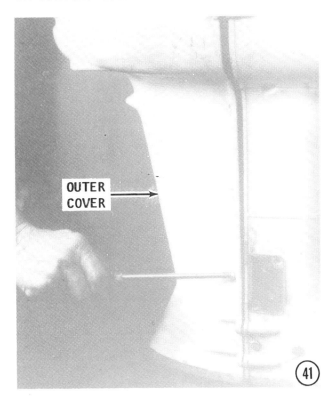

8-76 Lower Unit

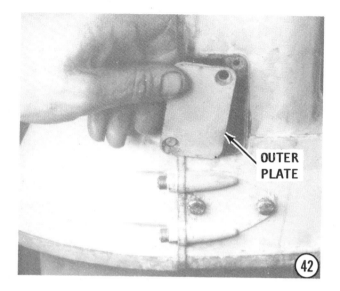

44- Coat the propeller shaft with an anti-corrosion grease. Install the propeller with the drive pin holes aligned. Install the washer and drive pin. Slide the propeller cap into place and secure it with the cotter pin.

If the unit being serviced has the propeller using the rear-type shear pin arrangement: Install the shear pin and then coat the propeller shaft with anti-corrosion grease. Install the propeller, propeller nut, and then the cotter pin.

45- Perform a functional check of the completed work by mounting the engine in a test tank, in a body of water, or with a flush attachment connected to the lower unit. If the flush attachment is used, **NEVER** operate the engine above an idle speed, because

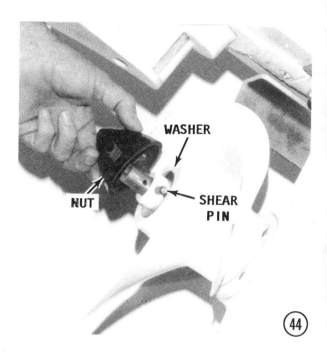

the no-load condition on the propeller would allow the engine to **RUNAWAY** resulting in serious damage or destruction of the engine.

CAUTION: Water must circulate through the lower unit to the engine any time the engine is run to prevent damage to the water pump in the lower unit. Just five seconds without water will damage the water pump.

Start the engine and observe the mist of water from the idle relief port in the exhaust housing. The water pump installation work is verified. Shift the engine into the three gears and check for smoothness of operation and satisfactory performance.

9
HAND STARTER

9-1 INTRODUCTION

Three different type hand starters may be installed on the Johnson/Evinrude outboard engines covered in this manual. Each type starter may have two, and sometimes three, models. All will be covered in this chapter with detailed procedures and illustrations.

Type I

The first type is a cylinder with a pinion gear arrangement similar to an automotive starter motor. The unit is mounted on the side of the engine in a vertical position. As the starter rope is pulled, a nylon drive gear slides upward and engages the flywheel ring gear. Once the engine starts the drive gear automatically disengages. Service procedures for both models of this starter are presented in Sections 9-2 and 9-3.

Type II

One model of this type starter will be found on the outboards covered in this manual. The starter assembly is mounted on the port side of the engine and the drive gear works on an axis and engages the flywheel directly. As the rope is pulled, a swing

Starter installation on the 5 hp and 6 hp engines. This starter is similar in operation to an automotive type starter.

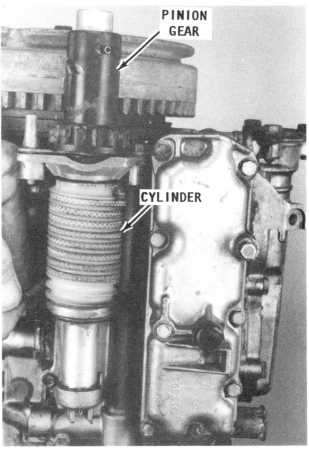

Starter installation on all 9.5 hp engines. Notice the difference with the unit in the left column, this page.

arm moves the drive gear upward to engage with the teeth of the flywheel ring gear. A coil spring winds and tightens as the rope unwinds. The spring then coils the rope around a pulley as the rope handle is returned to the control panel.

Service procedures for this hand starter are presented in Section 9-4.

Type III

This type starter is mounted atop the flywheel with three mounting legs attached to the powerhead. The mechanism consists of a starting rope and handle, a rewind spring, and an arrangement of pawls and corresponding clutch dogs. The complete assembly is contained in a housing secured by three legs extending to the powerhead.

When the starter rope is pulled, the pawls automatically engage the clutch dogs attached to the flyweel and the engine is cranked.

Three models of this type starter may be installed on various Johnson/Evinrude engines covered in this manual.

The first model contains pawls with return springs, a center cone, and a set of spring-release dogs that engage with the flywheel ring gear when the starter rope is pulled. Service procedures are presented in Section 9-5.

The second model does not have spring pawls or a center cone, but has friction-type dogs. Service procedures are given in Section 9-6.

The third model has only one nylon pawl which engages into cut-a-ways in the outer edge of the flywheel. Service procedures for this model are in Section 9-7.

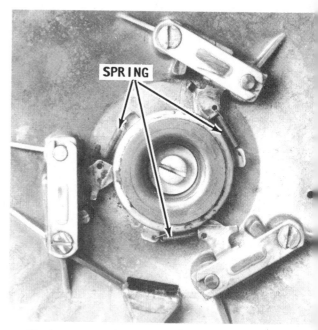

Starter mounted atop the flywheel. This unit has the three pawls with return springs. This starter model may be found on the 28 hp, 30 hp, 35 hp, and 40 hp engines.

OPERATION

Normally, very few problems are encountered with the hand starter. It is strictly a mechanical device to crank the engine for starting. The spring will last an incredibly long time, if used properly. The greatest enemy of the spring is the operator.

Swing arm type starter mounted on the port side on 3 hp and 4 hp engines.

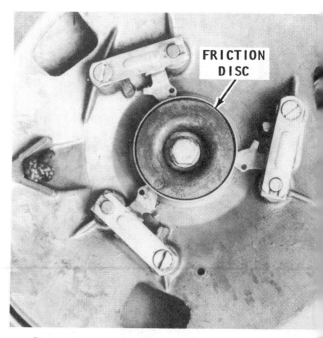

Starter mounted atop the flywheel. This model starter has the three pawls, with a friction disc and no return springs. This model may be found on the 28 hp, 33 hp, and 40 hp engines.

Three causes contribute to starter failure. Two may be prevented, the third cannot.

The most common problem is the result of the operator pulling the starter rope too far outward. If the operator places one hand on the engine and pulls the rope with the other hand, it is physically impossible, in this positon, to pull the rope too far. Problems develop when the operator uses both hands to pull on the rope, with no control on how far the rope can be extended. The rope may be broken or the knot released from the starter disc. In either case, the spring rewinds with tremendous speed and in almost all cases travels past its normal rewind position bending the end of the spring in reverse. Therefore, more maintenance work is involved than merely replacing the rope.

Another bad habit, while using the hand starter, is to release the grip on the rope when it is in the extended position, allowing the rope to freely rewind. The operator should **NEVER** release his grip, but hold onto the rope, and thus control the rewind. The owner should always be alert to any wear on the rope and replace it long before the possibility of breaking might occur. If the rope should break, the spring would rewind with incredible speed, the same as if the rope were released, causing damage to the spring and other starter parts.

The third cause of spring failure cannot be prevented -- age. As the outboard continues to perform year after year, the age of the spring steel will finally take its toll.

The rewind spring is made of spring steel. Depending on the model and the powerhead, from 6 feet to 12 feet of spring length is wound into about a 4 inch diameter. This places the spring under unbelievable tension, making it a highly **DANGEROUS** force. Therefore, any time the hand starter is serviced, especially during work on the spring, **SAFETY GLASSES** should be worn and the work performed with the utmost care.

Any time the rope is broken, the starter spring will rewind with incredible speed. Such action will cause the spring to rewind past its normal travel and the end of the spring will be bent back out of shape. Therefore, if the rope has been broken, the starter should be completely disassembled and the spring repaired or replaced.

9-2 TYPE I STARTER CYLINDER WITH PINION GEAR 5 HP AND 6 HP ENGINES

This gear-drive starter is a new design employing the principle of an automotive type starter motor. When the starter rope is pulled, the starter rotates, and a nylon pinion gear slides upward and engages the flywheel ring gear. The gear automatically disengages when the engine starts. The ratio between the pinion gear and the ring on the flywheel has been selected to provide maximum cranking speed with minimum pulling effort to ensure fast, easy engine start.

This single pawl starter may be installed on 3 hp to 5 hp engines.

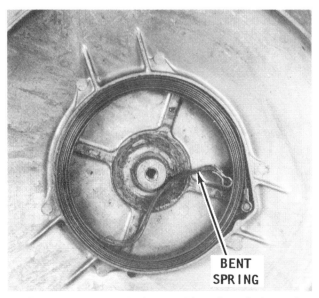

The starter rope broke on this unit and the spring rewound with such incredible speed it doubled back in reverse.

9-4 HAND STARTERS

STARTER ROPE REPLACEMENT

REMOVAL

1- Disconnect the high-tension leads from the spark plugs. Ground the high-tension leads. Pull the starter rope out until it is fully extended. Now, allow the rope to retract just a little, until the knot end on the spool is facing the port side of the engine. Lift the pinion gear to engage the flywheel ring gear. Hold the pinion gear engaged with the ring gear, and at the same time, slide the handles of a pair of pliers under the pinion gear to lock the pinion gear with the ring gear, as shown.

2- Remove the handle from the end of the starter rope. **OBSERVE** how the rope is wound onto the spool and how the rope is secured by a loop formed in a slot in the spool. Remove the rope from the starter spool.

INSTALLATION

Rope Purchase Instructions

The length and diameter of the starter rope required will vary depending on the size horsepower engine being serviced. Therefore, check the Hand Starter Rope Specifications in the Appendix, and then purchase a quality nylon piece of the proper length and diameter size. Only with the proper rope, will you be assured of efficient operation following installation.

Each end of the nylon rope should be "fused" by burning them slightly with a very small flame (a match flame will do) to melt the fibers together. After the end fibers have been "fused" and while they are still

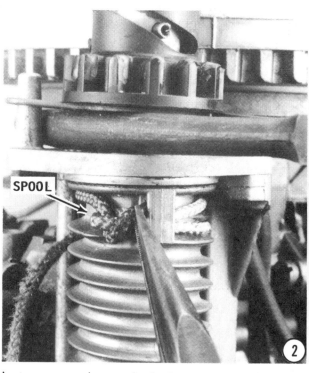

hot, use a piece of cloth as protection and pull the end out flat to prevent a "glob" from forming.

3- Feed one end of the new rope through the spool anchor, make a loop, and then thread the end back through the hole in the anchor, but **DO NOT** pull it tight at this time, leave a loop.

4- Bring the short end of the rope through the loop just formed.

5- Work the short end back through the anchor, as shown.

6- Now, pull both ends of the rope tight.

7- Feed the rope through the front engine cowling, and then install the starter rope handle. Pull and hold tension on the rope,

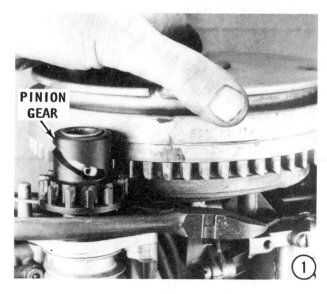

PINION GEAR 5 & 6 HP 9-5

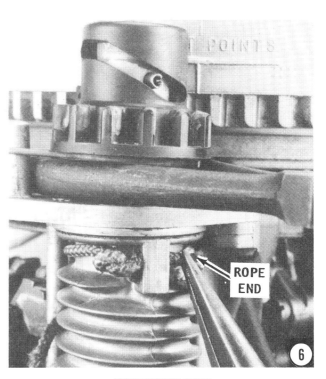

and at the same time remove the pliers from under the pinion gear. Allow the starter rope to rewind in a normal manner. After the rope is fully wound onto the starter spool, the rope handle should be up tight against the engine cowling. If the handle is not up tight, the rope was installed to long or the starter spring is weak and should be replaced.

STARTER REMOVAL

AUTHOR'S NOTE

For photographic clarity, the accompanying pictures were taken servicing a starter from a powerhead removed from the exhaust housing. The hood need only be removed to work on the starter.

1- Pull approximately 3/4 of the starter rope out, and then form a knot in the rope to prevent it from recoiling. Allow the rope to recoil until the knot is tight against the cowling. Remove the rope handle.

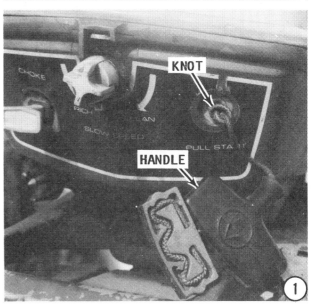

9-6 HAND STARTERS

2- The rope may be removed now, or later. To remove the rope now, first pull the rope all the way out. Slide the handles of a pair of pliers under the pinion gear to hold the gear engaged with the flywheel. Remove the rope from the starter spool.

3- Grasp the spool firmly. Remove the pliers and allow the spool to slip a little at a time until the spring is completely unwound.

4- Remove the two retaining bolts on top of the starter.

5- Loosen, but **DO NOT** remove the bolts on the bottom and on each side of the starter spool. When the bottom two bolts are loosened, the retainer will separate from the lower cap. Lift the starter from the engine housing. If the spring is still clipped into the lower retainer, release the spring by disengaging the spring tang from the retainer.

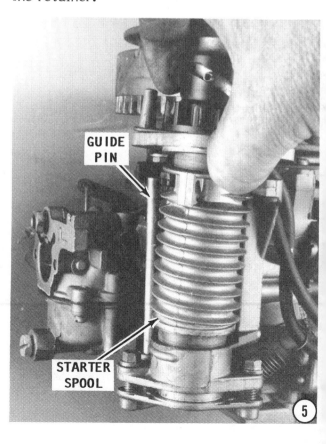

PINION GEAR 5 & 6 HP 9-7

DISASSEMBLING

6- Remove the pin in the pinion gear, and then remove the pinion gear from the collar. Slip the bearing head off the spool. Remove the retainer, installed under the pinion gear, from the starter.

7- Remove the spring retainer (the long one) from the center of the spool.

8- Pull the starter spring from the spool.

CLEANING AND INSPECTING

Wash all parts in solvent, and then dry them with compressed air.

Inspect and replace the main spring if it is damaged or worn. Check the bottom end of the spring very carefully to be sure the two tangs (one on the inner and the other on the outer spring) are in good condition with no sign of distortion.

Inspect the bushing in the bottom collar of the starter housing. This bushing was not removed in the disassembling procedures. Feel with a finger for any roughness, burrs, or other evidence of excessive wear or damage. If the bushing is in good condition, it need not be removed.

Check the rope condition. If the rope is frayed or shows any sign of weakness, it should be replaced. There will never be an easier time to replace the rope than while the starter is disassembled.

Inspect the teeth of the pinion gear. The teeth will show some signs of normal wear. A broken tooth or excessive wear on one side of the teeth is justification for replacement.

Inspect the groove through the pinion gear. This is the groove to accommodate the roll pin. Check to be sure the upper part of the pinion gear is not cracked or distorted.

DO NOT lubricate the pinion gear. Oil applied to the pinion gear will attract dirt causing the gear to bind on the spool. Lubricate the upper and lower spool bearing surfaces with just a drop of outboard lubricant. Apply outboard oil to the spring on the pinion gear. **DO NOT** oil the pinion gear bearing or the surfaces of the spring.

ASSEMBLING

GOOD NEWS

Two methods of assembling and installing this starter are presented. The first is the factory suggested procedure and begins with the following steps on this page.

An alternate method is also outlined which many professional mechanics feel is much simpler, easier, and quicker. The alternate method begins on Page 9-9 and includes steps 1A thru 5A.

After the starter is assembled and installed on the powerhead, continue the work with Step 7.

Factory Method

1- Install the spring retainer from the bottom side. Slide the spring onto the

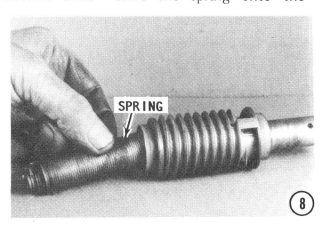

HAND STARTERS

spring retainer. Work the spring upward until the inner spring tang engages the slot on the bottom of the retainer.

2- Align the hole in the retainer with the hole in the spool sleeve.

3- Slip the bearing head and pinion gear down over the spool shaft. Install the roll pin through the pinion gear and sleeve.

4- If the lower bushing was removed, install a **NEW** bushing into the bottom collar on the engine.

HELPFUL WORD

As an assist to installation, first soak the bushing in hot water for about ten minutes, and then lubricate it with just a drop of outboard oil.

5- The tang on the outer spring must hook into the slot of the lower spring retainer plate. Pull on the outer spring to elongate the spring, and at the same time, lower the spring into the spring retainer plate and hook the tang into the slot in the plate. Rotate the spring **CLOCKWISE** to lock the tang in the plate. Hold upward and turn the spring **CLOCKWISE,** and at the same time tighten the two screws in the lower spring retainer plate.

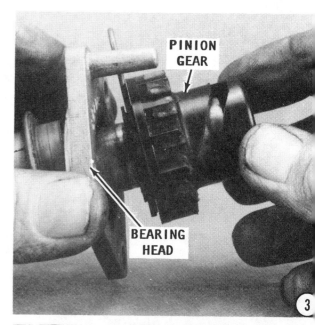

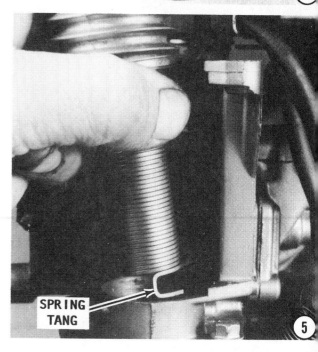

PINION GEAR 5 & 6 HP 9-9

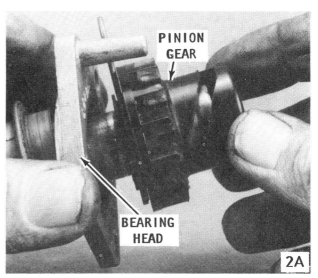

6- Place the starter assembly in position on the engine and start the two upper screws through the upper bearing support. Check to be sure the guide is in place in the bottom and top retainers. Tighten the two bottom retainer screws.

Alternate Assembling Method

1A- If the lower bushing was removed, install a **NEW** bushing into the bottom collar on the engine.

HELPFUL WORD

As an assist to installation, first soak the bushing in hot water for about ten minutes, and then lubricate it with just a dorp of outboard oil.

2A- Install the spring retainer from the bottom side of the spool. Slip the bearing head and pinion gear down the spool shaft. Align the holes and install the roll pin through the pinion gear and spool.

3A- Take the spring and lower it into the bottom retainer. Hook the outer spring into the retainer.

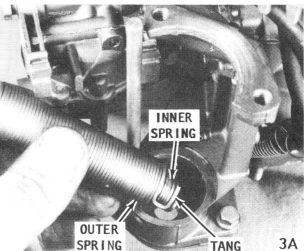

4A- Lower the spool assembly down over the spring and engage the spring retainer into the tang of the inner spring.

5A- Start the two upper screws through the upper bearing support. Check to be sure the guide is in place in the bottom and top retainers. Tighten the two bottom retainer screws.

7- Insert a large size screwdriver into the top of the spool, and then rotate the spool, by count, exactly 16-1/2 complete turns. Lift the pinion gear to engage the flywheel ring gear, and then slip a pair of plier handles under the pinion gear to hold it in mesh with the ring gear.

8- Feed one end of the new rope through the spool anchor, make a loop, and then thread the end back through the hole in the anchor, but **DO NOT** pull it tight at this time, leave a loop.

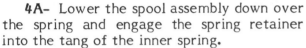

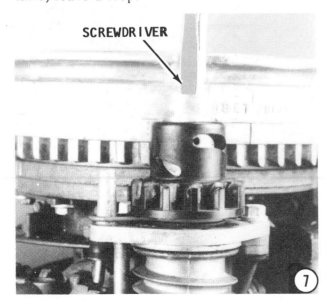

PINION GEAR ALL 9.5 HP 9-11

9- Bring the short end of the rope through the loop just formed.

10- Work the short end back through the anchor, as shown.

11- Now, pull both ends of the rope tight.

12- Feed the rope through the front engine cowling, and then install the starter rope handle. Hold tension on the rope with the handle and at the same time, remove the pliers from underneath the pinion gear. Allow the starter rope to wind onto the spool. After the rope has been wound onto the spool, the starter handle should be up tight against the engine cowling. If the handle is not up tight against the cowling the rope was installed too long and needs to be shortened.

9-3 TYPE I STARTER CYLINDER WITH PINION GEAR ALL 9.5 HP ENGINES

This gear-drive starter is a new design employing the principle of an automotive type starter motor. When the starter rope is pulled, the starter rotates, and a nylon pinion gear slides upward and engages the flywheel ring gear. The gear automatically disengages when the engine starts. The ratio between the pinion gear and the ring on the flywheel has been selected to provide maximum cranking speed with minimum pulling effort to ensure fast, easy engine start.

ROPE REMOVAL

1- Disconnect the high-tension leads from the spark plugs. Ground the high tension leads. Pull the starter rope out until it is fully extended. Now, allow the rope to retract just a little, until the knot end on the spool is facing the port side of the engine. Lift the pinion gear to engage the flywheel ring gear. Hold the pinion gear

9-12 HAND STARTERS

engaged with the ring gear, and at the same time, slide the handles of a pair of pliers under the pinion gear to lock the pinion gear with the ring gear, as shown.

2- Remove the handle from the end of the starter rope. **OBSERVE** how the rope is wound onto the spool and how the rope is secured by a knot. Pull the knot and the rope from the spool.

INSTALLATION

Rope Purchase Instructions

The length and diameter of the starter rope required will vary depending on the size horsepower engine being serviced. Therefore, check the Hand Starter Rope Specifications in the Appendix, and then purchase a quality nylon piece of the proper length and diameter size. Only with the proper rope, will you be assured of efficient operation following installation.

Each end of the nylon rope should be "fused" by burning them slightly with a very small flame (a match flame will do) to melt the fibers together. After the end fibers have been "fused" and while they are still hot, use a piece of cloth as protection and pull the end out flat to prevent a "glob" from forming.

3- Tie a figure 8 knot in one end of the new rope. Feed the rope through the spool anchor around the back of the spool and out the hole in the cowling. Install the handle on the the end of the rope.

4- Pull and hold tension on the rope, and at the same time remove the pliers from under the pinion gear. Allow the starter rope to rewind in a normal manner. After the rope is fully wound onto the starter spool, the rope handle should be up tight

against the engine cowling. If the handle is not up tight, the rope was installed to long or the starter spring is weak and should be replaced.

AUTHOR'S WORD

The engine exhaust shroud has been removed only for photographic clarity in the accompanying illustrations.

STARTER REMOVAL

1- Pull the starter rope out until it is fully extended. Now, allow the rope to retract just a little, until the knot end on the spool is facing the port side of the engine. Lift the pinion gear to engage the flywheel ring gear. Hold the pinion gear engaged with the ring gear, and at the same

PINION GEAR ALL 9.5 HP 9-13

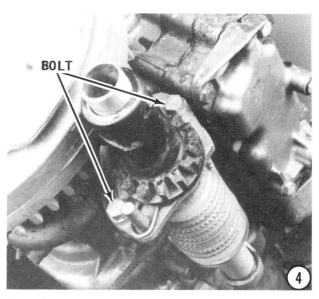

time, slide the handles of a pair of pliers under the pinion gear to lock the pinion gear with the ring gear, as shown.

2- Remove the handle from the end of the starter rope. **OBSERVE** how the rope is wound onto the spool and how the rope is secured by a knot. Pull the knot and the rope from the spool.

3- Grasp the spool firmly and remove the pliers from under the pinion gear. Now, allow the spool to slip a little at a time until the spring is completely unwound.

4- Remove the two retaining bolts on top of the starter. Lift the starter assembly from the powerhead.

OBSERVE

As the starter is removed, take special note of how the upper bearing retainer extends over a hole in the powerhead. This hole is a water passage. A gasket is installed under the retainer to form a seal for the water passage. This gasket may remain on the block or come with the starter retainer as the starter is removed. To ensure a good seal, the gasket should be discarded and replaced with a new one at time of installation.

5- Remove the roll pin extending through the pinion gear.

6- Remove the pinion gear, spring and bearing head from the spool.

7- Pull the main spring and upper spring retainer from the spool.

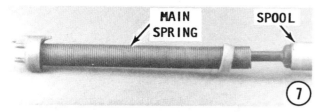

8- Notice the cap on the bottom of the spool secured with a set screw. Remove the set screw from the cap, and then pull the cap and bushing from the spring.

9- Remove the spring retainer and outer bearing from the spring.

CLEANING AND INSPECTING

Wash all parts in solvent, and then blow them dry with compressed air.

Inspect the main spring. Check the tab on the bottom end of the inner spring to be sure it is not bent or cracked.

Inspect the teeth of the pinion gear. The teeth will show some signs of normal wear. A broken tooth or excessive wear on one side of the teeth is justification for replacement.

Inspect the groove through the pinion gear. This is the groove to accommodate the roll pin. Check to be sure the upper part of the pinion gear is not cracked or distorted.

DO NOT lubricate the pinion gear, spring, or spool. Oil applied to these parts will attract dirt causing the gear to bind on the spool. Lubricate the upper and lower spool bearing surfaces with just a drop of outboard lubricant. Apply outboard oil to the spring on the pinion gear. **DO NOT** oil the pinion gear bearing or the surfaces of the spring.

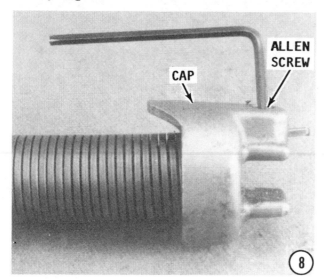

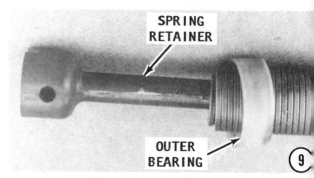

Check the rope condition. If the rope is frayed or shows any sign of weakness, it should be replaced. There will never be an easier time to replace the rope than while the starter is disassembled.

ASSEMBLING

1- Place the outer bearing, spring retainer, bushing and lower spring retainer onto the main spring in the order given. Guide the tab on the outer spring through the hole in the lower spring retainer.

2- Install the set screw securing the inner spring to the lower retainer. Set the completed spring assembly aside.

3- Slide the outer bearing down over the outer spring, as shown. Insert the spring retainer through the center of the inner spring from the **TOP**.

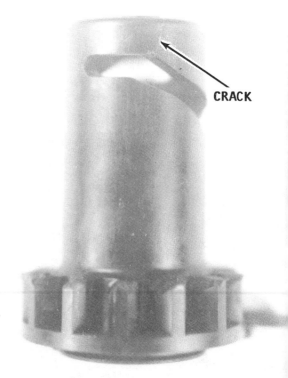

Pinion gear from a hand starter. Notice the crack on the top side, indicated by the callout. This unit must be replaced.

PINION GEAR ALL 9.5 HP

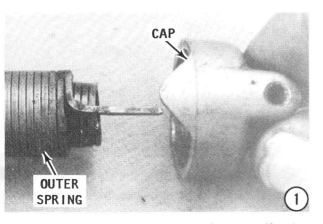

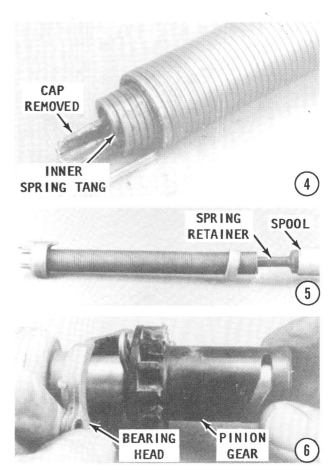

4- Rotate the inner spring until the "hook" on the end of the spring seats in the groove of the spring retainer.

5- Insert the assembled springs and retainer from Step 4, into the bottom of the spool.

6- Install the bearing head, spring, and pinion gear down over the spool shaft.

7- Align the hole in the upper spring retainer with the hole in the starter spool and with the slot in the pinion gear.

INSTALLATION

AUTHOR'S WORD

For the following illustrations, the engine exhaust shroud was removed, only for photographic clarity.

8- Cover both sides of a **NEW** gasket with Perfect Seal No. 4, and then place the gasket in position on the powerhead over the water passage.

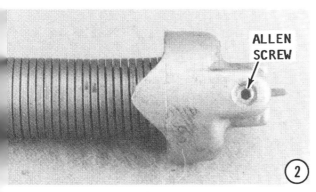

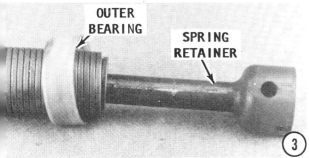

9-16 HAND STARTERS

9- Lower the spool assembly down into place on the side of the engine with the lower spring retainer indexed over the pin in the bottom portion of the housing.

10- Install the two bolts securing the bearing head to the powerhead.

Rope Purchase Instructions

The length and diameter of the starter rope required will vary depending on the size horsepower engine being serviced. Therefore, check the Hand Starter Rope Specifications in the Appendix, and then purchase a quality nylon piece of the proper length and diameter size. Only with the proper rope, will you be assured of efficient operation following installation.

Each end of the nylon rope should be "fused" by burning them slightly with a very small flame (a match flame will do) to melt the fibers together. After the end fibers have been "fused" and while they are still hot, use a piece of cloth as protection and pull the end out flat to prevent a "glob" from forming.

11- Insert a large screwdriver blade into the top of the spool, and then rotate the spool **COUNTERCLOCKWISE,** by count, 20-1/2 complete turns. After the required numbers of turns have been made, hold

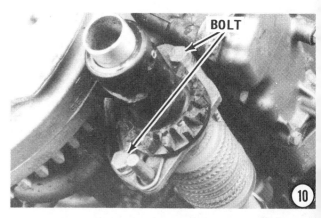

pressure on the screwdriver, and at the same time lift the pinion gear to engage with the flywheel ring gear, and insert the handles of a pair of pliers under the pinion gear to hold the gear engaged.

12- Tie a figure **8** knot in one end of the new rope. Feed the rope through the spool anchor around the back of the spool and out the hole in the cowling. Install the handle on the the end of the rope.

13- Pull and hold tension on the rope, and at the same time remove the pliers from under the pinion gear. Allow the starter rope to rewind in a normal manner. After the rope is fully wound onto the starter spool, the rope handle should be up tight against the engine cowling. If the handle is not up tight, the rope was installed too long or the starter spring is weak and should be replaced.

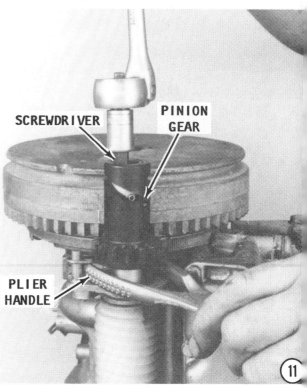

SWING ARM DRIVE GEAR 9-17

9-4 TYPE II STARTER
COIL SPRING WITH SWING ARM DRIVE GEAR
3 HP 1968
4 HP 1969-70

This type hand starter is a flat type mounted on the port side of the engine. As the rope is pulled, a swing arm moves the drive gear upward to engage with the teeth of the flywheel ring gear. A coil spring winds and tightens as the rope unwinds. The spring then coils the rope around a pulley as the rope handle is returned to the front of the engine cowling.

The coil spring consists of a lengthy piece of spring steel (approximately 12-feet) tightly wound inside a housing (the cup and stop assembly). Movement of the drive gear to the retracted position is accomplished through a second spring.

The starter must be disassembled to replace the rope.

WARNING
The rewind spring is under tremendous tension and is a potential hazard. Therefore, **SAFETY GLASSES** should be worn and extreme **CARE** exercised to follow the procedures carefully during disassembling and asembling work with the starter.

SAFETY WORDS
Work on the starter can be very dangerous. Because approximately 12-feet of spring steel is tightly wound into about a 4-inch housing, the spring is placed under tremendous tension -- a real tiger in a cage. If the spring should accidently be released, severe personal injury could result from being struck by the spring with force. Therefore, the service instructions **MUST**, and we say again **MUST**, be followed closely to prevent release of the spring at the wrong time. Such action would be a **BAD SCENE**, a very **BAD SCENE**, because serious personal injury could result.

The starter rope should **NEVER** be released from the extended position. Such action would allow the spring to wind with incredible speed resulting in serious damage to the starter mechanism.

REMOVAL

1- Remove the spark plugs and ground the high tension leads. Pull the starter rope out, and then tie a knot in the rope behind the handle. Allow the rope to rewind until the knot is against the engine cowling. Untie the knot in the end of the rope, and then remove the handle and the rubber bumper.

9-18 HAND STARTERS

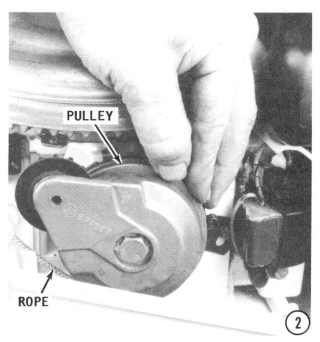

2- Remove the knot tied in the rope in Step 1. Allow the rope to **SLOWLY** wind into the starter. Before the rope end passes the cowling, firmly grasp the starter pulley, and then allow the starter to unwind.

3- Observe the back side of the starter. Notice the hook of the starter spring protruding out of a hole in the starter. Grasp the spring hook with a pair of needle-nose pliers, and then pull the spring out as far as possible, to relieve tension on the spring.

4- Remove the 3/8" bolt from the bracket between the starter and the exhaust housing.

5- Hold the starter together with one hand, and at the same time **LOOSEN** the large bolt from the center of the starter. **DO NOT** remove this bolt at this time. Remove the starter from the engine.

6- If the starter is only removed in order to accomplish other work, install a 3/8" x 16 nut onto the far side of the thru-bolt to hold the starter together and prevent the spring from escaping.

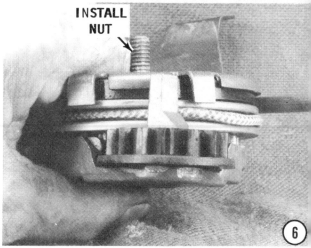

SWING ARM DRIVE GEAR

DISASSEMBLING

7- Remove the idler gear arm, idler gear, and the idler gear arm spring.

SAFETY WORDS

The next step could be dangerous. Removing the pulley from the cup **MUST** be done with care to prevent personal injury.

8- Lift the pulley **SLIGHTLY** and then use a screwdriver and work the spring free of the pulley. **DO NOT** allow the spring to be released from the cup. After the pulley has been removed, notice the position of the spring loop. Remove the rope from the pulley. Notice how the rope unwinds from the pulley **COUNTERCLOCKWISE.**

9- Remove the bushings from the idler gear arm. A bushing is installed on each side of the pulley.

10- Two different methods are suggested to remove the spring from the starter cup. One method involves pulling continuously on the end of the spring that contains the loop. The second method is to simply toss the cup a safe distance onto carpeting or a lawn, allowing the spring to be released instantly from the cup.

If this second method is used, be sure the spring will not cause a threat to any individual in the area when it is released.

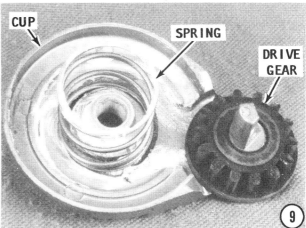

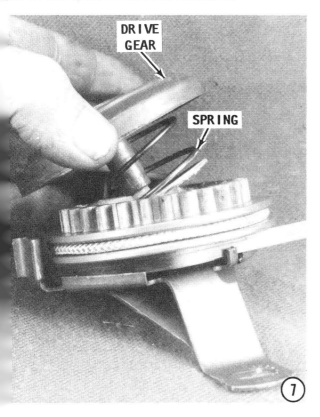

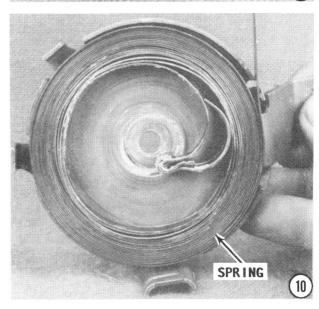

CLEANING AND INSPECTING

Wash all parts except the rope in solvent and then blow them dry with compressed air.

Remove any trace of corrosion and wipe all metal parts with an oil dampened cloth.

Inspect the starter spring end loops. Replace the spring if it is weak, corroded or cracked.

Inspect the rope. Replace the rope if it appears to be weak or frayed. If the rope is frayed, check the hole through which the rope passes for rough edges or burrs. Remove the rough edges or burrs with a file, and polish the surface until it is smooth.

Inspect the dog ears of the pulley gears to be sure they are not worn and are free of burrs. Check the idler gear for cracks and missing teeth

ASSEMBLING

Rope Purchase Instructions

The length and diameter of the starter rope required will vary depending on the size horsepower engine being serviced. Therefore, check the Hand Starter Rope Specifications in the Appendix, and then purchase a quality nylon piece of the proper length and diameter size. Only with the proper rope, will you be assured of efficient operation following installation.

Each end of the nylon rope should be "fused" by burning them slightly with a very

small flame (a match flame will do) to melt the fibers together. After the end fibers have been "fused" and while they are still hot, use a piece of cloth as protection and pull the end out flat to prevent a "glob" from forming.

1- Tie a figure 8 knot in one end of the rope.

2- Feed the other end of the rope through the pulley hole, and then pull the rope tight until the knot is seated in the pulley.

3- Wind the rope **CLOCKWISE** around the pulley. Use a piece of masking tape or rubber band to hold the rope in place in the pulley.

4- Coat the bushings with a light film of OMC Type A lubricant. Insert one bushing in the pulley and the other bushing in the idler gear arm.

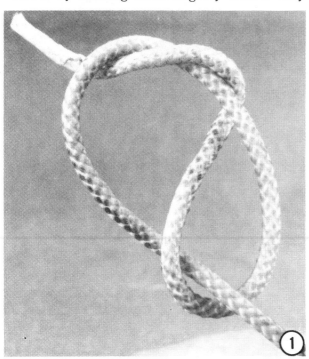

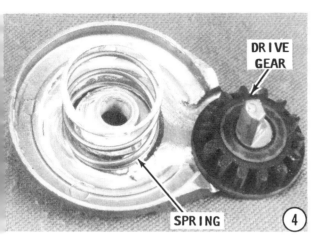

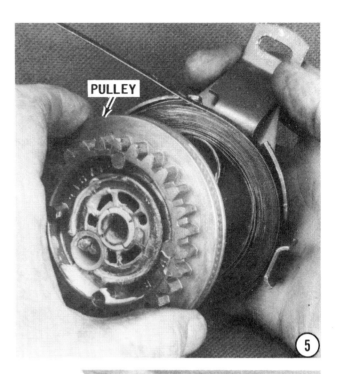

5- If the spring was **NOT** removed from the cup, lower the pulley down over the spring and insert the end of the spring into the spring anchor post of the pulley. If the spring **WAS** removed from the cup, hook the end of the spring into the pulley and then allow the spring to come out the slot in the cup. Turn the pulley and wind the spring into the cup until about 1/2 of the spring length has been wound. Hold the gear and allow it to back off slowly. **DO NOT** allow the spring to rewind quickly. The remainder of the spring will be installed later.

6- Assemble the idler gear with the shoulder against the idler gear arm. Install the idler gear arm and spring to the pulley and cup with the stop on the elder gear shaft located between the upper stop and the lower stop on the cup and stop assembly, as shown.

7- Hold the assembly together and install the assembly onto the engine. Thread the shoulder screw into place first. This screw will hold the starter assembly together.

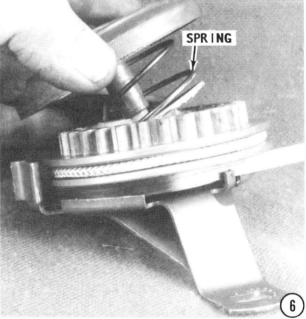

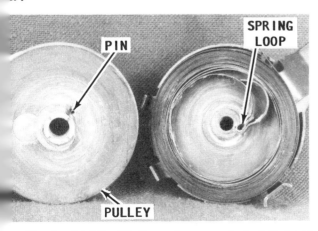

Housing (right) with the spring properly installed and the spring end bent toward the center. The pin in the pulley (left) must index into the loop on the spring end during installation.

9-22 HAND STARTERS

8- Install the screw through the idler arm and into the exhaust manifold. **DO NOT** tighten this screw at this time. Apply a light coating of OMC Type A lubricant to the portion of the spring extending out of the starter.

SPECIAL WORDS

Special tools are available to lock-in the starter to the flywheel. However, the tools are usually not available; they are expensive; and professional mechanics have developed an alternate method. The procedure will take time and patience, but it is the only way without the special tools. To work without the special tools proceed as follows:

9- Remove the rubber band or the masking tape from the coiled rope. Working from the back side of the engine, pull on the rope and the idler gear will engage with the flywheel ring gear. Continue to pull the rope, and at the same time, work the spring

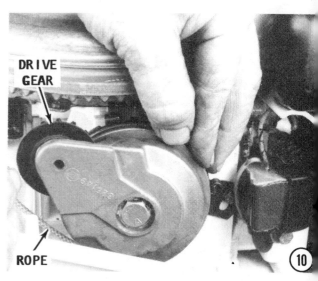

down into the cup. If the rope becomes fully extended, before the spring is installed into the cup, allow the rope to rewind onto the pulley as far as possible and then wind the rope around the pulley again. Now, pull on the rope again from the back side of the engine, and continue to work the spring into the cup until it is completely installed

10- Ease back on the rope until there is no spring tension on the starter. Thread the rope into the pulley **CLOCKWISE** around the starter. Two, or possibly more, loops may be required to accomplish the task. Use all of the rope in the pulley with the starter in the relaxed position. After all of the rope has been fed into the pulley, grab the end of the rope in front of the starter and pull it

out, then feed it through the cowling at the front of the engine. Continue to pull the rope until about two feet is extending out through the cowling. Tie a slip knot in the rope.

11- Install the rubber bumper and handle onto the rope. Tie a figure 8 knot in the end of the rope, and then pull the knot into the handle.

12- Untie the slip knot and ease the rope back into the starter. The starter handle must be up tight against the cowling when the rope is completely rewound on the starter pulley. If the rope is not tight against the cowling, remove the knot and handle from the rope, and then wind the rope around the pulley one complete turn. Tie another knot in the end of the rope as described earlier in this step and then check to be sure the handle is tight against the cowling when the rope is wound onto the pulley.

Starter Adjustment

13- Hold the idler gear arm stop against the cup stop. Fully engage the idler gear teeth with the teeth in the flywheel ring gear. Tighten the cup and stop assembly screw. Tighten the shoulder screw according to the specification listed in the Appendix.

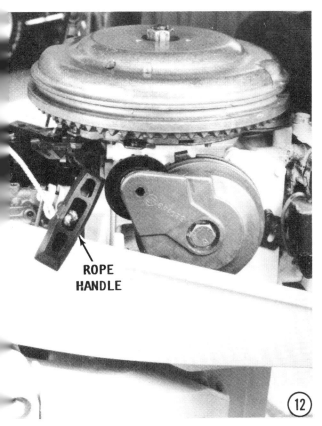

9-5 TYPE III MOUNTED ATOP FLYWHEEL MODEL WITH RETURN SPRINGS
28 HP 1962-63
30 HP 1956
35 HP 1957-59
40 HP 1960-63

This type starter is installed as original equipment by the manufacturer. However, if the starter pulley was damaged sometime in the past, and the hub replaced, the replacement kit would contain parts modifying the unit. The most noticeable change is the absence of the pawl return springs. The starter unit must then be serviced according to the procedures outlined in the next section -- 9-6.

Therefore, if the starter unit has not been modified, still has the pawl return springs, follow the procedures in this section. If the unit has been modified, no pawl return springs, follow the steps in the next section.

The starter rope should **NEVER** be released from the extended position. Such action would allow the spring to wind with incredible speed, resulting in serious damage to the starter mechanism.

Any time the rope is broken, the starter spring will rewind with incredible speed. Such action will cause the spring to rewind past its normal travel and the end of the spring will be bent back out of shape. Therefore, if the rope has been broken, the starter must be completely disassembled and the spring repaired or replaced.

9-24 HAND STARTERS

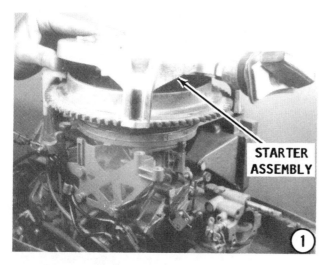

STARTER REMOVAL

1- Disconnect any linkage between the hand starter and the carburetor. Remove the compression release linkage between the starter and the engine head. Move the linkage out of the way. Remove the starter leg retaining screws, and then lift the complete starter from the engine.

2- Pull the rope out far enough to tie a knot in the rope. Allow the rope to rewind to the knot. Work the rope anchor out of the rubber covered handle, then remove the rope from the anchor. Remove the handle from the rope. Untie the knot in the rope, and then hold the disc pulley, but permit it to turn and thus allow the rope to wind back onto the pulley **SLOWLY**. Continue to allow the spring in the pulley to unwind **SLOWLY** until all tension has been released.

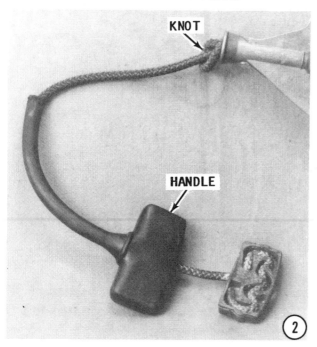

3- Remove the center bolt nut from the top side of the starter. Some models do not have a center bolt nut. On other models, the nut may have vibrated loose, but if the center bolt has threads showing, a nut **MUST** be installed during assembling.

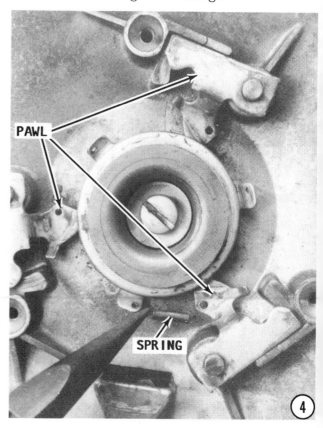

ATOP FLYWHEEL WITH SPRINGS

4- Lay the starter on its back on a work surface. Remove the three screws securing the pawl retainers to the pulley. Remove the three pawl springs, as shown. In the illustration, two springs have already been removed and the third is being removed.

5- Remove the three pawls from their retainers. Remove the center screw from the hub. Hold the pulley, and at the same time, lift the spindle and pin assembly, then the equalizer cup, and friction spring, from the pulley.

WARNING

The rewind spring is a potential hazard. The spring is under tremendous tension when it is wound -- a real tiger in a cage. If the spring should accidently be released, severe personal injury could result from being struck by the spring with force. Therefore, the following step **MUST** be performed with care to prevent personal injury to self and others in the area. If the spring should be accidently released at the wrong time, such action would be a **BAD SCENE**, a very **BAD SCENE**, because serious personal injury could result.

6- Lift the pulley straight up and at the same time work the spring free of the

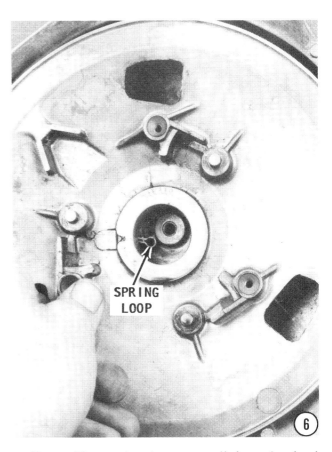

pulley. The spring has a small loop hooked into the pulley.

7- An alternate and safe method is to hold the pulley and the housing together tightly with the legs extending downward in the normal direction. Now, lower the complete assembly to the floor. When the legs

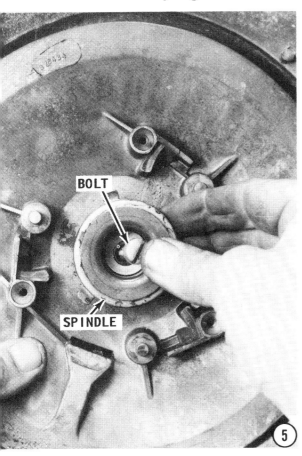

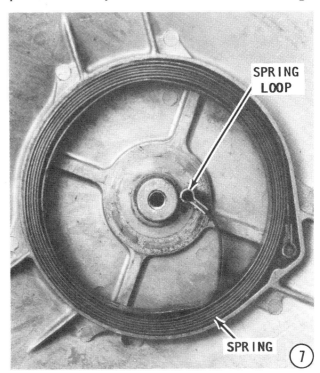

make contact with the floor, release the grip on the pulley. The pulley will fall and the spring will be released from the housing, but the three legs will contain the spring and prevent it from traveling across the room. If the spring was not released from the housing, the only safe method is to jar the three legs on the floor again to release the spring. Unwind the rope out of the pulley groove, and then pull it free.

CLEANING AND INSPECTING

If the rope was broken and the spring is bent backward, as shown in the accompanying illustration, it is a simple matter to bend the spring end back to its normal position. The next illustration clearly shows a spring end properly positioned in the housing.

Wash all parts except the rope in solvent and then blow them dry with compressed air.

Remove any trace of corrosion and wipe all metal parts with an oil dampened cloth.

Inspect the rope. Replace the rope if it appears to be weak or frayed. If the rope is

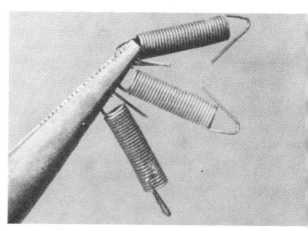
New pawl return springs ready for installation. used spring should appear much the same for satisfactory service.

frayed, check the hole through which the rope passes for rough edges or burrs. Remove the rough edges or burrs with a file and polish the surface until it is smooth.

Inspect the starter spring end loops. Replace the spring if is is weak, corroded or cracked. Check the spring pin located at the back side of the pulley to be sure it is straight and solid.

Check the inside surface of the housing and remove any burrs.

Check the condition of the pawl spring to be sure they are not stretched out of shape. The end of each spring should be bent back toward the coil of the spring. Inspect the pawls for wear and that the edges are not rounded.

Inspect the hub center locating pin to be sure it is straight and tight.

The rope on this unit broke, causing the spring to rewind with incredible speed. The end of the spring was bent back in the wrong direction.

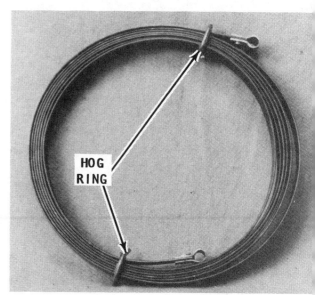
A new spring as it appears direct from the marine store. The hog rings must be CAREFULLY removed, as described in the text.

STARTER ASSEMBLING

GOOD WORDS

The accompanying illustration shows a new starter spring as it is purchased. Note how the spring is held wound with "hog rings". The spring **MUST** be released to its full extended position before it can be installed. Therefore, use care and remove the "hog rings" and allow the spring to unwind until it is a straight piece of spring steel.

SAFETY WORD

Wear a good pair of gloves while unwinding and installing the spring. The spring will develop tension and the edges of the spring steel are sharp. The gloves will prevent cuts on your hands and fingers.

A safe method involves one person removing the hog rings while an assistant holds the spring. After the rings have been removed, both persons work to unwrap the spring, one coil at-a-time.

1- Slide the spring onto the outer pin and then start the spring from the outside edge of the housing and insert it into the housing **COUNTERCLOCKWISE,** as shown in the accompanying illustration. Notice the small hump in the housing. This hump prevents the spring from being wound in the wrong direction. Work the first turn into the housing, and then hold the spring down with one hand and continue to wind the spring into the housing. Patience and time are required to work the spring completely into the housing. After the last portion is in place, bend the end of the spring towards the center of the housing. This position will allow the pulley pin to align with the loop in the end of the spring, when the pulley is installed.

2- Lower the pulley down over the top of the spring with the pulley pin indexing into the loop in the end of the spring. In the accompanying illustration, notice the call-out for the boss on the backside of the pulley. The pin is located directly under the boss. The boss can, therefore, be a guide during pulley installation.

TAKE NOTE

Observe closely the accompanying illustration for this step. Notice how the spring is on the inside diameter of the spindle and spring assembly and the friction spring is wound down on the shoulder of the spindle. During assembling the spring **MUST** remain down on the spindle shoulder. Also notice the pin protruding from the bottom of the spindle. This pin **MUST** drop into the hole in the starter housing. Observe into the housing and visually locate this hole.

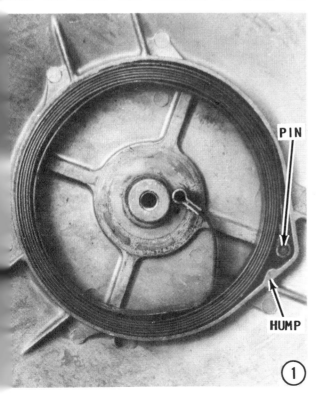

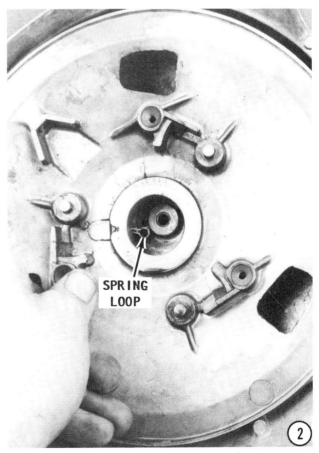

9-28 HAND STARTERS

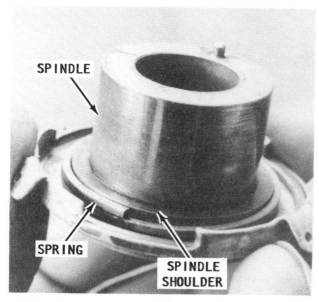

A spring properly installed and seated on the spindle shoulder.

3- Slide the equalizer cup onto the spindle. Work the spring around the outside diameter of the spindle. Lower the spindle assembly down through the pulley and index the pin into the hole in the housing. Attempt to rotate the large pulley. The pulley should turn freely if the spindle has been installed properly. Place the washer inside the spindle housing and then install the bolt through the washer into the housing. Tighten the bolt securely.

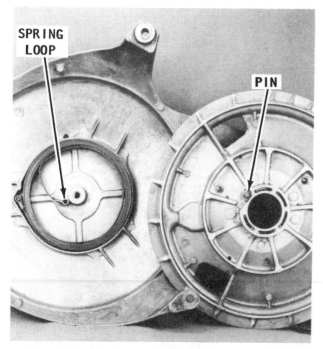

Housing (left) with the spring properly installed and the spring end bent toward the center. The pin in the pulley (right) must index into the loop on the spring end during installation.

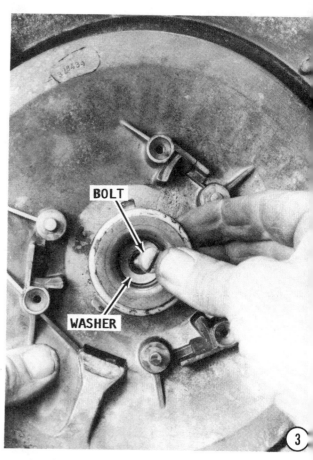

4- Turn the starter over and install the retaining nut onto the thru-bolt (if a nut is used). Tighten the nut securely. Check to be sure the pulley will rotate smoothly and does not bind on the spindle. If the spring on the spindle was allowed to slip out of place on the shoulder, then the spindle must be removed, the spring properly positioned

on the shoulder and the spindle installed again. Rotate the pulley slightly **COUNTERCLOCKWISE** and then release it to be sure there is proper engagement with the spring and the pulley has good spring tension.

5- Install the three pawls onto the three pins with the end of the pawls indexed in the three cutouts of the equalizer cup. Connect the three pawl springs to the equalizer cup.

6- Install the three retainers over the top of the pawls, and then secure them in place with the retaining screws.

ROPE INSTALLATION

Rope Purchase Instructions

The length and diameter of the starter rope required will vary depending on the size horsepower engine being serviced. Therefore, check the Hand Starter Rope Specifications in the Appendix, and then purchase a quality nylon piece of the proper length and diameter size. Only with the proper rope, will you be assured of efficient operation following installation.

Each end of the nylon rope should be "fused" by burning them slightly with a very small flame (a match flame will do) to melt

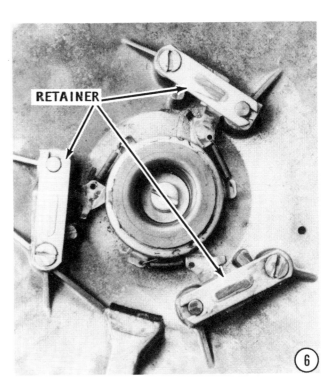

the fibers together. After the end fibers have been "fused" and while they are still hot, use a piece of cloth as protection and pull the end out flat to prevent a "glob" from forming.

7- Tie a figure 8 knot in one end of the starter rope. Set the rope aside, by handy, to be picked up with one hand.

8- Hold the housing with one hand and rotate the pulley three complete turns **COUNTERCLOCKWISE** with the other hand.

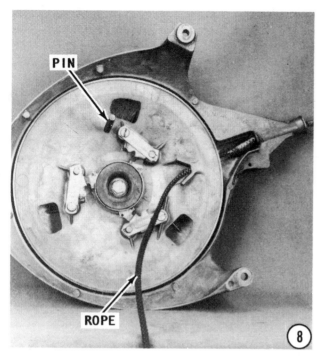

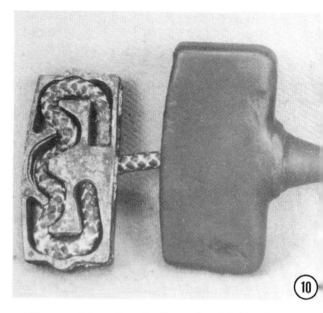

After three complete turns have been made, align the rope outlet in the pulley with the outlet in the housing. Insert a drift pin or other suitable tool through the hole in the pulley and the hole in the housing to hold the pulley in the desired position. Pick up the rope and feed the end through the pulley and the housing and out the other side of the housing. Pull the rope tight until the knot is seated against the pulley.

9- A special tool is manufactured by OMC to install the handle onto the rope. This tool has three prongs and is inserted through the handle and then attached to the rope and pulled back through the handle. If the special tool is not available take a stiff piece of wire; insert it through the handle; thread it through the rope; apply just a little oil to the rope; then pull the wire and rope through the handle.

10- Work the end of the rope into the handle anchor. Secure the rope in place by pushing the anchor into the rubber handle.

11- Lightly pull on the rope to relieve tension on the pin installed through the pulley and housing in Step 8. Maintain some tension on the rope, remove the pin, and allow the spring to **SLOWLY** wind the rope onto the pulley. Check the bolt through the spindle to be sure it is tight.

12- Lay the starter on its back and pull the rope with quick movements, and at the same time check the pawls to be sure they move towards the center of the pulley. Release the rope slowly and check to be sure the pawls return to their original position under the retainers.

STARTER INSTALLATION

13- Position the starter over the flywheel with the three legs aligned over the holes in the powerhead for the retaining

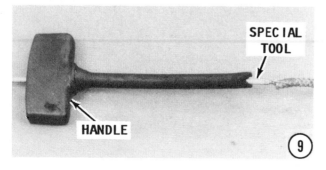

ATOP FLYWHEEL NO RETURN SPRINGS 9-31

9-6 TYPE III MOUNTED ATOP FLYWHEEL MODEL WITH NO RETURN SPRINGS
28 HP 1964
33 HP 1965-70
40 HP 1964-70

WARNING

As with other types of hand starters, the rewind spring is a potential hazard. The spring is under tremendous tension when it is wound -- a real tiger in a cage. If the spring should accidentally be released, severe personal injury could result from being struck by the spring with force. Therefore, the service instructions **MUST**, and we say again **MUST**, be followed closely to prevent release of the spring at the wrong time. Such action would be a **BAD SCENE**, a very **BAD SCENE**, because serious personal injury could result.

The starter rope should **NEVER** be released from the extended position. Such action would allow the spring to wind with incredible speed, resulting in serious damage to the starter mechanism.

Any time the rope is broken, the starter spring will rewind with incredible speed. Such action will cause the spring to rewind past its normal travel and the end of the spring will be bent back out of shape. Therefore, if the rope has been broken, the starter should be completely disassembled and the spring repaired or replaced.

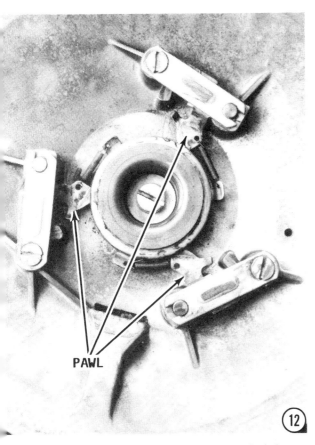

bolts. Install the retaining bolts and tighten them securely. Connect the linkage from the carburetor and the compression release.

9-32 HAND STARTERS

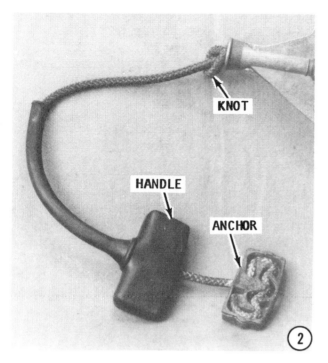

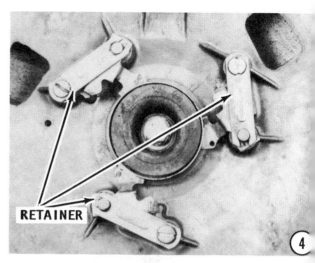

permit it to turn and thus allow the rope to wind back onto the pulley **SLOWLY**. Continue to allow the spring in the pulley to unwind **SLOWLY** until all tension has been released.

3- Remove the center bolt nut from the top side of the starter. Some models do not have a center bolt nut. On other models the nut may have vibrated loose, but if the center bolt has threads showing, a nut MUST be installed during assembling.

4- Lay the starter on its back on a work surface. Remove the three screws securing the pawl retainers to the pulley.

5- Remove the three pawls from their retainers.

6- Remove the center screw from the hub.

7- Hold the pulley, and at the same time, remove the spindle, the wavy washer, friction ring, and the nylon bushing from the center of the pulley.

STARTER REMOVAL

1- Disconnect any linkage between the starter and the carburetor. Move the linkage out of the way. Remove the starter leg retaining screws, and then lift the complete starter from the engine.

2- Pull the rope out far enough, and then tie a knot in the rope. Allow the rope to rewind to the knot. Work the rope anchor out of the rubber covered handle, then remove the rope from the anchor. Remove the handle from the rope. Untie the knot in the rope, and then hold the disc pulley, but

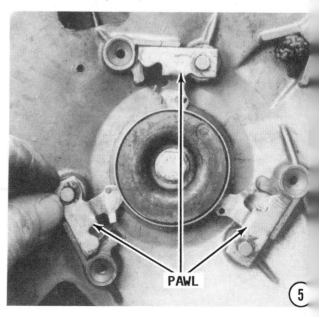

ATOP FLYWHEEL NO RETURN SPRINGS

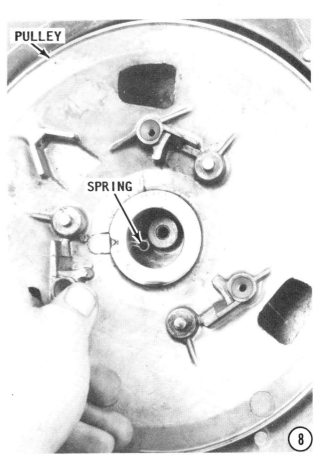

WARNING

The rewind spring is a potential hazard. The spring is under tremendous tension when it is wound -- a real tiger in a cage. If the spring should accidentally be released, severe personal injury could result from being struck by the spring with force. Therefore, the following step **MUST** be performed with care to prevent personal injury to self and others in the area. If the spring should be accidently released at the wrong time, such action would be a **BAD SCENE**, a very **BAD SCENE**, because serious personal injury could result.

8- Lift the pulley straight up and at the same time work the spring free of the pulley. The spring has a small loop hooked into the pulley.

9- An alternate and safe method is to hold the pulley and the housing together tightly and turn the complete assembly so the legs are facing downward. Now, lower the complete assembly to the floor. When the legs make contact with the floor, release the grip on the pulley. The pulley

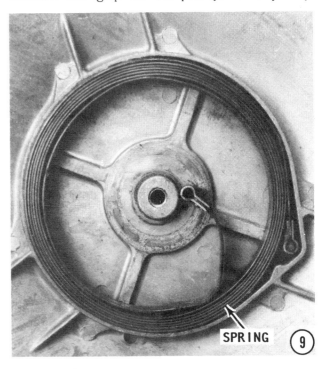

9-34 HAND STARTERS

will fall and the spring will be released from the housing, but the three legs will contain the spring and prevent it from traveling across the room. If the spring was not released from the housing, the only safe method is to jar the three legs on the floor to release the spring. Unwind the rope out of the pulley groove, and then pull it free.

CLEANING AND INSPECTING

If the rope was broken and the spring is bent backward, as shown in the accompanying illustration, it is a simple matter to bend the spring end back to its normal position. The next illustration clearly shows a spring end properly positioned in the housing.

Wash all parts except the rope in solvent and then blow them dry with compressed air.

Remove any trace of corrosion and wipe all metal parts with an oil dampened cloth.

Inspect the rope. Replace the rope if it appears to be weak or frayed. If the rope is frayed, check the hole through which the rope passes for rough edges or burrs. Remove the rough edges or burrs with a file, and polish the surface until it is smooth.

Inspect the starter spring end loops. Replace the spring if it is weak, corroded or cracked. Check the spring pin located at the back side of the pulley to be sure it is straight and solid.

Check the inside surface of the housing and remove any burrs.

Check the condition of the pawl springs to be sure they are not stretched out of shape. The end of each spring should be bent back toward the coil of the spring. Inspect the pawls for wear and that the edges are not rounded.

Inspect the hub center locating pin to be sure it is straight and tight.

STARTER ASSEMBLING

GOOD WORDS

The accompanying illustration shows a new starter spring as it is purchased. Note how the spring is held wound with "hog rings". The spring **MUST** be released to its full extended position before it can be installed. Therefore, use care and remove the "hog rings" and allow the spring to unwind until it is a straight piece of spring steel.

SAFETY WORD

Wear a good pair of gloves while unwinding and installing the spring. The spring will develop tension and the edges of the spring steel are sharp. The gloves will prevent cuts on your hands and fingers.

The rope on this unit broke, causing the spring to rewind with incredible speed. The end of the spring was bent back in the wrong direction.

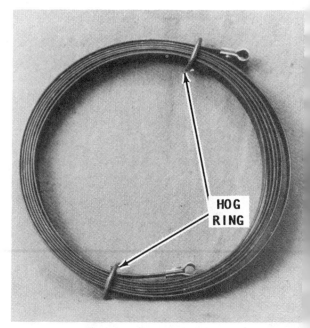

A new spring as it appears direct from the marine store. The hog rings must be CAREFULLY removed, as described in the text.

A safe method is for one person to remove the hog rings while an assistant holds the spring. After the rings have been removed, both persons work to unwrap the spring, one coil at-a-time.

1- Slide the spring onto the outer pin and then start the spring from the outside edge of the housing and insert it into the housing **COUNTERCLOCKWISE,** as shown in the accompanying illustration. Notice the small hump in the housing. This hump prevents the spring from being wound in the wrong direction. Work the first turn into the housing, and then hold the spring down with one hand and continue to wind the spring into the housing. Patience and time are required to work the spring completely into the housing. After the last portion is in place, bend the end of the spring towards the center of the housing. This position will allow the pulley pin to align with the loop in the end of the spring, when the pulley is installed.

2- Lower the pulley down over the top of the spring with the pulley pin indexing into the loop in the end of the spring. In the accompanying illustration, notice the callout for the boss on the backside of the pulley. The pin is located directly under the boss. The boss can, therefore, be a guide during pulley installation.

3- Coat the spindle with a thin film of OMC Type A lubricant. Place the wavy washer onto the spindle.

4- Slide the friction ring onto the spindle with the flats in the washer indexed with the flats on the spindle.

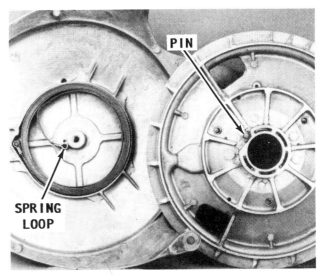

Housing (left) with the spring properly installed and the spring end bent toward the center. The pin in the pulley (right) must index into the loop on the spring end during installation.

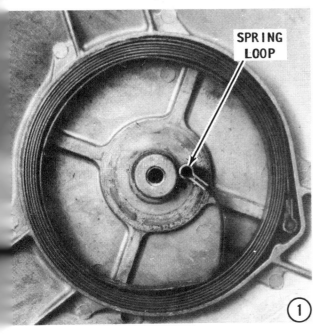

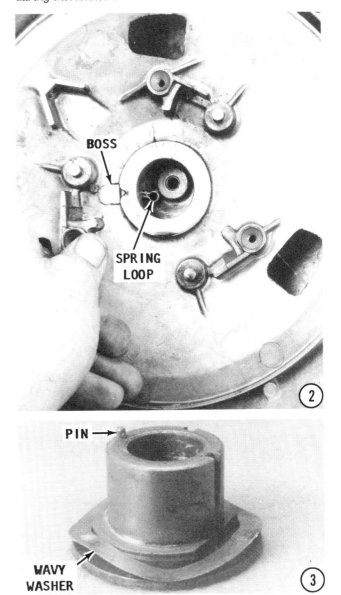

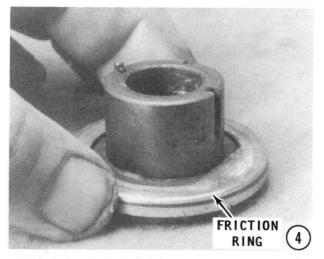

FRICTION RING 4

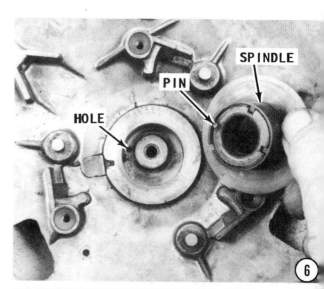

SPINDLE PIN HOLE 6

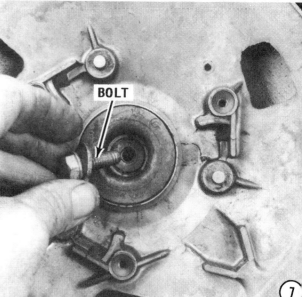

BOLT 7

5- Install the nylon bushing onto the spindle with the protrusions on the bushing indexed into the slots in the spindle. Notice the pin protruding from the bottom of the spindle. This pin **MUST** drop into the hole in the starter housing. Observe into the housing and visually locate this hole.

6- Lower the spindle assembly down through pulley and index the pin into the hole in the housing.

7- Place the washer inside the spindle housing and then install the bolt through the washer into the housing. Tighten the bolt securely.

8- Install the three pawls into their retainers with the tip of each pawl laying over the top of the nylon bushing.

9- Install the retainers securing the pawls to the pulley.

10- Turn the starter over and install the retaining nut onto the thru-bolt (if a nut is used). Tighten the nut securely. Check to be sure the pulley will rotate smoothly and does not bind on the spindle. Rotate the

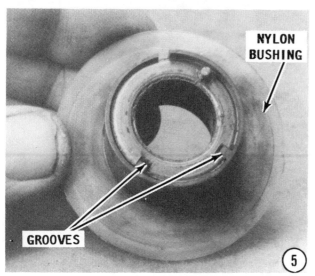

NYLON BUSHING

GROOVES 5

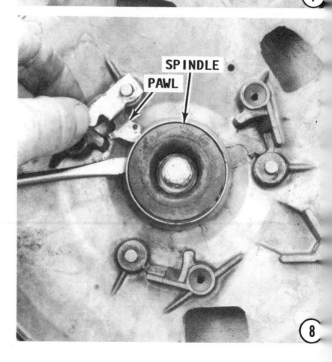

SPINDLE PAWL 8

ATOP FLYWHEEL NO RETURN SPRINGS 9-37

pulley slightly **COUNTERCLOCKWISE,** then release it to be sure the spring is properly engaged with the pulley and that the pulley has good spring tension.

ROPE INSTALLATION

Rope Purchase Instructions

The length and diameter of the starter rope required will vary depending on the size horsepower engine being serviced. Therefore, check the Hand Starter Rope Specifications in the Appendix, and then purchase a quality nylon piece of the proper length and diameter size. Only with the proper rope, will you be assured of efficient operation following installation.

Each end of the nylon rope should be "fused" by burning them slightly with a very small flame (a match flame will do) to melt the fibers together. After the end fibers have been "fused" and while they are still hot, use a piece of cloth as protection and pull the end out flat to prevent a "glob" from forming.

11- Tie a figure **8** knot in one end of the rope. Set the rope aside, but handy, to be picked up with one hand.

12- Hold the housing with one hand and rotate the pulley three complete turns **COUNTERCLOCKWISE** with the other hand. After three complete turns have been made, align the rope outlet in the pulley with the outlet in the housing. Insert a drift pin or other suitable tool through the hole in the pulley and the hole in the housing to hold the pulley in the desired position. Feed the the rope through the pulley and the housing

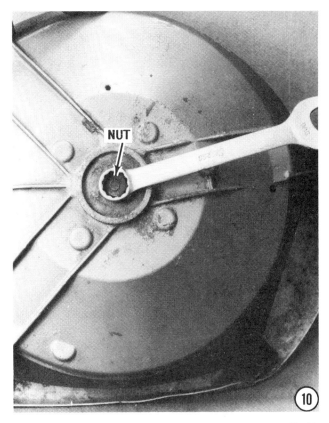

and out the other side of the housing. Pull the rope tight until the knot is seated against the pulley.

13- A special tool is manufactured by OMC to install the handle onto the rope. This tool has three prongs and is inserted through the handle and then attached to the rope and pulled back through the handle. If

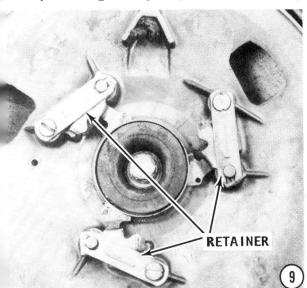

9-38 HAND STARTERS

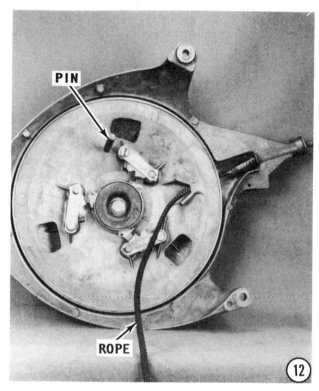

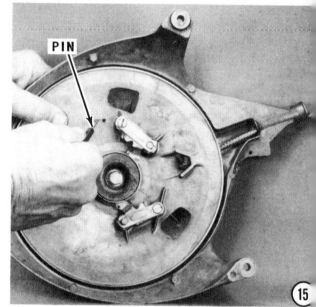

the special tool is not available take a stiff piece of wire; insert it through the handle; thread it through the rope; apply just a little oil to the rope; then pull the wire and rope through the handle.

14- Work the end of the rope into the handle anchor. Secure the rope in place by pushing the anchor into the rubber handle.

15- Lightly pull on the rope to relieve tension on the pin installed through the pulley and housing in Step 12. Maintain some tension on the rope, remove the pin, and allow the spring to **SLOWLY** wind the rope onto the pulley. Check the bolt through the spindle to be sure it is tight.

16- Lay the starter on its back and pull the rope with quick movements, and at the same time check the pawls to be sure they move towards the center of the pulley. Release the rope slowly and check to be sure the pawls return to their original position under the retainers.

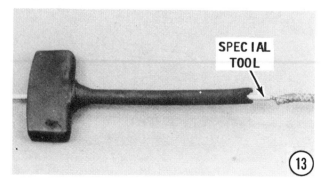

STARTER INSTALLATION

17- Position the starter over the flywheel with the three legs aligned over the holes in the powerhead for the retaining bolts. Install the retaining bolts and tighten them securely. Connect the linkage from the carburetor.

9-7 TYPE III MOUNTED ATOP FLYWHEEL MODEL WITH ONE NYLON PAWL
3 HP 1956-68
5.5 HP 1956-64
7.5 HP 1956-58
10 HP 1956-63
15 HP 1956
18 HP 1956-70
20 HP 1966-70
25 HP 1969-70

WARNING

As with other types of hand starters, the rewind spring is a potential hazard. The spring is under tremendous tension when it is wound -- a real tiger in a cage. If the spring should accidentally be released, severe personal injury could result from being struck by the spring with force. Therefore, the service instructions **MUST**, and we say again **MUST**, be followed closely to prevent release of the spring at the wrong time. Such action would be a **BAD SCENE,** a very **BAD SCENE,** because serious personal injury could result.

The starter rope should **NEVER** be released from the extended position. Such action would allow the spring to wind with incredible speed, resulting in serious damage to the starter mechanism.

Any time the rope is broken, the starter spring will rewind with incredible speed. Such action will cause the spring to rewind past its normal travel and the end of the spring will be bent back out of shape. Therefore, if the rope has been broken, the starter must be completely disassembled and the spring repaired or replaced.

Hand Starter Timing

Surprising as it may sound, this starter, mounted on top of the engine over the flywheel, can actually timed to the engine. This timing can best be described by using an example.

If two marks were made on the flywheel $180°$ apart, and matching marks made on the engine, then each time the engine was shut down, one set of marks on the flywheel would align very closely with one of the

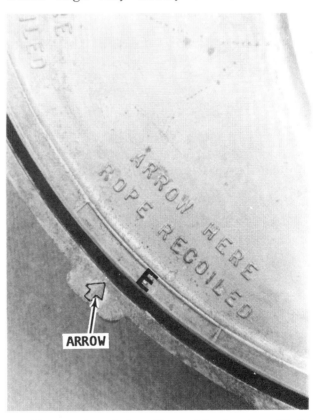

Closeup view of the pulley and the housing with the arrow on the housing aligned with the marks on the pulley. This alignment is necessary to "time" the starter with the engine.

marks on the engine. What is actually happening, is the engine is stopping with either the top piston at TDC (top dead center) or the bottom piston at the TDC position.

Now, assume the engine has been operating at idle speed and then suddenly stops for any number of reasons. The problem is corrected and the engine is once again ready to be started. Two notches are manufactured into the inside diameter of the flywheel. A single dog on the pulley engages with one of these dogs when the rope is pulled. Now, if it is necessary to pull an excessive amount of rope before the dog is able to engage the flywheel, full starting rotation power would not be available in the rope.

Therefore, the starter is timed to engage the starter with the flywheel after the rope has been pulled exactly the same amount each time. This distance is very short to allow as much rope pull as possible to rotate the crankshaft for fast start. This "timing" will assure an adequate amount available for starting **AND** that one of the pistons will return to TDC when the pull is completed, if the engine fails to start. If an excessive amount of pull is necessary before the flywheel begins to rotate, the starter was not assembled properly -- the arrow on the pulley was not aligned with the two marks on the starter housing when the spring is relaxed and the rope handle is retracted.

STARTER REMOVAL

1- Remove the attaching bolts securing the three legs of the starter housing to the powerhead. On some smaller horsepower engines, the starter housing is attached to the fuel tank with screws. Remove the hand starter and lay it on the bench with the pulley facing toward you.

DISASSEMBLING

2- Pull the rope out enough to tie a knot in the rope. Tie a knot, and then allow the rope to rewind to the knot. Work the rope anchor out of the rubber covered handle then remove the rope from the anchor. Remove the handle from the rope. Untie

The starter rope on this engine has been pulled much too far before the flywheel begins to rotate. This starter is, therefore, not timed properly with the engine, as described in the text.

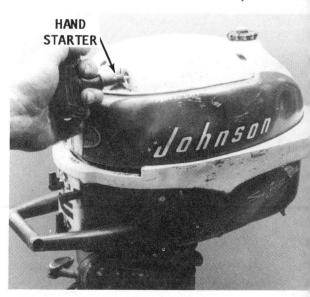

Starter installed atop the flywheel in the center the fuel tank.

ATOP FLYWHEEL WITH NYLON PAWL

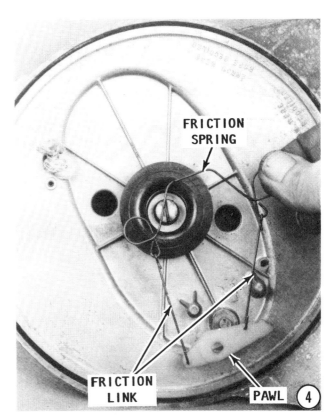

the knot in the rope, and then hold the disc pulley, but permit it to turn and thus allow the rope to wind back onto the pulley **SLOWLY**. Continue to allow the spring in the pulley to unwind **SLOWLY** until all tension has been released.

3- Remove the E-clip from the nylon pawl. Lift the pawl from the stud.

4- Remove the friction spring and friction link from the pawl.

5- Remove the bolt, lockwasher, and washer from the center of the pulley spindle. Lift the spindle out of the pulley, and

at the same time hold the pulley firmly together with the housing.

6- Lift the pulley straight up and at the same time work the spring free of the pulley. The spring has a small loop hooked into the pulley. An alternate and safe method is to hold the pulley and the housing together tightly and turn the complete assembly with the legs extending downward in the normal manner. Now, lower the complete assembly to the floor. When the legs

make contact with the floor, release your grip. The pulley will fall and the spring will be released from the housing, but the three legs will contain the spring and prevent it from traveling across the room. If the spring was not released from the housing, the only safe method is to again make contact with the three legs on the floor and jar the spring free.

7- Unwind the rope out of the pulley groove. Notice the pin next to the knot in the rope and how the rope feeds **BEHIND** the pin. Pull the knot and the rope out far enough to untie the knot, and then pull the rope free of the pulley.

CLEANING AND INSPECTING

If the rope was broken and the spring is bent backward, as shown in the accompanying illustration, it is a simple matter to bend the spring end back to its normal position. The next illustration clearly shows a spring end properly positioned in the housing.

Wash all parts except the rope in solvent and then blow them dry with compressed air.

Remove any trace of corrosion and wipe all metal parts with an oil dampened cloth.

Inspect the rope. Replace the rope if it appears to be weak or frayed. If the rope is frayed, check the hole through which the rope passes for rough edges or burrs. Remove the rough edges or burrs with a file, and polish the surface until it is smooth.

Inspect the starter spring end loops. Replace the spring if it is weak, corroded or cracked. Check the spring pin located at

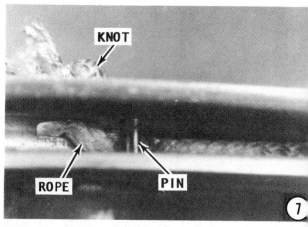

the back side of the pulley to be sure it is straight and solid.

Check the inside surface of the housing and remove any burrs.

Check the condition of the pawl spring to be sure it is not stretched out of shape.

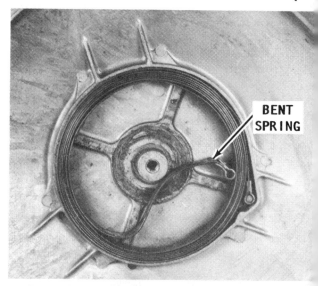

The rope on this unit broke, causing the spring to rewind with incredible speed. The end of the spring was bent back in the wrong direction.

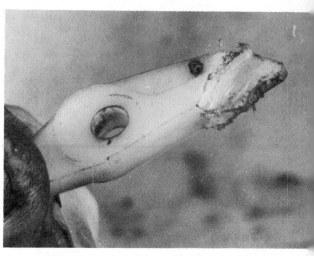

Damaged pawl unfit for further service.

ATOP FLYWHEEL WITH NYLON PAWL

e end of the spring should be bent back ward the coil of the spring. Inspect the wl for wear and that the edges are not nded.

Check the friction spring and link to be e they are not distorted.

Inspect the spindle. The spindle must be aight and tight.

ARTER ASSEMBLING

PE INSTALLATION

pe Purchase Instructions

The length and diameter of the starter pe required will vary depending on the e horsepower engine being serviced. erefore, check the Hand Starter Rope ecifications in the Appendix, and then rchase a quality nylon piece of the proper gth and diameter size. Only with the pper rope, will you be assured of efficient eration following installation.

Each end of the nylon rope should be sed" by burning them slightly with a very all flame (a match flame will do) to melt e fibers together. After the end fibers ve been "fused" and while they are still t, use a piece of cloth as protection and ll the end out flat to prevent a "glob" m forming.

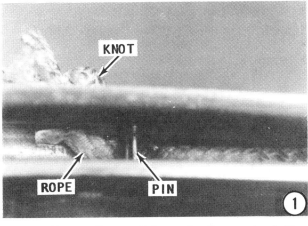

1- Tie a figure **8** knot in the end of the rope. Insert one end of the new rope through the hole in the pulley and housing and on the back side of the pin, as shown. Continue to wrap the remainder of the rope **COUNTERCLOCKWISE** around the pulley.

SAFETY WORD

Wear a good pair of gloves while installing the spring. The spring will develop tension and the edges of the spring steel are sharp. The gloves will prevent cuts on your hands and fingers.

2- Slide the spring onto the outer pin and then start the spring from the outside edge of the housing and insert it into the housing **COUNTERCLOCKWISE**, as shown in the accompanying illustration. Notice the small hump in the housing. This hump prevents the spring from being wound in the wrong direction. Work the first turn into the housing, and then hold the spring down

Knot tied in the end of the starter rope to prevent e rope from being pulled from the starter.

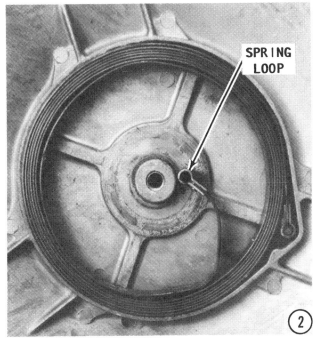

with one hand and continue to wind the spring into the housing. Patience and time are required to work the spring completely into the housing. After the last portion is in place, bend the end of the spring towards the center of the housing. This position will allow the pulley pin to align with the loop in the end of the spring, when the pulley is installed.

Alternate Method

An alternate method of installing a **NEW** spring into the housing with less risk of personal injury is presented with accompanying illustrations.

1A- Remove **ONLY** the hog ring next to the end of the outside wrap of the spring.

2A- Pull on the outside end of the spring. As the spring is pulled, the inside diameter will get smaller and smaller, as shown. When the diameter is a bit smaller than the inside diameter of the starter housing, wrap the entire free end of the spring around the coiled portion.

3A- **CAREFULLY** lower the coiled spring into the starter housing with the loop on the free end of the spring indexed over the peg in the housing and the spring feeding **COUNTERCLOCKWISE**, as shown. Remove the second hog ring **WITHOUT** allowing the spring to escape from the housing. An easy and safe method is to cut the hog ring with a pair of "dikes". Bend the inside end of the spring toward the center of the housing to permit the pin in the pulley to index into the loop.

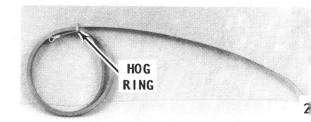

3- Lower the pulley down over the t of the spring with the pulley pin indexi into the loop in the end of the spring. In t accompanying illustration, notice the ca out for the boss on the backside of t pulley. The pin is located directly under t boss. The boss can, therefore, be a gui during pulley installation.

4- Lower the spindle assembly do through pulley. Place the washer and loc washer inside the spindle housing and th install the bolt through the washer into t housing. Tighten the bolt securely. Che

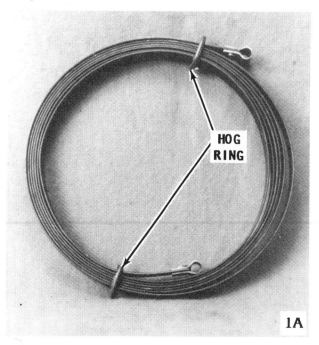

ATOP FLYWHEEL WITH NYLON PAWL 9-45

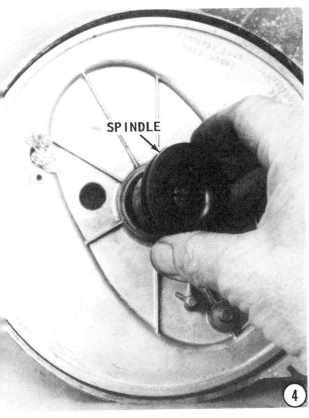

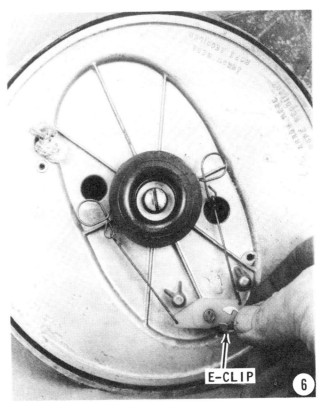

to be sure the pulley will rotate smoothly and does not bind on the spindle. Rotate the pulley slightly **COUNTERCLOCKWISE** and then release it to be sure there is proper engagement with the spring and the pulley has good spring tension.

5- Install the friction spring and link and the nylon pawl onto the starter hub. The friction spring fits into a groove in the spindle. Pull just a little on the pawl and set it over the stud on the flywheel pulley.

6- Snap the E-clip over the top of the pawl to secure it in place.

7- Rotate the pulley three complete revolutions, and then work the rope out through the hole in the pulley and the housing. Pull on the rope until a couple of feet are exposed. Tie a knot in the rope and allow the rope to rewind until the knot is tight against the housing. Work the end of

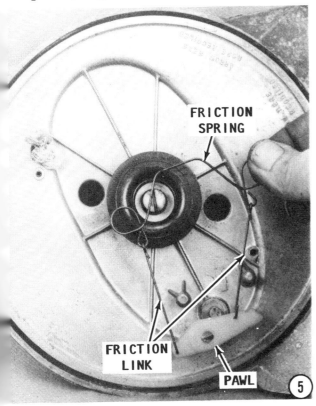

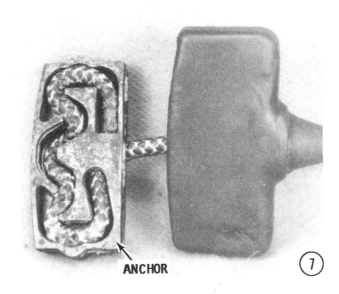

9-46 HAND STARTERS

the rope into the handle anchor. Secure the rope in place by pushing the anchor into the rubber handle.

8- Pull on the rope enough to untie the knot, and then allow the rope to slowly recoil into the pulley.

9- Lay the starter on its back with the pulley facing toward you. Notice the imprint on the pulley: **ARROW HERE ROPE-RECOILED**. Also notice the arrow on the housing. On Johnson engines, when the rope is fully coiled (the starter pulley completely wound) the arrow must fall between the marks on the pulley. Further up on the pulley you will notice the letter E (for Evinrude). On the Evinrude engines, the arrow must fall between the two marks on the pulley. If the arrow is not properly aligned the starter rope is not the proper length. It is either too long or too short. The arrow must align properly for the starter to be timed with the engine. See the beginning of this section.

10- With the starter still on its back, pull the rope with quick movements, and at the same time check the pawl to be sure it moves toward the center of the pulley. Release the rope slowly and check to be sure the pawl returns to its original position.

STARTER INSTALLATION

11- Position the starter over the flywheel with the three legs aligned over the holes in the powerhead for the retaining bolts. Install the retaining bolts and tighten them securely.

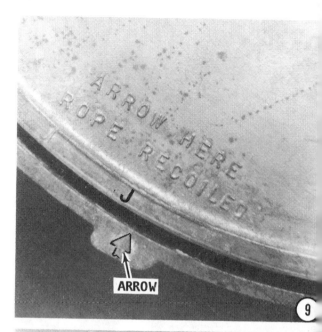

10
MAINTENANCE

10-1 INTRODUCTION

The authors estimate 75% of engine repair work can be directly or indirectly attributed to lack of proper care for the engine. This is especially true of care during the off-season period. There is no way on this green earth for a mechanical engine, particularily an outboard motor, to be left sitting idle for an extended period of time, say for six months, and then be ready for instant satisfactory service.

Imagine, if you will, leaving your automobile for six months, and then expecting to turn the key, have it roar to life, and be able to drive off in the same manner as a daily occurrence.

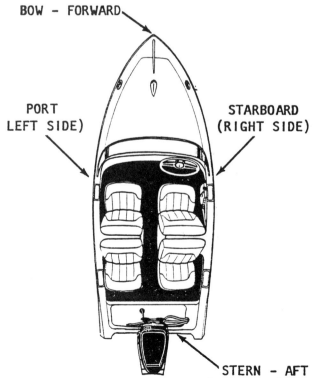

Common terminology used throughout the world for reference designation on boats. These are the terms used in this book.

It is critical for an outboard engine to be run at least once a month, preferably, in the water. At the same time, the shift mechanism should be operated through the full range several times and the steering operated from hard-over to hard-over.

Only through a regular maintenance program can the owner expect to receive long life and satisfactory performance at minimum cost.

Many times, if an outboard is not performing properly, the owner will "nurse" it through the season with good intentions of working on the unit once it is no longer being used. As with many New Year's resolutions, the good intentions are not completed and the outboard may lie for many months before the work is begun or the unit is taken to the marine shop for repair.

Imagine, if you will, the cause of the problem being a blown head gasket. And let us assume water has found its way into a cylinder. This water, allowed to remain over a long period of time, will do considerably more damage than it would have if the unit had been disassembled and the repair work performed immediately. **THEREFORE**, if an outboard is not functioning properly, **DO NOT** stow it away with promises to get at it when you get time, because the work and expense will only get worse, the longer corrective action is postponed. In the example of the blown head gasket, a relatively simple and inexpensive repair job could very well develop into major overhaul and rebuild work.

Outboards On Sail Boats

Owners of sail boats pride themselves in their ability to use the wind to clear a harbor or for movement from Port A to Port B, or maybe just for a day sail on a lake. The outboard is carried only as a last resort -- in case the wind fails completely, or in an emergency situation.

As a result, the outboard is stowed below, usually in a very poorly ventilated area, and subjected to moisture, stale air --in short, an excellent enviroment for "sweating" and corrosion.

If the dyed-in-the-wool "rag bagger" could just take the time about once every month or two, to pull out his outboard, clean it up, and give it a short run, not only would he have "peace of mind" knowing it **WILL** start in an emergency, but also his maintenance cost will be drastically reduced.

Chapter Coverage

The material presented in this chapter is divided into five general areas.

1- General information every boat owner should know.

2- Maintenance tasks that should be performed periodically to keep the boat operating at minimum cost.

3- Care necessary to maintain the appearance of the boat and to give the owner that "Pride of Ownership" look.

4- Winter storage practices to minimize damage during the off-season when the boat is not in use.

5- Preseason preparation work that should be performed to ensure satisfactory performance the first time it is put in service.

In nautical terms, the front of the boat is the **bow** and the direction is **forward**; the rear is the **stern** and the direction is **aft**; the right side, when facing forward, is the **starboard** side; and the left side is the **port** side. All directional references in this manual use this terminology. Therefore, the direction from which an item is viewed is of no consequence, because **starboard** and **port NEVER** change no matter where the individual is located or in which direction he may be looking.

10-2 ENGINE SERIAL NUMBERS

The engine serial numbers are the manufacturer's key to engine changes. These numbers identify the year of manufacture, the qualified horsepower rating, and the parts book identification. If any correspondence or parts are required, the engine model number **MUST** be used or proper identification is not possible. The accompanying illustrations will be very helpful in locating the engine identification tag for the various models.

ONE MORE WORD

The model number establishes the year in which the engine was produced and not neccessarily the year of first installation.

On some model engines, the serial number and model number were stamped on a plate mounted between the two swivel brackets underneath the hood.

On other model engines, the plate is mounted on the port side of the engine on the front or side of the swivel bracket. The hp and rpm range will also be found on the plate.

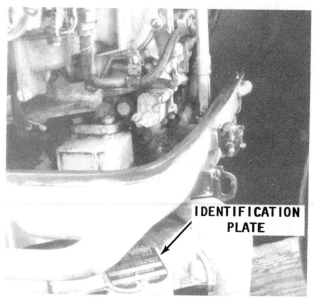

Manufacturer's identification plate installed between the transom brackets.

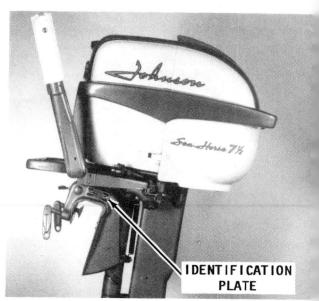

Manufacturer's identification plate installed on the port side of the transom bracket.

FIBERGLASS AND ALUMINUM HULLS 10-3

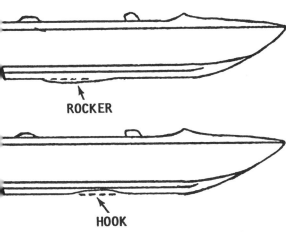

Simple drawing to illustrate two types of possible damage to the hull. Such injury to the boat will affect the boat's performance and subtract from the owner's enjoyment.

10-3 FIBERGLASS HULLS

Fiberglass reinforced plastic hulls are tough, durable, and highly resistant to impact. However, like any other material they can be damaged. One of the advantages of this type of construction is the relative ease with which it may be repaired. Because of the break characteristics, and the simple techniques used in restoration, these hulls have gained popularity throughout the world. From the most congested urban marina, to isolated lakes in wilderness areas, to the severe cold of far off northern seas, and in sunny tropic remote rivers of primitive islands or continents, fiberglass boats can be found performing their daily task with a minimum of maintenance.

A fiberglass hull has almost no internal stresses. Therefore, when the hull is broken or stove-in, it retains its true form. It will not dent to take an out-of-shape set. When

An aluminum boat ready for an engine. The owner of this type boat will probably carry it atop his vehicle and be saved the expense and trouble of trailering to the water.

the hull sustains a severe blow, the impact will be either absorbed by deflection of the laminated panel or the blow will result in a definite, localized break. In addition to hull damage, bulkheads, stringers, and other stiffening structures attached to the hull, may also be affected and therefore, should be checked. Repairs are usually confined to the general area of the rupture.

10-4 ALUMINUM HULLS

Aluminum boats have become popular in recent years because they are so lightweight and may be carried with ease atop an automobile or other vehicle. These aluminum craft are available in sizes ranging from small 8-foot prams to twin-hulled pontoon houseboats or swimming "rafts" in excess of 30 feet. Naturally, the large units cannot be carried atop a vehicle.

A new fiberglass boat and trailer outfit ready for an owner and a power package.

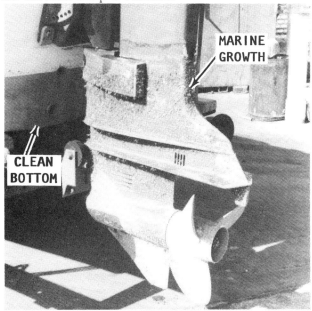

A boat and outboard used in salt water. Notice the marine growth on the lower unit and the anti-fouling bottom paint on the hull which prevented the marine growth.

10-4 MAINTENANCE

One of the advantages of an aluminum hull is the easy maintenance program required, and the ability of the material to resist corrosion.

As an added protection against marine growth, the below the waterline area may be painted with an anti-fouling paint. Bottom paint sold for use on a wooden or fiberglass hull is **NOT** suitable. At the time of purchase, check to be sure the paint contains the chemical properties required for an aluminum surface. The label should clearly indicate the intended use is specifically for aluminum.

If the aluminum hull does not have anti-fouling paint but requires cleaning to remove marine growth, one method is to rub the hull with a gunny sack just as soon as the boat is removed from the water and while it is still wet. The roughness of the sack is fairly effective in cleaning the surface of marine growth, including crustaceans (barnacles for instance) that have attached themselves to the hull. As soon as the rubdown has been completed the hull should be washed with high-pressure fresh water.

If the rubdown and wash was not accomplished immediately after the boat was removed from the water and the hull was allowed to dry, it will be necessary to cover the hull with wet blankets, gunny sacks or other suitable material and to continue soaking the covering until the growth is loosened. An easy alternate method, of course, is to return the boat to the water, if possible, and then too pull it out after it has been allowed to soak.

If an aluminum boat should strike an underwater object resulting in damage to the hull and a leak develops, the only emergency action possible is to make an attempt to reduce the amount of water being taken on by stuffing any type of available material into the opening until the boat is returned to shore. The aluminum cannot be repaired while it is wet. Repair of a damaged hull must be performed by a shop equipped for heliarc welding and other aluminum work.

Styrofoam blocks are installed under the seats of all aluminum boats. The foam blocks are designed for flotation to prevent the boat from sinking even if it should fill with water. Once each season, the wooden seat should be removed and the foam allowed to dry. Some manufacturers enclose the foam blocks in plastic bags prior to installation to protect them from moisture and loss of their flotation ability. New blocks may be purchased in a wide range of sizes. If new blocks are obtained, make an attempt to enclose the block in some form of plastic covering, then seal the package before installing it under the seat.

10-5 BELOW WATERLINE SERVICE

A foul bottom can seriously affect boat performance. This is one reason why racers, large and small, both powerboat and sail, are constantly giving attention to the condition of the hull below the waterline.

In areas where marine growth is prevalent, a coating of vinyl, anti-fouling bottom paint should be applied. If growth has developed on the bottom, it can be removed with a solution of muriatic acid applied with a brush or swab and then rinsed with clean water. **ALWAYS** use rubber gloves when working with muriatic acid and **TAKE EXTRA CARE** to keep it away from your face and hands. The **FUMES ARE TOXIC**. Therefore, work in a well-ventilated area or if outside, keep your face on the windward side of the work.

Barnacles have a nasty habit of making their home on the bottom of boats which have not been treated with anti-fouling paint. Actually they will not harm the fiberglass hull, but can develop into a major nuisance.

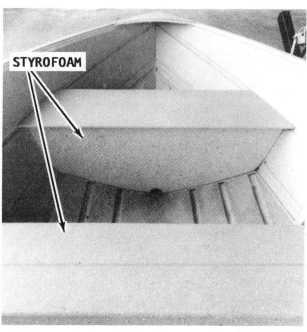

Aluminum boat with the wooden seat removed exposing the Styrofoam blocks for flotation. The seat should be removed at least once each season and the blocks thoroughly dried.

SUBMERGED ENGINE 10-5

If barnacles or other crustaceans have attached themselves to the hull, extra work will be required to bring the bottom back to a satisfactory condition. First, if practical, put the boat into a body of fresh water and allow it to remain for a few days. A large percentage of the growth can be removed in this manner. If this remedy is not possible, wash the bottom thoroughly with a high-pressure fresh water source and use a scraper. Small particles of hard shell may still hold fast. These can be removed with sandpaper.

10-6 SUBMERGED ENGINE SERVICE

A submerged engine is always the result of an unforeseen accident. Once the engine is recovered, special care and service procedures **MUST** be closely followed in order to return the unit to satisfactory performance.

NEVER, again we say **NEVER** allow an engine that has been submerged to stand more than a couple hours before following the procedures outlined in this section and making every effort to get it running. Such delay will result in serious internal damage. If all efforts fail and the engine cannot be started after the following procedures have been performed, the engine should be disassembled, cleaned, assembled, using new gaskets, seals, and O-rings, and then started as soon as possible.

Submerged engine treatment is divided into three unique problem areas: submersion in salt water; submerged engine while running; and a submerged engine in fresh water, including special instructions.

The most critical of these three circumstances is the engine submerged in salt water, with submersion while running a close second.

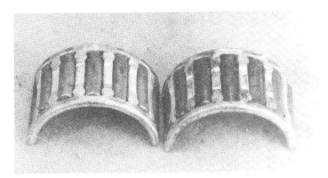

Rod bearing and cages badly damaged by salt water corrosion.

Salt Water Submersion

NEVER attempt to start the engine after it has been recovered. This action will only result in additional parts being damaged and the cost of restoring the engine increased considerably. If the engine was submerged in salt water the complete unit **MUST** be disassembled, cleaned, and assembled with new gaskets, O-rings, and seals. The corrosive effect of salt water can only be eliminated by the complete job being properly performed.

Cleaner to restore an engine recovered from fresh water after it has been submerged.

Crankshaft from a submerged two-cylinder engine recovered from salt water. In a very short time the crank was severely damaged by corrosion.

10-6 MAINTENANCE

Submerged While Running
Special Instructions

If the engine was running when it was submerged, the chances of internal engine damage is greatly increased. After the engine has been recovered, remove the spark plugs and attempt to rotate the flywheel with the rewind starter. On larger horsepower engines without a rewind starter, use a socket wrench on the flywheel nut. If the attempt to rotate the flywheel fails, the chances of serious internal damage, such as, bent connecting rod, bent crankshaft, or damaged cylinder, is greatly increased. If all attempts to rotate the flywheel fail, the powerhead must be completely disassembled.

Submerged Engine -- Fresh Water
SPECIAL WORD

As an aid to performing the restoration work, the following steps are numbered and should be followed in sequence. However, illustrations are not included with the procedural steps because the work involved is general in nature.

1- Recover the engine as quickly as possible.

2- Remove the hood and the spark plugs.

3- Remove the carburetor. To rebuild the carburetor, see Chapter 4.

4- Flush the outside of the engine with fresh water to remove silt, mud, sand, weeds, and other debris. **DO NOT** attempt to start the engine if sand has entered the powerhead. Such action will only result in serious damage to powerhead components. Sand in the powerhead means the unit must be disassembled.

5- Remove as much water as possible from the powerhead. Most of the water can be eliminated by first holding the engine in a horizontal position with the spark plug holes **DOWN**, and then cranking the engine with the rewind starter or with a socket wrench on the flywheel nut.

6- Alcohol will absorb water. Therefore, pour alcohol into the carburetor throat and again crank the engine.

7- Lay the engine in a horizontal position, and then roll it over until the spark plug openings are facing **UPWARD**. Pour alcohol into the spark plug openings and again crank the engine.

8- Roll the engine in the horizontal position until the spark plug openings are again facing **DOWN**. Pour engine oil into the carburetor throat and, at the same time, crank the engine to distribute oil throughout the crankcase.

9- With the engine still in a horizontal position, roll it over until the spark plug holes are again facing **UPWARD**. Pour approximately one teaspoon of engine oil into each spark plug opening. Crank the engine to distribute the oil in the cylinders.

10- Install the spark plugs and tighten them to the torque value given in the Appendix. Connect the high-tension leads to the spark plugs.

11- Install the carburetor onto the engine with a **NEW** gasket on the intake manifold.

12- Mount the engine in a test tank or body of water.

CAUTION: Water must circulate through the lower unit to the engine any time the engine is run to prevent damage to the water pump in the lower unit. Just five seconds without water will damage the water pump.

Obtain **FRESH** fuel and attempt to start the engine. If the engine will start, allow it to run for approximately an hour to eliminate any water remaining in the engine.

Rust preventative to be sprayed inside the engine in preparation for storage, as explained in the text.

13- If the engine fails to start, determine the cause, electrical or fuel, correct the problem, and again attempt to get it running. **NEVER** allow an engine to remain unstarted for more than a couple hours without following the procedures in this section and attempting to start it. If attempts to start the engine fail, the unit should be disassembled, cleaned, assembled, using new gaskets, seals, and O-rings, just as soon as possible.

10-7 WINTER STORAGE

Taking extra time to store the boat properly at the end of each season, will increase the chances of satisfactory service for the next season. **REMEMBER**, idleness is the greatest enemy of an outboard motor. The unit should be run on a monthly basis. The boat steering and shifting mechanism should also be worked through complete cycles several times each month. The owner who spends a small amount of time involved in such maintenance will be rewarded by satisfactory performance, and greatly reduced maintenance expense for parts and labor.

ALWAYS remove the drain plug and position the boat with the bow higher than the stern. This will allow any rain water and melted snow to drain from the boat and prevent "trailer sinking". This term is used to describe a boat that has filled with rain water and ruined the interior, because the plug was not removed or the bow was not high enough to allow the water to drain properly.

Proper storage for the engine involves adequate protection of the unit from physical damage, rust, corrosion, and dirt.

The following steps provide an adequate maintenance program for storing the unit at the end of a season.

1- Remove the hood. Start the engine and allow it to warm to operating temperature.

CAUTION: Water must circulate through the lower unit to the engine any time the engine is run to prevent damage to the water pump in the lower unit. Just five seconds without water will damage the water pump.

Disconnect the fuel line from the engine and allow the unit to run at **LOW** rpm and, at the same time, inject about 4 ounces of rust preventative spray through each carburetor throat. Allow the engine to run until it shuts down from lack of fuel, indicating the carburetor/s are dry of fuel.

2- Drain the fuel tank and the fuel lines. Pour approximately one qt. of benzol (benzene) into the fuel tank, and then rinse the tank and pickup filter with the benzol. Drain the tank. Store the fuel tank in a cool dry area with the vent **OPEN** to allow air to circulate through the tank. **DO NOT** store the fuel tank on bare concrete. Place the tank to allow air to circulate around it. If

Engine mounted in a test tank. The engine can be safely operated at idle speeds in preparation for winter storage, as explained in the text.

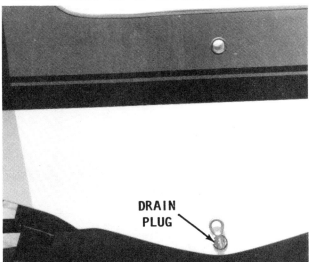

Drain plug removed from the transom to allow rain and melted snow to drain from the boat. Failure to remove this plug during long periods of storage can cause "boku" problems.

10-8 MAINTENANCE

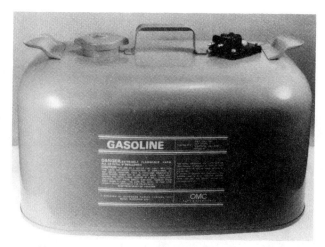

Standard OMC fuel tank. During periods of storage the tank should be empty and the cap "cracked" open to allow the tank to "breathe".

the fuel tank containing fuel is to be stored for more than a month, a commercial additive such as Sta-Bil should be added to the fuel. This type of additive will maintain the fuel in a "fresh" condition for up to a full year.

3- Clean the carburetor fuel filter/s with benzol, see Chapter 4, Carburetor Repair Section.

OMC fuel conditioner added to the fuel will keep it fresh for up to one full year.

Rust preventative to be used when preparing the engine for long periods of non-use and storage.

Chemical additives, such as Sta-Bil and the OMC fuel conditioner at the top of the page will prevent fuel from "souring" for up to twelve months.

4- Drain, and then fill the lower unit with OMC Lower Unit Gear Lubricant, as outlined in Section 10-8.

5- Lubricate the throttle and shift linkage. Lubricate the swivel pin and the tilt tube with Multipurpose Lubricant, or equivalent.

Clean the engine thoroughly. Coat the powerhead with Corrosion and Rust Preventative spray. Install the hood and then apply a thin film of fresh engine oil to all painted surfaces.

Remove the propeller. Apply Perfect Seal or a waterproof sealer to the propeller shaft, and then install the propeller back in position.

FINAL WORDS: Be sure all drain holes in the gear housing are open and free of obstruction. Check to be sure the **FLUSH** plug has been removed to allow all water to drain. Trapped water could freeze, expand, and cause expensive castings to crack.

ALWAYS store the engine off the boat with the lower unit below the powerhead to prevent any water from being trapped inside. The ideal storage position for an outboard is to hang it with the lower unit down. If hanging is not practical, lay the outboard on its back. This will place the lower unit below the powerhead.

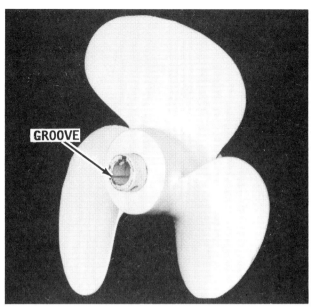

Propeller with grooves worn into the hub.

10-8 LOWER UNIT SERVICE

PROPELLER

The propeller should be checked regularly to be sure all the blades are in good condition. If any of the blades become bent or nicked, such damage will set up vibrations in the motor. Remove and inspect the propeller. Use a file to trim nicks and burrs. **TAKE CARE** not to remove any more material than is absolutely necessary. For a complete check, take the propeller to your marine dealer where the proper equipment and knowledgeable mechanics are available to perform a proper job at modest cost.

Inspect the propeller shaft to be sure it is still true and not bent. If the shaft is not perfectly true, it should be replaced.

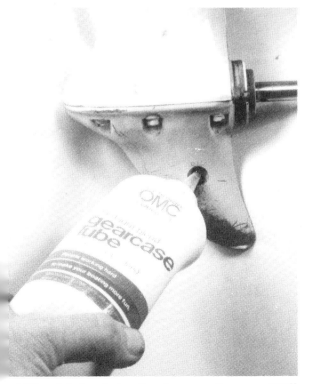

Adding lubricant to the lower unit on a small horsepower engine. The lubricant must always be added through the drain plug after the upper vent plug has been removed.

Sea weed entangled in the propeller. The propeller should be removed frequently and any foreign material, especially fish line, removed before it can cause damage.

10-10 MAINTENANCE

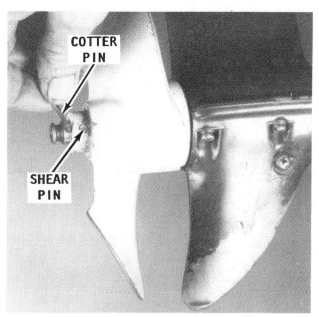

Propeller installation with a shear pin and cotter pin extending through the propeller shaft. A rubber cap covers both pins.

Propeller Removal

On some model engines, the shear pin is installed behind the propeller. First, pull the cotter key, and then remove the propeller nut, drive pin, and washer. Because the drive pin is not a tight fit, the propeller is able to move on the pin and cause burrs on the hole. These burrs may make removing the propeller difficult. To overcome this

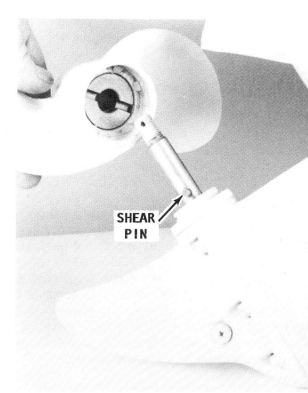

Propeller with a slot for the shear pin. The pin inserted first, then the propeller is installed onto t propeller shaft and over the shear pin.

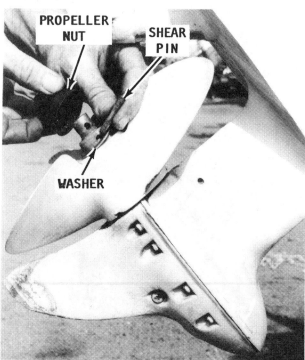

Propeller installation with a washer and shear pin. The cotter pin is installed through the propeller nut. This type of installation is the most popular.

OMC anti-corrosion lubricant that should be applie to the propeller shaft before the propeller is installe Such lubricant on the shaft will not only fight corrosio but assist in propeller removal.

problem, the propeller hub has two grooves running the full length of the hub. Hold the shaft from turning, and then rotate the propeller 1/4 turn to position the grooves over the drive pin holes. The propeller can then be pulled straight off the shaft. After the propeller has been removed, file the drive pin holes on both sides of the shaft to remove the burrs.

Installation

On some model engines, the shear pin is installed behind the propeller. On such units the propeller shaft should be coated with Perfect Seal, and then the propeller installed. After the propeller is on the shaft, install the washer, shear pin, propeller nut, and finally a **NEW** cotter pin.

Draining Lower Unit

Remove the **VENT** plug just above the anti-cavitation plate **FIRST,** and then the **FILL** plug from the gear housing. **NEVER** remove the vent or filler plug when the drive unit is hot. Expanded lubricant would be released through the plug hole.

CRITICAL WORD

The Phillips screw securing the shift fork in place is located very close to the vent screw. If the wrong screw is removed, **BAD NEWS, VERY BAD NEWS.** The lower unit will have to be disassembled in order to return the shift fork to its proper location.

Allow the gear lubricant to drain into the container. As the lubricant drains, catch some with your fingers, from time-to-time, and rub it between your thumb and finger to determine if any metal particles are present. If metal is detected in the lubricant, the unit must be completely disassembled, inspected, and the damaged parts replaced.

If the lubricant appears milky brown, or if large amounts of lubricant must be added to bring the lubricant up to the full mark, a thorough check should be made to determine the cause of the loss.

Filling Lower Unit

Add only OMC lower unit lubricant. Lubricant, for engines covered in this manual, is as follows: Use **ONLY** Type C, now known as Premium Blend Gearcase Lube, in all electric shift models. Use either Premium Blend Lube or the OMC Hi-Vis Gearcase Lube for all other engines. **NEVER** use regular automotive-type grease in the lower unit because it expands and foams too much. Lower units do not have provisions to accommodate such expansion.

The gearcase lubricant should be changed twice each year or season. If the lubricant is purchased in a large container, say the one-gallon size, a considerable savings can be realized. What is not used this

Draining lubricant from a lower unit. TAKE CARE not to remove the Phillips screw by mistake. This screw secures the shift fork in position.

Lubricant being drained from a lower unit without shift capabilities. Therefore, the Phillips screw mentioned in the illustration in the left column is not present.

10-12 MAINTENANCE

season will be used in the next or the one after. A small inexpensive pump can be be purchased to move the lubricant from the large container to the lower unit.

Position the drive unit approximately vertical and without a list to either port or starboard. Insert the lubricant tube into the **FILL/DRAIN** hole at the bottom plug hole, and inject lubricant until the excess begins to come out the **VENT** hole. Install the **VENT** and **FILL** plugs with **NEW** gaskets.

After the lower plug has been installed, remove the vent plug again and using a squirt-type oil can, add lubricant through this vent hole. A squirt-type oil can must be used to allow the trapped air in the lower unit to escape at the same time the final lubricant is added. Once the unit is completely full, install and tighten the vent plug.

Check to be sure the vent and drain plug gaskets are properly positioned to prevent water from entering the housing.

See the Appendix for lower unit capacities.

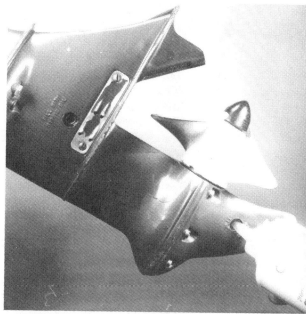
Filling the lower unit on a small horsepower engine through the lower hole after the vent plug has been removed.

The two types of lubricant used for Johnson/Evinrude engines. The text clearly identifies the lubricant to be used on the different model engines covered in this manual.

Filling a non-shifting lower unit with lubricant.

Using a squirt can to "top off" the lubricant in the lower unit, as explained in the text.

10-9 BATTERY STORAGE

Remove the batteries from the boat and keep them charged during the storage period. Clean the batteries thoroughly of any dirt or corrosion, and then charge them to full specific gravity reading. After they are fully charged, store them in a clean cool dry place where they will not be damaged or knocked over.

NEVER store the battery with anything on top of it or cover the battery in such a manner as to prevent air from circulating around the filler caps. All batteries, both new and old, will discharge during periods of storage, more so if they are hot than if they remain cool. Therefore, the electrolyte level and the specific gravity should be checked at regular intervals. A drop in the specific gravity reading is cause to charge them back to a full reading.

In cold climates, **EXERCISE CARE** in selecting the battery storage area. A fully-charged battery will freeze at about 60

Today, numerous type spark plugs are available for service. **ALWAYS** check with the local OMC dealer to be sure you are purchasing the proper plugs for the engine being serviced.

degrees below zero. A discharged battery, almost dead, will have ice forming at about 19 degrees above zero.

10-10 PRESEASON PREPARATION

Satisfactory performance and maximum enjoyment can be realized if a little time is spent in preparing the engine for service at the beginning of the season. Assuming the unit has been properly stored, as outlined in Section 10-6, a minimum amount of work is required to prepare the engine for use.

The following steps outline an adequate and logical sequence of tasks to be performed before using the engine the first time in a new season.

1- Lubricate the engine according to the manufacturer's recommendations. Remove, clean, inspect, adjust, and install the spark

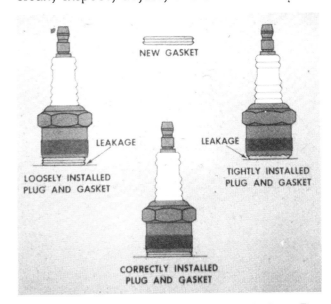

A check of the electrolyte in the battery should be on the maintenance schedule for any boat. A hydrometer reading of 1.300 or in the green band, indicates the battery is in satisfactory condition. If the reading is 1.150 or in the red band, the battery needs to be charged. Observe the six safety points given in the text, when using a hydrometer.

Correct and incorrect spark plug installation. The plugs **MUST** be installed properly and tightened to the proper torque value for satisfactory performance.

10-14 MAINTENANCE

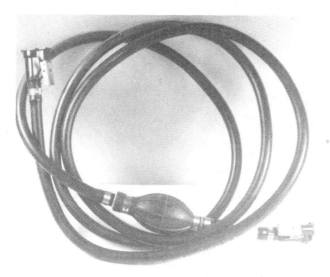

*Typical fuel hose with squeeze bulb. The hose and bulb must remain flexible. The O-rings **MUST** prevent fuel leakage.*

plugs with new gaskets if they require gaskets. Make a thorough check of the ignition system. This check should include: the points, coil, condenser, condition of the wiring, and the battery electrolyte level and charge.

2- If a built-in fuel tank is installed, take time to check the tank and all of the fuel lines, fittings, couplings, valves, and the flexible tank fill and vent. Turn on the fuel supply valve at the tank. If the fuel was not drained at the end of the previous season, make a careful inspection for gum formation. If a six-gallon fuel tank is used,

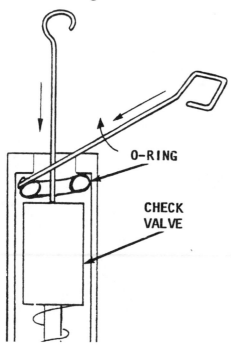

Method of removing an O-ring from a connector. The connector is also replaceable.

Adding OMC oil to the fuel. Only a high grade oil should be added to the fuel to ensure proper lubrication.

take the same action. When gasoline is allowed to stand for long periods of time, particularly in the presence of copper, gummy deposits form. This gum can clog the filters, lines, and passageways in the carburetor. See Chapter 4, Fuel System Service.

3- Check the oil level in the lower unit by first removing the vent screw on the port side just above the anti-cavitation plate.

OMC lubricants for Johnson/Evinrude outboards.

PRE-SEASON PREPARATION   10-15

sert a short piece of wire into the hole and
check the level. Fill the lower unit according
to procedures outlined in Section 10-8.

4- Close all water drains. Check and
replace any defective water hoses. Check
to be sure the connections do not leak.
Replace any spring-type hose clamps, if
they have lost their tension, or if they have
distorted the water hose, with band-type
clamps.

5- The engine can be run with the lower
unit in water to flush it. If this is not
practical, a flush attachment may be used.
This unit is attached to the water pick-up in
the lower unit. Attach a garden hose, turn
on the water, allow the water to flow into
the engine for awhile, and then run the
engine.

CAUTION: Water must circulate through
the lower unit to the engine any time the engine
is run to prevent damage to the water
pump in the lower unit. Just five seconds
without water will damage the water pump.

Check the idle exhaust port for water
discharge. At idle speed, only a fine water
mist will be visible. Check for leaks.
Check operation of the thermostat. After
the engine has reached operating temperature,
tighten the cylinder head bolts to the
torque value given in the Specifications in
the Appendix.

6- Check the electrolyte level in the
batteries and the voltage for a full charge.
Clean and inspect the battery terminals and

The battery should be located near the engine and well secured to prevent even the slightest amount of movement. The battery, including the terminals, **MUST** *be kept clean for maximum performance.*

cable connections. **TAKE TIME** to check
the polarity, if a new battery is being installed.
Cover the cable connections with
grease or special protective compound as a
prevention to corrosion formation. Check
all electrical wiring and grounding circuits.

Using a squirt can to "top off" the lubricant in a lower unit, as described in the text.

An engine mounted in a test tank can be safely operated at idle speed for adjustment purposes.

7- Check all electrical parts on the engine and electrical fixture or connections in the lower portions of the hull inside the boat to be sure they are not of a type that could cause ignition of an explosive atmosphere. Rubber caps help keep spark insulators clean and reduce the possibility of arcing. Starters, generators, distributors, alternators, electric fuel pumps, voltage regulators, and high-tension wiring harnesses should be of a marine type that cannot cause an explosive mixture to ignite.

ONE FINAL WORD

Before putting the boat in the water, **TAKE TIME** to **VERIFY** the drain plugs are installed. Countless number of boating excursions have had a very sad beginning because the boat was eased into the water only to have water begin filling the inside.

Keep your gas tank full, the fuel pump pumping, the spark plugs sparking, and the pistons, well, keep them working too.

Clarence & Howard

An OMC degreaser widely used for cleaning the boat and engine.

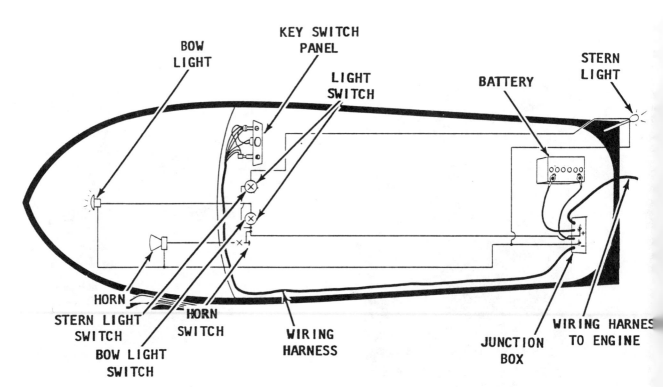

Principle electrical points requiring a careful check at the start of each season.

APPENDIX

METRIC CONVERSION CHART

LINEAR
inches	X 25.4	= millimetres (mm)
feet	X 0.3048	= metres (m)
yards	X 0.9144	= metres (m)
miles	X 1.6093	= kilometres (km)
inches	X 2.54	= centimetres (cm)

AREA
inches^2	X 645.16	= millimetres^2 (mm^2)
inches^2	X 6.452	= centimetres^2 (cm^2)
feet^2	X 0.0929	= metres^2 (m^2)
yards^2	X 0.8361	= metres^2 (m^2)
acres	X 0.4047	= hectares (10^4 m^2) (ha)
miles^2	X 2.590	= kilometres^2 (km^2)

VOLUME
inches^3	X 16387	= millimetres^3 (mm^3)
inches^3	X 16.387	= centimetres^3 (cm^3)
inches^3	X 0.01639	= litres (l)
quarts	X 0.94635	= litres (l)
gallons	X 3.7854	= litres (l)
feet^3	X 28.317	= litres (l)
feet^3	X 0.02832	= metres^3 (m^3)
fluid oz	X 23.57	= millilitres (ml)
yards^3	X 0.7646	= metres^3 (m^3)

MASS
ounces (av)	X 28.35	= grams (g)
pounds (av)	X 0.4536	= kilograms (kg)
tons (2000 lb)	X 907.18	= kilograms (kg)
tons (2000 lb)	X 0.90718	= metric tons (t)

FORCE
ounces - f (av)	X 0.278	= newtons (N)
pounds - f (av)	X 4.448	= newtons (N)
kilograms - f	X 9.807	= newtons (N)

ACCELERATION
feet/sec^2	X 0.3048	= metres/sec^2 (m/S^2)
inches/sec^2	X 0.0254	= metres/sec^2 (m/s^2)

ENERGY OR WORK (watt-second - joule - newton-metre)
foot-pounds	X 1.3558	= joules (j)
calories	X 4.187	= joules (j)
Btu	X 1055	= joules (j)
watt-hours	X 3500	= joules (j)
kilowatt - hrs	X 3.600	= megajoules (MJ)

FUEL ECONOMY AND FUEL CONSUMPTION
miles/gal	X 0.42514	= kilometres/litre (km/l)

Note:
235.2/(mi/gal) = litres/100km
235.2/(litres/100 km) = mi/gal

LIGHT
footcandles	X 10.76	= lumens/metre^2 (lm/m^2)

PRESSURE OR STRESS (newton/sq metre - pascal)
inches HG (60°F)	X 3.377	= kilopascals (kPa)
pounds/sq in	X 6.895	= kilopascals (kPa)
inches H2O (60°F)	X 0.2488	= kilopascals (kPa)
bars	X 100	= kilopascals (kPa)
pounds/sq ft	X 47.88	= pascals (Pa)

POWER
horsepower	X 0.746	= kilowatts (kW)
ft-lbf/min	X 0.0226	= watts (W)

TORQUE
pound-inches	X 0.11299	= newton-metres (N·m)
pound-feet	X 1.3558	= newton-metres (N·m)

VELOCITY
miles/hour	X 1.6093	= kilometres/hour (km/h)
feet/sec	X 0.3048	= metres/sec (m/s)
kilometres/hr	X 0.27778	= metres/sec (m/s)
miles/hour	X 0.4470	= metres/sec (m/s)

TEMPERATURE

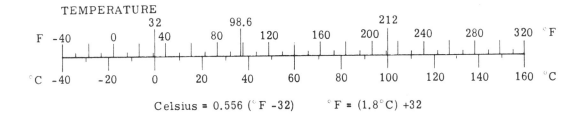

Celsius = 0.556 (°F -32)   °F = (1.8°C) +32

DRILL SIZE CONVERSION CHART

SHOWING MILLIMETER SIZES, FRACTIONAL AND DECIMAL INCH SIZES AND NUMBER DRILL SIZES

Milli-Meter	Dec. Equiv.	Frac-tional	Num-ber	Milli-Meter	Dec. Equiv.	Frac-tional	Num-ber	Milli-Meter	Dec. Equiv.	Frac-tional	Num-ber	Milli-Meter	Dec. Equiv.	Frac-tional	Num-ber	Milli-Meter	Dec. Equiv.	Frac-tional
.1	.0039			1.75	.0689			....	.1570		22	6.8	.2677			10.72	.4219	27/6
.15	.0059			....	.0700		50	4.0	.1575			6.9	.2716			11.0	.4330	
.2	.0079			1.8	.0709			....	.1590		21	....	.2720		I	11.11	.4375	7/16
.25	.0098			1.85	.0728			....	.1610		20	7.0	.2756			11.5	.4528	
.3	.0118			....	.0730		49	4.1	.1614			....	.2770		J	11.51	.4531	29/6
....	.0135		80	1.9	.0748			4.2	.1654			7.1	.2795			11.91	.4687	15/3
.35	.0138			....	.0760		48	....	.1660		19	....	.2811		K	12.0	.4724	
....	.0415		79	1.95	.0767			4.25	.1673			7.14	.2812	9/32	....	12.30	.4843	31/6
.39	.0156	1/64	....	1.98	.0781	5/64	....	4.3	.1693			7.2	.2835			12.5	.4921	
.4	.0157			....	.0785		47	....	.1695		18	7.25	.2854			12.7	.5000	1/2
....	.0160		78	2.0	.0787			4.37	.1719	11/64	....	7.3	.2874			13.0	.5118	
.45	.0177			2.05	.0807			....	.1730		17	....	.2900		L	13.10	.5156	33/6
....	.0180		77	....	.0810		46	4.4	.1732			7.4	.2913			13.49	.5312	17/3
.5	.0197			....	.0820		45	....	.1770		16	....	.2950		M	13.5	.5315	
....	.0200		76	2.1	.0827			4.5	.1771			7.5	.2953			13.89	.5469	35/6
....	.0210		75	2.15	.0846			....	.1800		15	7.54	.2968	19/64	....	14.0	.5512	
.55	.0217			....	.0860		44	4.6	.1811			7.6	.2992			14.29	.5624	9/16
....	.0225		74	2.2	.0866			....	.1820		14	....	.3020		N	14.5	.5709	
.6	.0236			2.25	.0855			4.7	.1850		13	7.7	.3031			14.68	.5781	37/6
....	.0240		73	....	.0890		43	4.75	.1870			7.75	.3051			15.0	.5906	
....	.0250		72	2.3	.0905			4.76	.1875	3/16	....	7.8	.3071			15.08	.5937	19/3
.65	.0256			2.35	.0925			4.8	.1890		12	7.9	.3110			15.48	.6094	39/6
....	.0260		71	....	.0935		42	....	.1910		11	7.94	.3125	5/16	....	15.5	.6102	
....	.0280		70	2.38	.0937	3/32	....	4.9	.1929			8.0	.3150			15.88	.6250	5/8
.7	.0276			2.4	.0945			....	.1935		10	....	.3160		O	16.0	.6299	
....	.0292		69	....	.0960		41	....	.1960		9	8.1	.3189			16.27	.6406	41/6
.75	.0295			2.45	.0964			5.0	.1968			8.2	.3228			16.5	.6496	
....	.0310		68	....	.0980		40	....	.1990		8	....	.3230		P	16.67	.6562	21/3
.79	.0312	1/32	....	2.5	.0984			5.1	.2008			8.25	.3248			17.0	.6693	
.8	.0315			....	.0995		39	....	.2010		7	8.3	.3268			17.06	.6719	43/6
....	.0320		67	....	.1015		38	5.16	.2031	13/64	....	8.33	.3281	21/64	....	17.46	.6875	11/1
....	.0330		66	2.6	.1024			....	.2040		6	8.4	.3307			17.5	.6890	
.85	.0335			....	.1040		37	5.2	.2047			....	.3320		Q	17.86	.7031	45/6
....	.0350		65	2.7	.1063			....	.2055		5	8.5	.3346			18.0	.7087	
.9	.0354			....	.1065		36	5.25	.2067			8.6	.3386			18.26	.7187	23/3
....	.0360		64	2.75	.1082			5.3	.2086			....	.3390		R	18.5	.7283	
....	.0370		63	2.78	.1094	7/64	....	....	.2090		4	8.7	.3425			18.65	.7344	47/6
.95	.0374			....	.1100		35	5.4	.2126			8.73	.3437	11/32	....	19.0	.7480	
....	.0380		62	2.8	.1102			....	.2130		3	8.75	.3445			19.05	.7500	3/4
....	.0390		61	....	.1110		34	5.5	.2165			8.8	.3465			19.45	.7656	49/6
1.0	.0394			....	.1130		33	5.56	.2187	7/32	....	....	.3480		S	19.5	.7677	
....	.0400		60	2.9	.1141			5.6	.2205			8.9	.3504			19.84	.7812	25/3
....	.0410		59	....	.1160		32	....	.2210		2	9.0	.3543			20.0	.7874	
1.05	.0413			3.0	.1181			5.7	.2244			....	.3580		T	20.24	.7969	51/6
....	.0420		58	....	.1200		31	5.75	.2263			9.1	.3583			20.5	.8071	
....	.0430		57	3.1	.1220			....	.2280		1	9.13	.3594	23/64	....	20.64	.8125	13/1
1.1	.0433			3.18	.1250	1/8	....	5.8	.2283			9.2	.3622			21.0	.8268	
1.15	.0452			3.2	.1260			5.9	.2323			9.25	.3641			21.04	.8218	53/6
....	.0465		56	3.25	.1279			....	.2340		A	9.3	.3661			21.43	.8437	27/3
1.19	.0469	3/64	....	....	.1285		30	5.95	.2344	15/64	....	....	.3680		U	21.5	.8465	
1.2	.0472			3.3	.1299			6.0	.2362			9.4	.3701			21.83	.8594	55/
1.25	.0492			3.4	.1338			....	.2380		B	9.5	.3740			22.0	.8661	
1.3	.0512			....	.1360		29	6.1	.2401			9.53	.3750	3/8	....	22.23	.8750	7/
....	.0520		55	3.5	.1378			....	.2420		C	....	.3770		V	22.5	.8858	
1.35	.0513			....	.1405		28	6.2	.2441			9.6	.3780			22.62	.8906	57/
....	.0550		54	3.57	.1406	9/64	....	6.25	.2460		D	9.7	.3819			23.0	.9055	
1.4	.0551			3.6	.1417			6.3	.2480			9.75	.3838			23.02	.9062	29/
1.45	.0570			....	.1440		27	6.35	.2500	1/4	E	9.8	.3858			23.42	.9219	59/
1.5	.0591			3.7	.1457			6.4	.2520			....	.3860		W	23.5	.9252	
....	.0595		53	....	.1470		26	6.5	.2559			9.9	.3839			23.81	.9375	15/
1.55	.0610			3.75	.1476			....	.2570		F	9.92	.3906	25/64	....	24.0	.9449	
1.59	.0625	1/16	....	....	.1495		25	6.6	.2598			10.0	.3937			24.21	.9531	61/
1.6	.0629			3.8	.1496			....	.2610		G	....	.3970		X	24.5	.9646	
....	.0635		52	....	.1520		24	6.7	.2638			....	.4040		Y	24.61	.9687	31/
1.65	.0649			3.9	.1535			6.75	.2657	16/64	....	10.32	.4062	13/32	....	25.0	.9843	
1.7	.0669			....	.1540		23	6.75	.2657			....	.4130		Z	25.03	.9844	63/
....	.0670		51	3.97	.1562	5/32	....	....	.2660		H	10.5	.4134			25.4	1.0000	

TORQUE SPECIFICATIONS

All torque values given in **Inch/pounds**.
Divide by 12 for foot/pounds.

HP	YEAR	FLYWHEEL NUT In./Lbs	CONNECTING ROD BOLT In./Lbs	CYLINDER HEAD BOLTS In./Lbs	CRANKCASE TO CYL BOLTS UPPER LOWER In./Lbs	CRANKCASE TO CYL BOLTS CENTER In./Lbs
1.5	1968-70	264-300	60-66	60-80	60-80	60-80
3.0	1956-68	480-540	60-96	60-84	60-84	60-84
4.0	1969-70	360-480	60-66	60-80	60-80	60-80
5.0	1965-68	480-540	60-66	60-80	60-80	60-80
5.5	1956-64	480-540	60-66	60-84	60-84	60-84
6.0	1965-70	480-540	60-66	60-80	60-80	60-80
7.5	1956-58	480-540	60-66	60-84	60-84	60-84
9.5	1965-70	480-540	90-100	96-120	120-145	120-145
10	1956-63	480-540	180-186	96-120	120-144	120-144
15	1956	480-540	180-186	96-120	120-144	120-144
18	1957-70	480-540	180-186	96-120	120-144	120-144
20	1966-70	480-540	180-186	96-120	110-130	110-130
25	1969-70	480-540	180-186	96-120	120-144	120-144
28	1962-64	1200-1260	348-372	168-192	162-168	162-168
30	1956	720-780	180-186	216-240	144-168	162-168
33	1965-70	1200-1260	348-372	168-192	150-170	162-168
35	1957-59	720-780	216-222	216-240	162-168	162-168
40	1960-70	1200-1260	348-372	168-192	150-170	162-168

STANDARD BOLTS AND NUTS
Torque value for special bolts may vary.

Bolt Size	In./Lbs	Ft/Lbs	Newton Meters
No. 6	7-10	--	0.8-1.2
No. 8	15-22	--	1.6-2.4
No. 10	25-35	2-3	2.8-4.0
No. 12	35-40	3-4	4.0-4.6
1/4"	60-80	5-7	7-9
5/16"	120-140	10-12	14-16
3/8"	220-240	18-20	24-27
7/16"	340-360	28-30	38-40

POWERHEAD SPECIFICATIONS

HP	YEAR	STROKE	BORE STD	BORE OVERSIZE	BORE OVERSIZE	PISTON TO CYL CLEARANCE MAX	MIN
1.5	1968	1.37	1.56	0.020	---	0.0025	0.0013
1.5	1969	1.37	1.56	0.020	---	0.0055	0.0043
1.5	1970	1.37	1.56	0.020	---	0.0055	0.0043
3.0	1956-67	1.37	1.56	0.020	---	0.0013	0.002
3.0	1968	1.37	1.56	0.020	---	0.0025	0.0013
4.0	1969	1.37	1.56	0.020	---	0.0049	0.0014
4.0	1970	1.37	1.56	0.020	---	0.008	0.0020
5.0	1965-68	1.50	1.94	0.020	---	0.003	0.0018
5.5	1956-64	1.50	1.94	0.020	---	0.0025	0.0013
6.0	1965-70	1.50	1.94	0.020	---	0.003	0.0018
7.5	1956-57	1.75	2.125	0.000	---	0.0015	0.0030
9.5	1968	1.81	2.31	0.020	---	0.0045	0.0030
9.5	1969-70	1.81	2.32	0.020	---	0.0050	0.0035
10	1956-63	1.875	2.37	N/A	---	0.0035	0.0030
15	1956	2.50	2.37	N/A	---	0.0035	0.0030
18	1957-64	2.25	2.50	0.025	0.040	0.0035	0.0040
18	1965-68	2.25	2.50	0.020	0.040	0.0045	0.0030
18-20	1969-1970	2.25	2.50	0.020	0.040	0.0047	0.0032
20	1966-68	2.25	2.50	0.020	0.040	0.0045	0.0030
25	1969-70	2.25	2.50	0.020	0.040	0.0045	0.0033
28	1962-64	2.75	2.875	0.020	0.040	0.025	0.0040
30	1956	2.75	2.875	0.020	0.040	0.025	0.0040
33	1965-70	2.75	3.06	0.020	0.040	0.045	0.0030
35	1957-59	2.75	3.06	0.020	0.040	0.0035	0.0040
40	1960-61	2.75	3.19	N/A	0.025	0.0035	0.0040
40	1962-68	2.75	3.19	N/A	0.025	0.0045	0.0030
40	1969-70	2.75	3.19	N/A	0.025	0.0045	0.0030

APPENDIX A-5

POWERHEAD SPECIFICATIONS

HP	YEAR	PISTON RING GROOVE		WIDTH OF RING		PISTON RING END GAP	
		MAX	MIN	MAX	MIN	MAX	MIN
1.5	1968	0.0035	0.0010	0.0935	0.0925	0.015	0.005
1.5	1969	0.0035	0.0010	0.0625	0.0615	0.015	0.005
1.5	9170	0.0040	0.0020	0.0625	0.0615	0.015	0.005
3.0	1956-67	0.0035	0.0010	---	---	0.015	0.005
3.0	1968	0.0035	0.0010	0.0935	0.0925	0.015	0.005
4.0	1969	0.0035	0.0010	0.0625	0.0615	0.015	0.005
4.0	1970	0.0040	0.0020	0.0625	0.0615	0.015	0.005
5.0	1965-68	0.0035	0.0010	0.0935	0.0925	0.015	0.005
5.5	1956-64	0.0035	0.0010	0.0935	0.0925	0.015	0.005
6.0	1965-70	0.0035	0.0010	0.0935	0.0925	0.015	0.005
7.5	1956-57	0.0035	0.0010	---	---	0.015	0.005
9.5	1968	0.0035	0.0010	0.0935	0.0925	0.017	0.007
9.5	1969-70	0.0035	0.0010	0.0935	0.0925	0.017	0.007
10	1956-63	0.0015	0.0030	---	---	0.017	0.007
15	1956	0.0015	0.0030	---	---	0.017	0.007
18	1957-64	0.0015	0.0030	0.0935	0.0925	0.017	0.007
18	1965-68	0.0035	0.0010	0.0935	0.0925	0.017	0.007
18-20	1969-70	0.0040	0.0020	0.0625	0.0615	0.017	0.007
20	1966-68	0.0035	0.0010	0.0625	0.0615	0.017	0.007
25	1969-70	0.0040	0.0020	0.0625	0.0615	0.017	0.007
28	1962064	0.0045	0.0070	---	---	0.017	0.007
30	1956	0.0045	0.0070	---	---	0.017	0.007
33	1965-70	0.0045	0.0070	0.0935	0.0925	0.017	0.007
35	1957-59	0.0065	0.0050	---	---	0.017	0.007
40	1960-61	0.0065	0.0050	---	---	0.017	0.007
40	1962-68	0.0045	0.0070	0.0935	0.0925	0.017	0.007
40	1969-70	0.0045	0.0020	0.0935	0.0925	0.017	0.007

TUNE-UP SPECIFICATIONS

SEE GENERAL AND SPECIAL NOTES APPENDIX PAGE A-10

JOHNSON MODEL	EVINRUDE MODEL	NO. CYL	HP	CU IN DISPL	W O T RPM	SPARK PLUG TYPE	SHIFT REMOVAL NOTE	PRIMARY PICKUP LOCATION	PRIMARY P/U ADJUSTMENT NOTE
1956	**1956**					**1956**			
JW-12	Lightwin, Ductwin	2	3	5.28	4000	J6J	A	7	1
CD-13	Fisherman	2	5.5	8.84	4000	J6J	B, C	7	1
AD-10	Fleetwin	2	7.5	12.40	4000	J6J	B, C	7	1
QD-17	Sportman	2	10	16.60	4000	J6J	D	7	1
FD, FDE-10	Fastwin	2	15	19.94	4000	J6J	D	7	1
RD, RDE, RJE-18	Bigtwin Lark	2	30	35.70	4000	J6J	D	9	1
1957	**1957**					**1957**			
JW-13	3022-3024	2	3	5.28	4000	J6J	A	7	1
CD-14	5514	2	5.5	8.84	4000	J6J	B, C	7	1
AD-11	7522	2	7.5	12.4	4000	J6J	B, C	7	1
QD-18	10014	2	10	16.6	4000	J6J	D	7	1
FD, FDE-11	15020-15922	2	18	22.0	4500	J4J	D	7	1
RD, RDE, RJE-19	25028-25930-25532	2	35	40.5	4500	J4J	D	9	1
1958	**1958**					**1958**			
JW-14	3026-3028	2	3.0	5.28	4000	J6J	A	7	1
CD-15	5516	2	5.5	8.84	4000	J6J	B, C	7	1
AD-12	7524	2	7.5	12.4	4000	J6J	B, C	7	1
QD-19	10016	2	10.0	16.6	4000	J6J	D	7	1
FD, FDE-12	15024-15926	2	18	22.0	4500	J4J	D	7	1
RD, RDE, RDS-19C	25034-25936-35514	2	35	40.5	4500	J4J	D	9	3
1959	**1959**					**1959**			
JW-15	3030-3032	2	3.0	5.28	4000	J6J	A	7	1
CD-16	5518	2	5.5	8.84	4000	J6J	B, C	7	1
QD-20	10018	2	10.0	16.6	4000	J4J	D	7	1
FD-13	15028	2	18.0	22.0	4500	J4J	D	7	1
RD, RDS-21	35012-35516	2	35.0	40.5	4500	J4J	D	9	3

NOTE: See Appendix Page A-10 for primary pickup location, primary pickup adjustment note, and shift removal note called out in this table.

TUNE-UP SPECIFICATIONS

SEE GENERAL AND SPECIAL NOTES APPENDIX PAGE A-10

APPENDIX A-7

JOHNSON MODEL	EVINRUDE MODEL	NO. CYL	HP	CU IN DISPL	W O T RPM	SPARK PLUG TYPE	SHIFT REMOVAL NOTE	PRIMARY PICKUP LOCATION	PRIMARY P/U ADJUSTMENT NOTE
1960	**1960**								
JW-16	3034-3036	2	3.0	5.28	4000	J6J	A	7	1
CD-17	5520	2	5.5	8.84	4000	J6J	B,C	7	1
QD-21	10020	2	10.0	16.6	4500	J4J	D	7	1
FD-14	15032	2	18	22.0	4500	J4J	D	7	1
RD, RDS-22	35018-35520	2	40	43.9	4500	J4J	D	9	3
1961	**1961**								
JW-17	3038-3040	2	3.0	5.28	4000	J6J	A	7	1
CD-18	5522	2	5.5	8.84	4000	J6J	B,C	7	1
QD-22	10032	2	10	16.6	4500	J4J	D	7	1
FD-15	15034	2	18	22	4500	J4J	D	7	1
RD, RDS-23	35022-35524	2	40	43.9	4500	J4J	D	9	3
1962	**1962**								
JW-17	3042-3044	2	3	5.28	4000	J6J	A	7	1
CD-19	5524	2	5.5	8.84	4000	J6J	B,C	7	1
QD-23	10024	2	10	16.6	4500	J4J	D	7	1
FD-16	15036	2	18	22	4500	J4J	D	7	1
RX-10	28202	2	28	35.7	4500	J4J	D	9	3
RD, RDS, RK-24	35028-30; 35932	2	40	43.9	4500	J4J	D	9	3
1963	**1963**								
JW-18	3302-3312	2	3	5.28	4000	J6J	A	7	1
CD-20	5302	2	5.5	8.84	4000	J6J	B,C	7	1
QD-24	10302	2	10	16.6	4500	J4J	D	7	1
FD-17	18302	2	18	22	4500	J4J	D	7	1
RX-11	28302	2	28	35.7	4500	J4J	D	9	3
RD, RDS, RK-25	40302-352; 40362-372	2	40	43.9	4500	J4J	D	9	4

NOTE: See Appendix Page A-10 for primary pickup location, primary pickup adjustment note, and shift removal note called out in this table.

A-8 APPENDIX

TUNE-UP SPECIFICATIONS
SEE GENERAL AND SPECIAL NOTES APPENDIX PAGE A-10

JOHNSON MODEL	EVINRUDE MODEL	NO. CYL	HP	CU IN DISPL	W O T RPM	SPARK PLUG TYPE	SHIFT REMOVAL NOTE	PRIMARY PICKUP LOCATION	PRIMARY P/U ADJUSTMENT NOTE
1964	**1964**								
JW, JH-19	3402-12; 3432	2	3	5.28	4000	J6J	A	7	1
CD-21	5402	2	5.5	8.84	4000	J6J	B, C	7	1
MQ-10	9422	2	9.5	15.2	4500	J4J	E	7	2
FD-18	18402	2	18	22	4500	J4J	D	7	1
RX-12	28402	2	28	35.7	4500	J4J	D	9	3
RD, RDS, Rk-26	40402-52; 40462-72	2	40	43.9	4500	J4J	D	9	4
1965	**1965**								
JW, JH-20B	3502-12; 3532	2	3	5.28	4000	J6J	A	7	1
LD-10S	5502	2	5	8.84	4000	J6J	A, B	7	1
CD-22M	6502	2	6	8.84	4500	J6J	B, E	7	1
MQ-11C	9522	2	9.5	15.2	4500	J4J	E	7	2
FD-19DL	18502	2	18	22	4500	J4J	D	7	1
RX, RXW-13BE	33502-52	2	33	40.5	4500	J4J	D	9	3
RD, RDS, RK, AM-27	40502-52; 40562-72	2	40	43.9	4500	J4J	D	9	4
1966	**1966**								
JW, JWL, JH, JHL-21	3602-03; 3612-32; 3633	2	3	5.28	4000	J6J	A	7	1
LD, LDL-11	5602-03	2	5	8.84	4000	J6J	A, B	7	1
CD, CDL-23	6602-03	2	6	8.84	4500	J6J	B, E	7	1
MQ, MQL-12	9622-23	2	9.5	15.2	4500	J4J	E	7	2
	18602-03	2	18	22	4500	J4J	D	7	1
FD, FDL-20	---	2	20	22	4500	J4J	D	7	1
RX, RXL, RXE, RXEL	33602-03; 33652-53	2	33	40.5	4500	J4J	D	9	3
RD, RDL-28	40602-03; 40652-53; 40672-73	2	40	43.9	4500	J4J	D	9	4
1967	**1967**								
JW, JWF, JWF/C, JH, JHF, JHF/C, JHL-22	3612-3716; 3732-33; 3706-07; 3702-03; 3736-37	2	3	5.28	40000	J6J	A	6	1
LD, LDL-12	5702-03	2	5	8.84	4000	J6J	A, B	6	1
CD, CDL-24	6702-03	2	6	8.84	4500	J6J	B, E	6	1
MQ, MQL-13	9722-23	2	9.5	15.2	4500	J4J	E	7	2

NOTE: See Appendix Page A-10 for primary pickup location, primary pickup adjust-

TUNE-UP SPECIFICATIONS

SEE GENERAL AND SPECIAL NOTES APPENDIX PAGE A-10

JOHNSON MODEL	EVINRUDE MODEL	NO. CYL	HP	CU IN DISPL	W O T RPM	SPARK PLUG TYPE	SHIFT REMOVAL NOTE	PRIMARY PICKUP LOCATION	PRIMARY P/U ADJUSTMENT NOTE
1967 (Continued)	**1967 (Continued)**								
---	18702-03	2	18	22	4500	J4J	D	8	1
FD, FDL-21	---	2	20	22	4500	J4J	D	8	1
RX, RXL, RXE, RXEL-15	33702-03; 33752-53	2	33	40.5	4500	J4J	D	9	3
RD, RDL, RDS, RDSL, RK, RKL-29	40702-03; 40752-53; 40772-73	2	40	43.9	4500	J4J	D	9	4
1968	**1968**								
SC-10	1802	1	1.5	2.64	4000	J6J	A	5	1
JH, JHL, JHF, JHF/C, JW, JWF, JWF/C-23	3806-07; 3802-03; 3832-33; 3836-37	2	3	5.28	4000	J6J	A	6	1
LD, LDL-13	5802-03	2	5	8.84	4000	J6J	A, B	6	1
CD, CDL-25	6802-03	2	6	8.84	4500	J6J	B, E	6	1
MQ, MQL-14	9822-23	2	9.5	15.2	4500	J4J	E	7	2
---	18802-03	2	18	22.0	4500	J4J	D	8	1
FD, FDL-22	---	2	20	22.0	4500	J4J	D	8	1
RX, RXL, RXE, RXEL-16	33802-03; 33852-53	2	33	40.5	4500	J4J	D	9	3
RD, RDL, RDS, RDSL-30	40802-03; 40852-53; 40872-73	2	40	43.9	4500	J4J	D	9	4
1969	**1969**								
1R-69	1902	1	1.5	2.67	4000	J6J	A	5	1
4R, 4W, 4WF-69	4902; 4906; 4936	2	4	5.28	4500	J6J	A	6	1
6R, 6RL-69	6902-03	2	6	8.84	4500	J6J	B, E	6	1
9R, 9RL-69	9922-23	2	9.5	15.2	4500	J4J	E	7	2
---	18902-03	2	18	22.0	4500	J4J	D	8	4
20R, 20RL-69	---	2	20	22.0	4500	J4J	D	8	4
25R, 25RL-69	25902-03	2	25	22.0	5500	J4J	D	8	4
33R, 33RL, 33E, 33EL-69	33902-03; 33952-53	2	33	40.5	4500	J4J	D	9	3
40R, 40RL, 40E, 40EL, 40ES, 40ESL-69	40902-03; 40952-53; 40972-73	2	40	43.9	4500	J4J	D	9	4

NOTE: See Appendix Page A-10 for primary pickup location, primary pickup adjustment note, and shift removal note called out in this table.

TUNE-UP SPECIFICATIONS

JOHNSON MODEL	EVINRUDE MODEL	NO. CYL	HP	CU IN DISPL	W O T RPM	SPARK PLUG TYPE	SHIFT REMOVAL NOTE	PRIMARY PICKUP LOCATION	PRIMARY P/U ADJUSTMENT NOTE
1970	**1970**					**1970**			
1R-70	1002	1	1.5	2.64	4000	J6J	A	5	1
4R, 4W-70	4006; 4036	2	4	5.28	4500	J6J	A	6	1
6R, 6RL-70	6002-03	2	6	8.84	4500	J6J	B, E	6	1
9R, 9RL-70	9022-23	2	9.5	15.2	4500	J4J	E	7	2
---	18002-03	2	18	22.0	4500	J4J	D	8	4
20R, 20RL-70	---	2	20	22.0	4500	J4J	D	8	4
25R, 25RL-70	25002-03	2	25	22.0	5500	J4J	D	8	4
33R, 33RL, 33E, 33EL-70	33002-03; 33052-53	2	33	40.5	4500	J4J	D	9	3
40R, 40RL, 40E, 40EL, 40ES, 40ESL-70	40002-03; 40052-54; 40072-73	2	40	43.9	4500	J4J	D	9	4

GENERAL NOTES

All engines -- flywheel magneto.
Top cylinder -- No. 1.
Bottom cylinder -- No. 2.
All points -- .020".
All spark plugs -- .030".
Timing **NOT** adjustable.

SPECIAL NOTES

Primary Pickup Adjustment

1- Loosen two screws under armature plate and move primary pickup in or out to meet follower.

2- Loosen center screw on throttle lever and move lever in or out to match line on cam (9.5 HP Model only).

3- Loosen eccentric lock screw on throttle shaft and turn eccentric to move roller to meet armature cam.

4- Loosen clamp on throttle shaft and move roller to meet armature cam.

Primary Pickup Location
5- Port side of mark.
6- Starboard side of mark.
7- Center of mark.
8- Between marks.
9- Mark at pointer.

Shift Removal
A- No shift rod.
B- Pin in upper end of driveshaft.
C- Remove powerhead.
D- Remove window.
E- Drop lower unit.

GEAR OIL CAPACITIES

ENGINE SIZE	MODEL	CAPACITY OUNCES
1.5 hp	1968-70	.75
3.0 hp	1956-70	2.9
4.0 hp	1969-70	1.3
4.0W hp	1969-70	3.4
5.0 hp	1965-68	2.9
5.5 hp	1956-64	8.5
6.0 hp	1965-70	8.5
7.5 hp	1956-58	8.5
9.5 hp	1964-70	9.7
10 hp	1956-63	10.0
15 hp	1956	8.3
18 hp	1956-70	8.3
20 hp	1966-68	8.3
25 hp	1969-70	8.3
28 hp	1962-65	13.9
30 hp	1956	13.9
33 hp	1965-68	13.9
35 hp	1957-59	13.9
40 hp	1969-70	13.9
40E hp	1962-70	15.0

STARTER MOTOR SPECIFICATIONS

Model Number	Brush Spring Tension (Ounces)	Armature End Play (Inches)	Volts	Max. Amperes	Min. RPM	Volts	Max. Amperes	Min. Lbs. Ft.
MDO-0	42-66	0.005 min.	10.0	38	10,000	4.0	170	1.5
MDO-1	42-66	.010-.035	10.0	38	10,000	4.0	170	1.5
MDW-0	42-66	.010-.035	10.0	26	8,500	4.0	160	2.1
MDW-1	42-66	.005 min.	10.0	26	8,500	4.0	160	2.1

Pinion position, 1-25/32 ± 1/16 inches from face of mounting flange to edge of pinion.

MDO-4002M	Use Test	MDO-0	CCW Rotation
MDO-4003M	Use Test	MDO-1	CW Rotation
MDW-4001M	Use Test	MDW-0	CW Rotation
MDW-4002M	Use Test	MDW-1	CCW Rotation

REGULATOR SPECIFICATIONS

Part No.	VRU-6101A
System Voltage	12
Ground Polarity	Negative
Armature Air Gap	
Circuit Breaker	.031-.034 in.
Voltage Reg.	.048-.052 in.
Current Regulator	.048-.052 in.
Current Regulator	
Setting Amps	9.0-11.0

CB Shunt Winding 107 to 121 ohms
VR Winding 43.7 to 49.3 ohms

Circuit Breaker	
Close Volts	12.6-13.6
Open Amps	
Discharge	3.0-5.0

Regulator (Hot) Operating Voltages
Tolerance ± .4 Volt

50°F	15.2 Volts
80°F	15.0 Volts
110°F	14.8 Volts
140°F	14.6 Volts

These figures are for a unit in normal operation while charging at 1/2 rated output or with 1/4 ohm fixed resistor in series with the battery.

GENERATOR SPECIFICATIONS

Generator	GJG-4001M
	GJG-4002M
Rot. D.E.	C
Ground Polarity	Negative
Brush Spring	
Tension	12-24 oz.
Field Coil Draw	
Volts	10.0
Amps	1.7-1.9
Monitoring Draw	
Volts	10.0
Amps	5.0-6.0
Generator Output	
Volts	15.0
Max. Amps	10.0
Max. RPM	7,000

CONDENSER SPECIFICATIONS

ENGINE SIZE	MODEL	CONDENSER OMC PART NO.	MFD CAPACITY
1.5 hp	All	580321	18-22
3.0 hp	All	580321	18-22
4.0 hp	All	580321	18-22
5.0 hp	All	580321	18-22
5.5 hp	All	580321	18-22
6.0 hp	All	580321	18-22
7.5 hp	All	580321	18-22
9.5 hp	All	580321	18-22
10 hp	All	580321	18-22
15 hp	All	580321	18-22
18 hp	1957-61	580321	18-22
30 hp	All	580321	18-22
35 hp	All	580321	18-22
40 hp	1960-61	580321	18-22
18 hp	1962-72	580419	25-29
20 hp	All	580419	25-29
25 hp	All	580419	25-29
28 hp	All	580419	25-29
33 hp	All	580419	25-29
40 hp	All	580419	25-29

STARTER ROPE SPECIFICATIONS

ENGINE SIZE	MODEL	DIAMETER INCHES	LENGTH INCHES
1.5 hp	1968-70	382712 Note 2	
3.0 hp	1956-61	5/32	65-1/4
3.0 hp	1962-67	5/32	71-1/2
3.0 hp	1968	0.130	56
4.0 hp	1969-70	0.130	64
5.0 hp	1965-67	0.130	64
5.0 hp	1968	0.130	56
5.5 hp	1956-61	5/32	70
5.5 hp	1962-64	5/32	71-1/2
6.0 hp	1965-67	0.130	64
6.0 hp	1968-70	0.130	56
7.5 hp	1956-58	5/32	70
9.5 hp	1964-70	5/32	71-1/2
10 hp	1956-61	7/32	70-3/16
10 hp	1962-63	7/32	75-3/4
15 hp	1956	7/32	70-3/16
18 hp	1957-61	7/32	70-3/16
18 hp	1962-70	7/32	75-3/4
20 hp	1966-70	7/32	75-3/4
25 hp	1969-70	7/32	75-3/4
28 hp	1962-64	7/32	75-3/4
30 hp	1956	7/32	69-3/4
33 hp	1965-70	7/32	75-3/4
35 hp	1957-59	7/32	73-3/4
40 hp	1960-70	7/32	75-3/4

Notes

1- Purchase a good grade of nylon rope.
2- Order OMC part number. Rope includes handle.

APPENDIX A-15

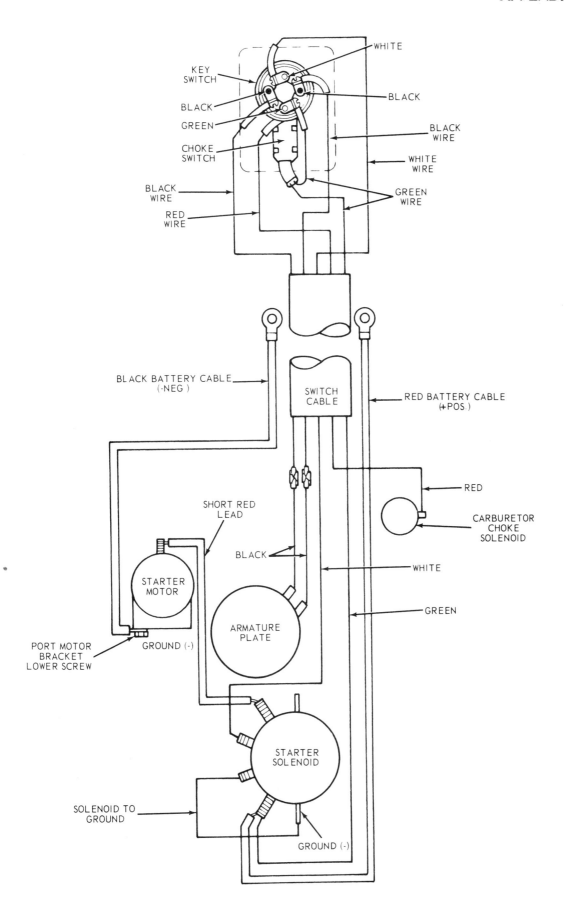

Wire Identification — 20 hp and 25 hp — 1971-72

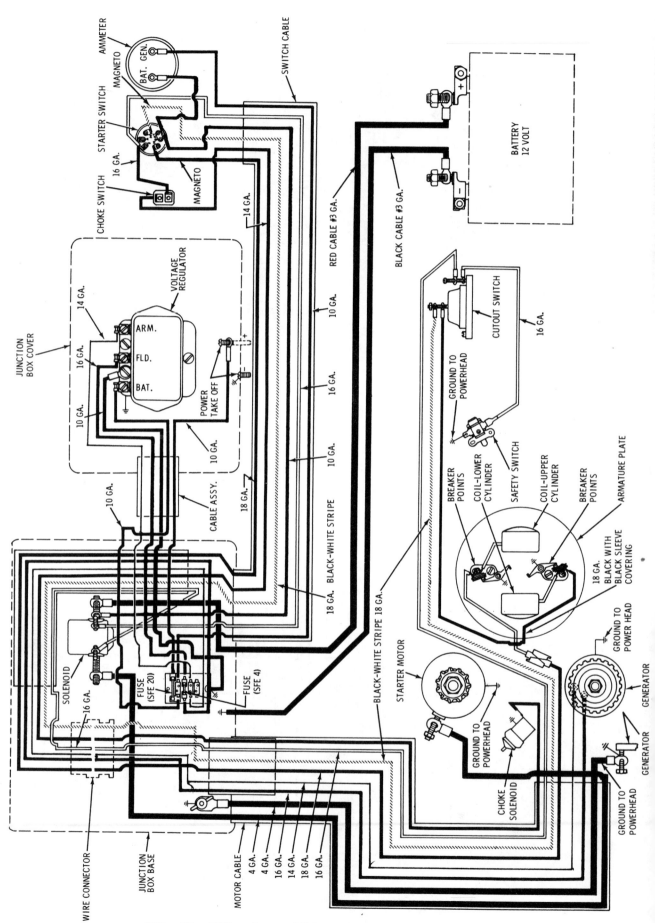

Wire Identification — 33 hp with Generator — 1965-67

APPENDIX A-17

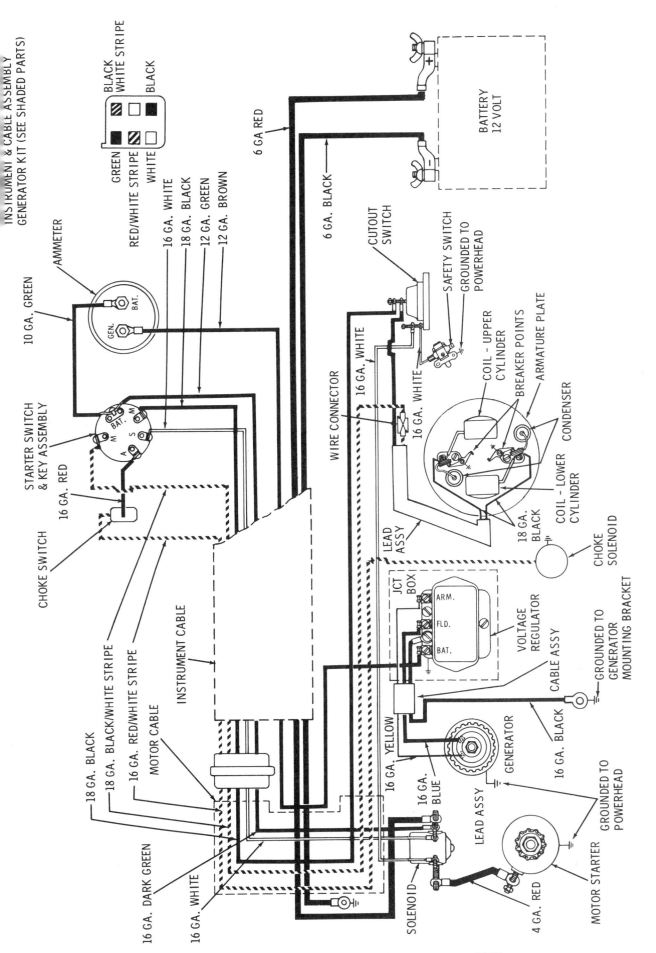

Wire Identification — 33 hp with Generator — 1968

A-18 APPENDIX

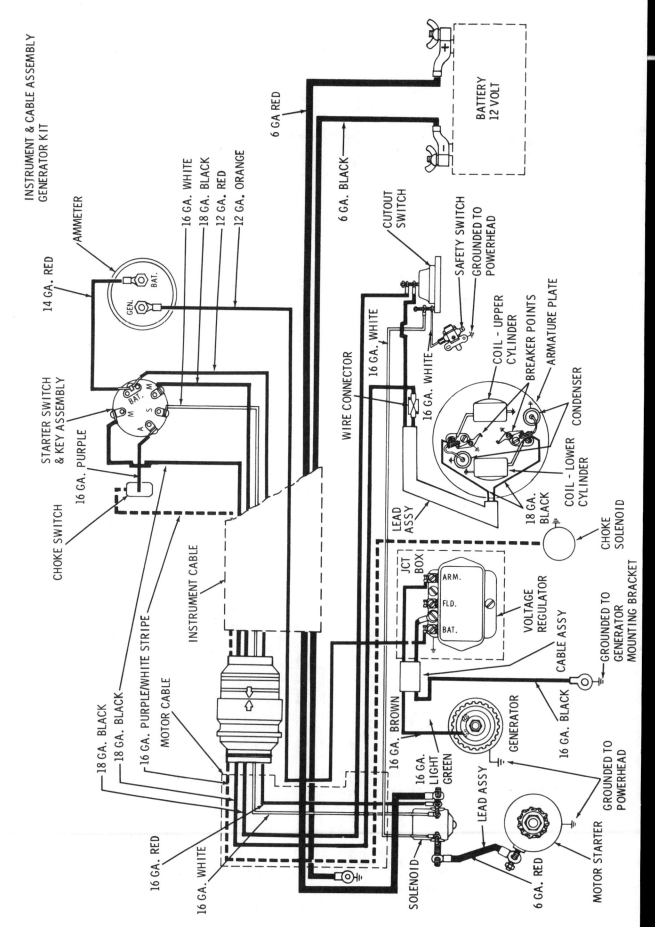

Wire Identification — 33 hp with Generator — 1969-70

APPENDIX A-19

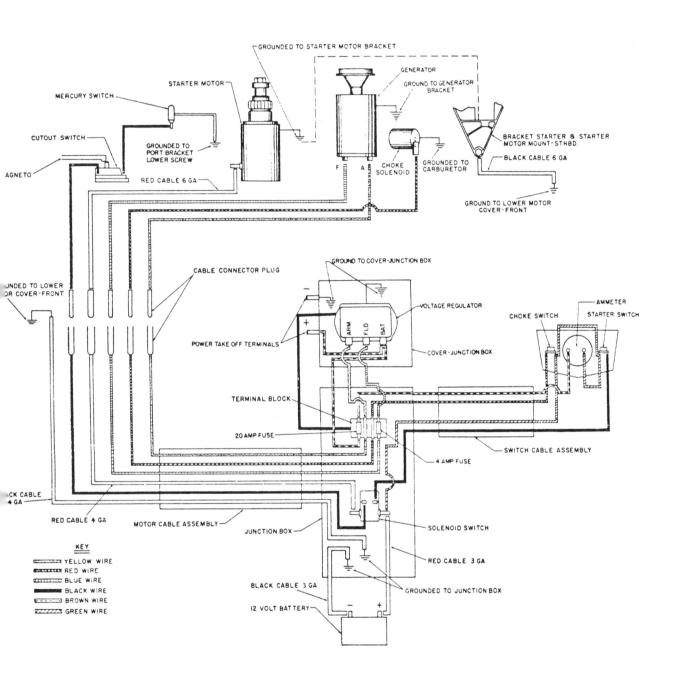

Wire Identification — 35 hp — 1957-59

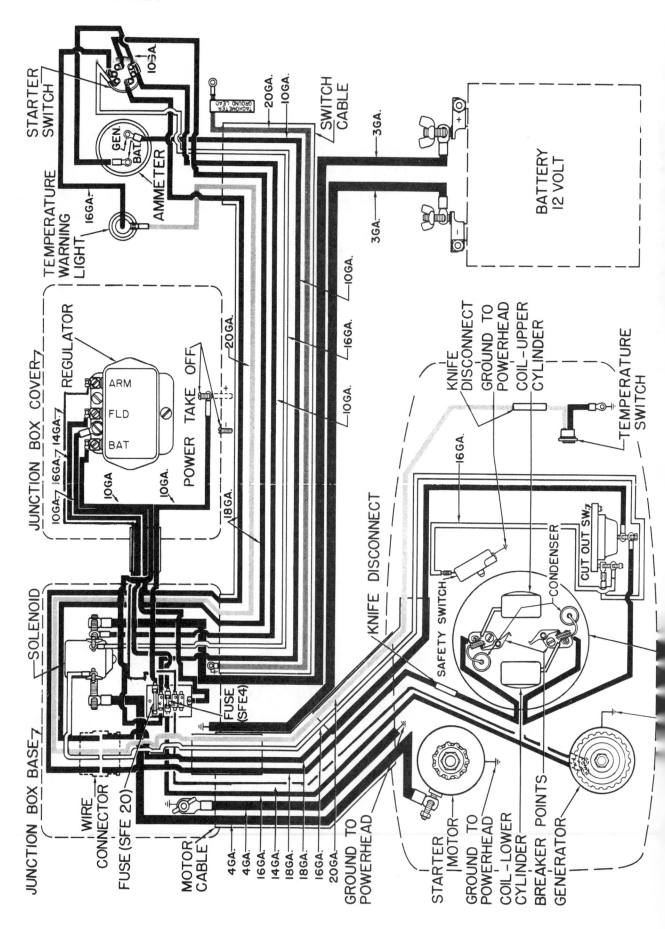

Wire Identification — 40 hp Standard Shift with Generator — 1960-66

APPENDIX A-21

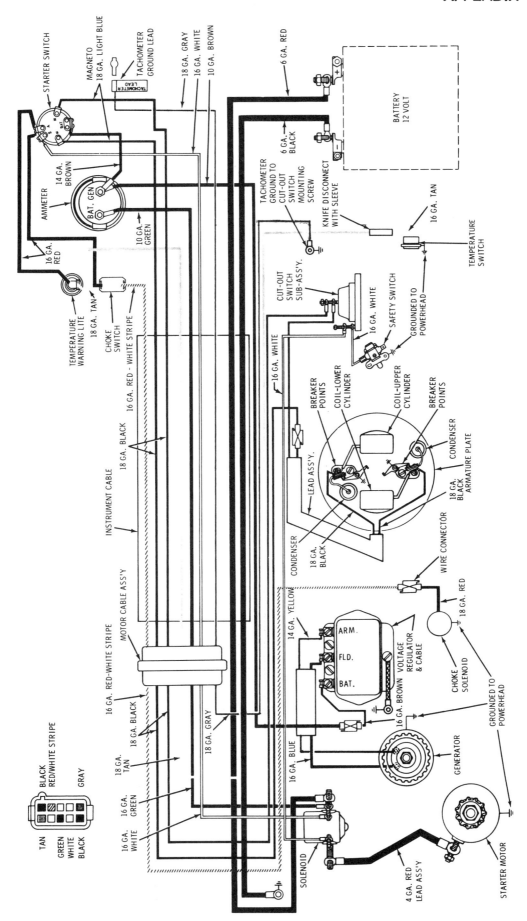

Wire Identification — 40 hp Standard Shift with Generator — 1967-68

A-22 APPENDIX

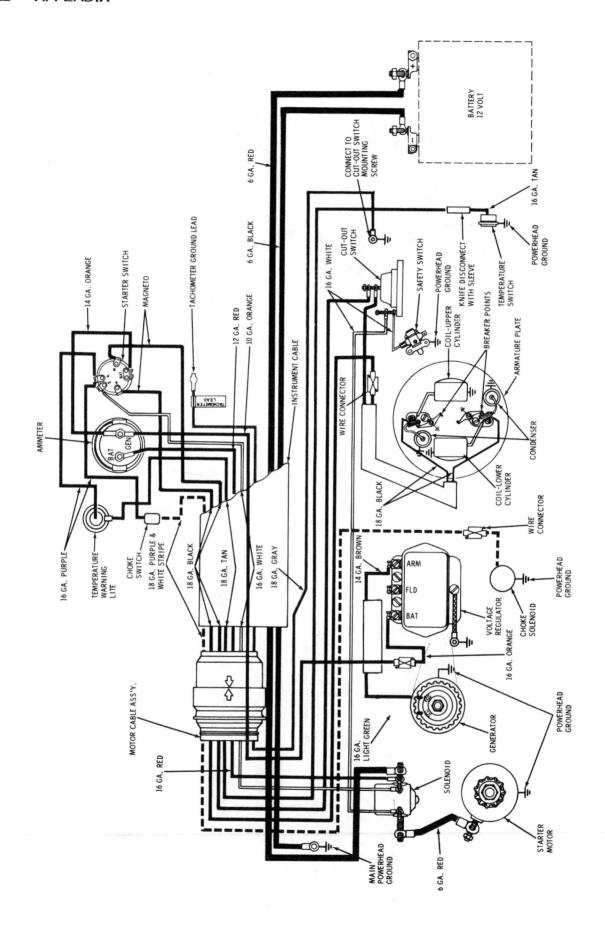

Wire Identification — 40 hp Standard Shift With Generator — 1969-70

APPENDIX A-23

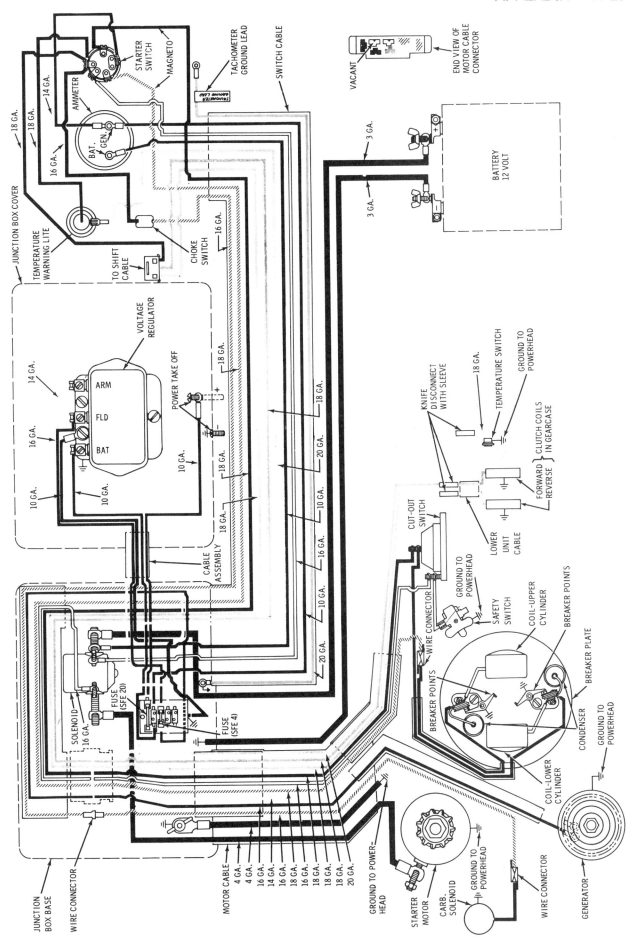

Wire Identification — 40 hp Electric Shift with Generator — 1961-66

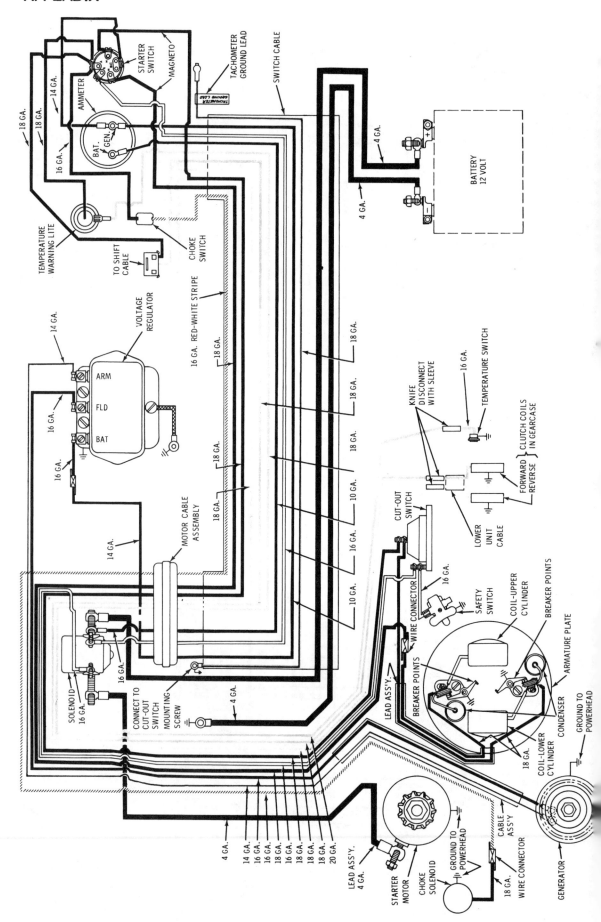

Wire Identification — 40 hp Electric Shift with Generator — 1967-68

APPENDIX A-25

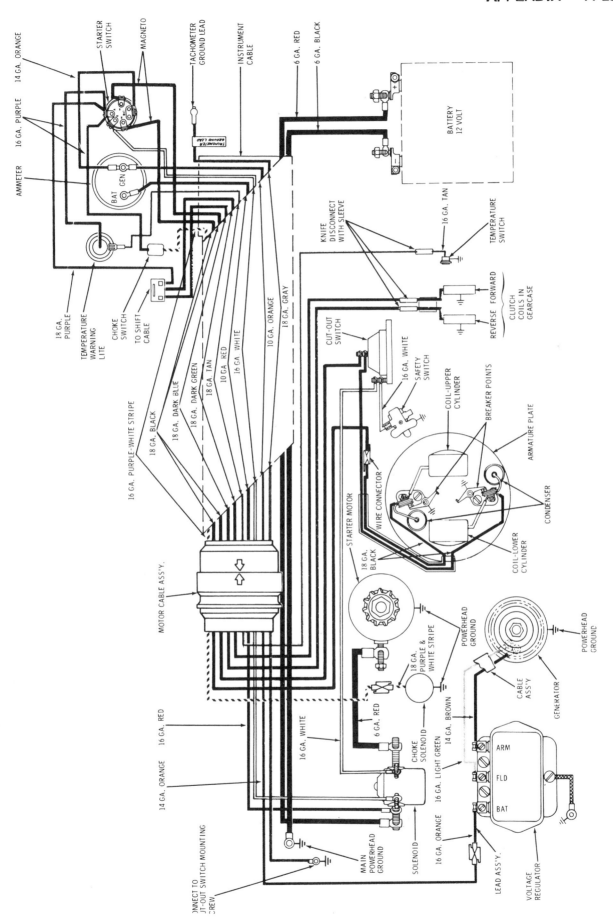

Wire Identification — 40 hp Electric Shift with Generator — 1969-70

NOTES & NUMBERS

NOTES & NUMBERS

SELOC PUBLISHING'S FULL-LINE MASTER LIST

ISBN	PART NO		TITLE/DESCRIPTION	YEARS
OUTBOARDS				
089330018-7	018-7	1000	Chrysler Outboards, All Engines	1962-8
089330055-1	055-1	1100	Force Outboards, All Engines	1984-9
089330048-9	048-9	1200	Honda Outboards, All Engines	1978-9
089330007-1	007-1	1300	Johnson/Evinrude Outboards, 1-2 Cyl	1956-7
089330008-X	008-X	1302	Johnson/Evinrude Outboards, 1-2 Cyl	1971-8
089330026-8	026-8	1304	Johnson/Evinrude Outboards, 1-2 Cyl	1990-9
089330009-8	009-8	1306	Johnson/Evinrude Outboards, 3-4 Cyl	1958-7
089330010-1	010-1	1308	Johnson/Evinrude Outboards, 3, 4 & 6 Cyl	1973-9
089330040-3	040-3	1310	Johnson/Evinrude Outboards - All V Engines	1992-9
089330063-2		1310	Johnson/Evinrude Outboards - All V Engines	1992-0
089330052-7		1312	Johnson/Evinrude Outboards, All In-line engines/2 & 4 Stroke	1996-0
089330015-2	015-2	1400	Mariner Outboards, 1-2 Cyl	1977-8
089330016-0	016-0	1402	Mariner Outboards, 3, 4 & 6 Cyl	1977-8
089330012-8	012-8	1404	Mercury Outboards, 1-2 Cyl	1965-9
089330013-6	013-6	1406	Mercury Outboards, 3-4 Cyl	1965-8
089330014-4	014-4	1408	Mercury Outboards, 6 Cyl	1965-8
089330051-9	051-9	1416	Mercury/Mariner Outboards, All Engines	1990-0
089330050-0	050-0	1600	Suzuki Outboards, All Engines	1988-9
089330021-7	021-7	1700	Yamaha Outboards, 1-2 Cyl	1984-9
089330022-5	022-5	1702	Yamaha Outboards, 3 Cyl	1984-9
089330023-3	023-3	1704	Yamaha Outboards, 4 & 6 Cyl	1984-9
089330047-0	047-0	1706	Yamaha Outboards, All Engines	1992-9
STERN DRIVES				
089330029-2	029-2	3000	Marine Jet Drive	1961-9
089330005-5	005-5	3200	Mercruiser Stern Drive	1964-9
089330053-5		3206	Mercruiser Stern Drive - All	1992-0
089330004-7	004-8	3400	OMC Stern Drive	1964-8
089330025-X	025-X	3402	OMC Cobra Stern Drive	1985-9
089330056-X	025-X	3402	OMC Cobra Stern Drive	1985-9
089330011-X	011-X	3600	Volvo/Penta Stern Drives	1968-9
089330038-1	038-1	3602	Volvo/Penta Stern Drives	1992-9
089330041-1	041-1	3604	Volvo/Penta Stern Drives	1992-9
089330057-8	041-1	3604	Volvo/Penta Stern Drives	1992-0
INBOARDS				
089330049-7	049-7	7400	Yanmar Inboards	1975-9
PERSONAL WATERCRAFT				
089330032-2	032-2	9200	Kawasaki	1973-9
089330042-X	042-X	9202	Kawasaki	1992-9
089330045-4	045-4	9400	Polaris	1992-9
089330033-0	033-0	9000	Sea-Doo/Bombardier	1988-9
089330043-8	043-8	9002	Sea-Doo/Bombardier	1992-9
089330034-9	034-9	9600	Yamaha	1987-9
089330044-6	044-6	9602	Yamaha	1992-9
Seloc-On-Line (Internet Access)				
089330075-6		5000	One Mfg/Model - Subscription per user 3 years	1990-0
		5002	Master Pack -Case of 6 CD's including POP Display	
PROFESSIONAL TECHNICIANS MANUALS				
089330060-8		4500	Labor Guide - Johnson and Evinrude	1980-0
089330061-6		4550	Labor Guide - Yamaha	1980-0
089330062-4		4600	Labor Guide - Mercury	1980-0
BRIGGS AND STRATTON				
096376610-4	610-4	610	Briggs & Stratton 4 Stroke Horizontal Crankshaft	1950
096376611-2	611-2	611	Briggs & Stratton 4 Stroke Vertical Crankshaft	1950
096376612-0	612-0	612	Briggs & Stratton 4 Stroke Overhead Crankshaft	1950